W9-CMG-161

THE CICHLID FISHES

THE CICHLID FISHES

Nature's Grand Experiment in Evolution

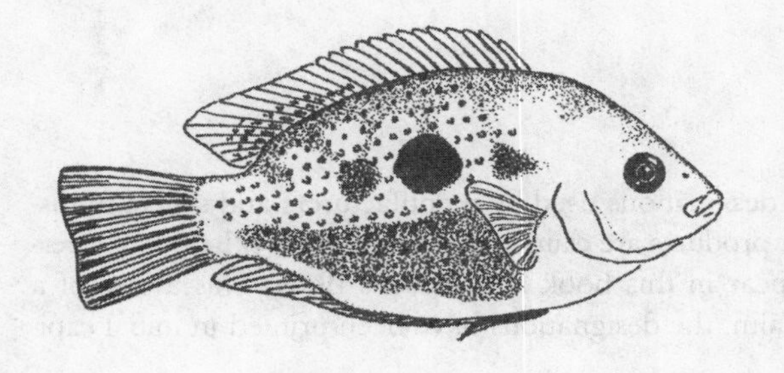

GEORGE W. BARLOW

PERSEUS PUBLISHING
Cambridge, Massachusetts

A CIP record for this book is available from the Library of Congress

 Printed in the United States of America.
Perseus Publishing is a member of the Perseus Books Group

Text design by Heather Hutchison
Set in 10.5-point Baskerville

First printing, November 2000

Find Perseus Publishing on the World Wide Web at
http://www.perseuspublishing.com

1 2 3 4 5 6 7 8 9 10–03 02 01 00

Dedicated to the memory of
Gerardus Pieter Baerends
30 March 1916–1 September 1999
Pioneer in the study of cichlid fishes
Scholar, source of inspiration, and friend

CONTENTS

FOREWORD

I have known and admired George Barlow ever since 1950 when he was a star on the water polo team at the University of California at Los Angeles and I was a beginning graduate student there. After graduation he was a Coast Guard officer during the Korean War for two years and then joined me and others in graduate work under Boyd W. Walker in the UCLA Zoology Department. When not ichthyologizing we often opposed each other at handball or went snorkeling in local kelp beds. I finished my dissertation and left before he did, but we have been friends ever since.

I have known and admired cichlids for an even longer time. They and other readily available aquarium fish turned me into an ichthyologist at a tender age, and most of my field and laboratory work in the past fifty years has been on fish, mostly of coastal marine habitats. But the almost entirely freshwater cichlids have always been a cherished aspect of my fish world. As a teenager I was deeply involved with the sex life and parental behaviors of the firemouth, still known by that name in English, although the Latin *Cichlasoma meeki* has changed.

I expected this new book by George Barlow would be outstanding, and my expectations have been met. Its two main messages are admirably presented. The first is that its author's thinking and long history of field and laboratory research, and the cichlid research of other dedicated biologists on the many conceptual challenges posed by this group of fishes, provide many fundamentally important findings and examples of research strategies designed to answer nagging questions on the biology of this special group. The second message is that there is an enormous list of important questions that, as yet, have no answers or, even worse, have too many answers. There is obviously more than one possible answer to many of the questions raised in this book, and research is needed to decide which if any of these previously suggested possibilities is actually valid. Hardly a page can be found that does not suggest a project that could be worthy of some years of study by a professional researcher or graduate student.

Here are a few suggestions for the special attention of those who think they do not have time to read the whole book. Everyone should know about the subtle and intricate mechanisms of cichlid jaws (Chapters 2 and 12), especially someone who still thinks of evolution as a story of steady progress from amoeba to man. The fish-to-mammal evolution of jaws is clearly a matter of gross simplification, and the evolution of primitive fish jaws to those of cichlids is one of complex and sophisticated engineering. I also think that everyone should know about the sex-ratio adaptations of anemone fish (Chapter 4). In many places, the book gives delightful examples of clever experimental tests of evolutionary ideas, especially Chapters 7 and 8. There is a superb treatment of the evolution of parental roles in Chapters 10 and 11, and of the amazing diversity of behavior patterns in the cichlids. Chapter 12 is a critical discussion of speciation and the lessons that cichlids can teach us about this phenomenon. I have always thought of myself as well informed about fishes and about cichlids in particular, but I sure learned a lot from this book.

A vital lesson that every chapter makes clear is that cichlid taxonomy presents a great mass of problems in urgent need of resolution. We cannot hope for reliable progress on questions about the ecology and evolution of cichlids unless we can improve our ability to identify evolving lineages and analyze the nature and timing of their divergences. Much of the needed research can be a continuation of the concepts and techniques of museum taxonomy as practiced over the past century. Today we can also take advantage of newer kinds of data. We not only can deal with morphology observable by dissection or through a microscope but can compare the morphology of molecules distinguishable by modern techniques. These molecular data provide independent evidence that can be critically compared with patterns shown by the diversity of traditional taxonomic characters.

Obviously no two scientists, if they really are scientists, would agree on all issues, and as I started reading I expected that even George Barlow would present some ideas that I would find seriously objectionable. I was wrong. He does not claim, as others have, that the special jaws of cichlids are the basic reason for the success of the group. Such claims are often made, but seldom with the required comparisons with the random phylogenies easily produced on a computer. Barlow's treatment of the cost of sex and of sexual selection is notably free of the misleading terms and implications so often seen. His use of the species concept is free of typological fallacies and uncritical acceptance of sympatric speciation proposals.

I would find it difficult to identify a category of biologist that would not benefit from this book. So many of the principles (and urgent questions) are so clearly illustrated by cichlid fishes and so clearly presented here. I would be hard put to identify organisms with problems that could not be better understood with prior familiarity with cichlid examples of the same phenomena. Compare parental motivations of cichlids (Chapter 10) with similar human motivations. Cichlid parents seem to show greater love for their offspring as they grow and approach independence. Anthropological observations suggest that the same is true for our species too. Do you notice any other human parallels as you read this book? They are well worth recognizing, not, of course, as basic principles that must be as applicable to us as to cichlids, but as possibilities worth considering and investigating.

George C. Williams
Department of Ecology and Evolution
State University of New York–Stony Brook

PREFACE

I've never met an animal, or a plant for that matter, that wasn't interesting, but some stand out as special. Cichlid fishes are right up there. In sheer number of species, they are one of the most successful of all families of vertebrate animals. The extent and speed with which they have evolved in some African lakes has made them the darlings of evolutionary biologists, and the attention is well deserved. That aspect of their biology fascinates me, but what captivates me even more is the complexity of their social lives and their devotion to family, not to mention the sheer beauty of many species.

I became hooked on fishes as a boy, and cichlids were the fishes that enchanted me most. I could sit for hours before an aquarium, watching a pair guiding their school of babies around, protecting them from the other fishes, who had their hearts set on eating the tasty small fry. I went on to become a professional ichthyologist, studying many kinds of fishes, but cichlids have remained central to my work. Much of my research has been dedicated to their social behavior. Despite my involvement with those remarkable fishes, I had no intention of writing a book about them, though the germ of the idea had been planted some time in the past.

Perhaps twenty years ago, one of my scientific idols, the late Dutch scientist Gerardus Baerends, said to me, "You should write the next book on cichlids." I took that as high praise because he meant it. In 1950 he and his wife, Josina Baerends-van Roon, had written the work that propelled the scientific study of cichlids center stage among behavioral biologists. It stands as a landmark in the field. At the time Baerends made his suggestion, I was frantically trying to do research and publish scientific reports because that is what life in the modern university demands, and like the cichlids, I had a family to nurture.

I was, nonetheless, tempted by Baerends's suggestion, even when I stopped to consider that undertaking such a book would have been much more difficult than when they published their monograph. The literature had grown enormously, and the growth was accelerating and has continued to expand. For the past ten years, the electronic database called BIOSIS lists 3,213 articles on cichlids, and that is only a portion of what is out there. Of course, that is nothing compared with the publications on medically important species such as the white rat, but the cichlid literature is diverse and challenging.

Two events pushed me over the edge. The first was an article about cichlid fishes that appeared in the science section of the *New York Times*. Based on a visit to my research laboratory and to that of my former Ph.D. student Axel Myer, it alerted the public to some of the fascinating aspects of the biology of cichlids. When the article was published I got a number of phone calls, including one from a major magazine that shall remain unnamed, asking if I had photographs of cichlids engaging in oral sex. Among the people who phoned me was Jack Repchek, then an editor for Addison-Wesley. He wanted me to write a book on these amazing fishes. The rest is, as they say, history.

I was encouraged to produce an unreferenced book virtually off the top of my head, using lots of photographs, and fast. I could have done that, but something within wouldn't let me. My psyche had been shaped by too many years of writing careful scientific papers. The goal that took shape in my mind was a book that could reach a wide audience while retaining scientific integrity. It would be directed toward the enquiring lay person, the kind of individual who has a strong bent for natural history and thus one who enjoys learning about the engaging lives of animals.

Among those readers, I was hoping to capture the imagination of students, both young and mature. I also wanted this book to be the one I searched for as a teenager but could never find. To that end, I have tried to write in a direct fashion, minimizing the use of scientific terminology (unfortunately, for the sake of economy, the reader will have to become familiar with a few such terms).

I realized from the outset that a large portion of the readers of this book would be cichlid-fish hobbyists. I know many of them and appreciate their drive to collect and breed cichlids, especially species new to the hobby. One thing that further distinguishes them is the passion behind their desire to know and name the many different kinds of cichlids. That is one reason why I provide what will seem a plethora of fish names in this book, although the hobbyist will be disappointed in the way I use the same small set of species to illustrate so many points. The nonhobbyist can skim over many of the names and still follow the development of the cichlid chronicle. The nearly obsessive mind set of the hobbyist is also part of the reason I have stuffed the book so full of information about cichlids.

While writing, the slightly unnerving thought kept coming back to me that many of my scientific colleagues and advanced, critical students would read the book. I have already been asked by several of them when the book will be published because they want to read it. And they will, because cichlids are such evolutionary celebrities, and no such book has ever been written about them, nor about any other family of fishes.

If you are acquainted with scientists, you realize that they cherish references to the literature from which the information came. (Sorry, but most of what you read will be from the published research of other scientists; my contribution is small.) To keep my colleagues happy, I initially supplied abundant citations to original research. Also, as the writing progressed I shifted my focus ever more to the scientists while continuing to write in a popular style. The result was the worst of both worlds–an overly long book that would please neither the general reader nor the more professionally inclined.

Therefore, I vigorously revised the first draft, cutting out 150 pages of manuscript. Extraneous material was reduced to a minimum, and examples were limited to those that best made the case. I was especially anguished by the need to reduce the number of citations. Where knowledge seemed general, and even where multiple citations seemed appropriate to me, I took a deep breath and expunged many of them. You'll often have to take my word for it when I make pronouncements about general biological findings.

To keep the prose flowing, I use small superscript numbers as keys to the literature cited. These are listed at the end of the book, chapter by chapter. Those numbers are linked to a list of authors and years, which in turn direct you to the full citations in the References. I provide a slim Glossary, as well, to help if you forget the meaning of certain terms.

My aim in this book is tell the story of how one family of fishes has evolved the highest level of parental care known for any kind of fish–a level of care that rivals that found in birds and mammals and other animals we think of as good parents. The foundation of this complex family structure rests squarely on one of the most difficult behavioral tasks known to higher vertebrates: Two individuals of the opposite sex live in peaceful cooperation; that is not easy, and I hope to explain to you why when I take up the theoretical basis of the conflict of the sexes. But that is only part of the story, and I don't want to get ahead of myself.

I also wish to persuade you, the reader, that one needs to understand many aspects of the biology of an animal to grasp how natural selection can produce such remarkable adaptations. If I have accomplished this, you should be able to digest the closing chapters on the stunning evolutionary radiations of cichlids in African lakes and how that most likely came to pass.

I should also remind you that I had to be selective in what I covered. I have emphasized the behavioral and ecological aspects of cichlid biology. I had to leave out many fascinating topics. I only briefly touch on bits of physiology, for instance, when treating plasticity of sex determination. Other topics, such as sensory systems, digestion, osmoregulation, and metabolism are totally ignored, and genetics is touched only lightly. Even within behavior and ecology, I chose not to treat feeding behavior in depth, although it is a crucial aspect of speciation. To have done so would have made an already long book much longer. The information provided should be sufficient, however, for the reader to appreciate the relevance of feeding behavior and its morphological basis.

Many illustrations have been borrowed from the literature, and I thank the authors from whose publications those illustrations came. Ad Konings and Paul Loiselle kindly made available several photographs from their film libraries, and a few more were provided by Juan Miguel Artigas, Jeffrey Baylis, Wayne Leibel, Michael Taborsky, Ole Seehausen, and Uli Schliewen. Having done a bit of fish photography myself (two of the photos are mine), I appreciate the skill and patience that went into catching those images on film. Karen Klitz and Randy Reiserer prepared several original black-line drawings. Melanie Stiassny generously volunteered a new diagram of the way cichlid pharyngeal jaws are suspended by muscle in the skull. Many thanks to all.

Over the years, my many Ph.D. and postdoctoral students have educated me in numerous ways. I wish to acknowledge the grace with which they shared their ideas with me, so much so that I am sure that I don't even realize it when I present their views as my own. I thank Jürg Lamprecht for sharing with me his unpublished observations and for sharing his thoughts with me. I am also indebted to Alison Galvani and Ron Coleman for reading several of the early chapters and offering suggestions for their improvement. An extra thanks to Ron for suggesting the title of the book. Richard Stratton, a skilled and experienced aquarist, read the entire manuscript. I am especially grateful to George C. Williams, my dear friend and renowned evolutionary biologist for writing the Foreword (thanks, Nafni). I am beholden to all for their time and for their helpful comments. The responsibility for those errors that inevitably persist, however, are mine.

Fittingly, the last person to whom I owe thanks is my editor, Amanda Cook. She firmly reigned in my inclination to keep on writing, citing yet another reference, adding more and more indispensable tidbits about those fascinating cichlids, and plunging into the convoluted theories of how they evolved. She taught me that less can be more, for which you, dear reader, should also be grateful.

George W. Barlow
Berkeley, California

INTRODUCTION

How many times have friends, freshly back from a vacation in Hawaii, told me excitedly about their first experience snorkeling among vibrant coral-reef fishes? That is a stirring experience, and I still remember vividly my first encounter there: a kaleidoscope of fishes swimming effortlessly through the water, coming close enough for me to see the movement of their fins, the glint of an eye guardedly looking back at the intruding snorkeler.

I asked Jack Randall, the ultimate authority on tropical marine fishes, how many fish species are found in Hawaii. Restricting the census to the inshore fishes down to a depth of 200 meters, he came up with 567 species. Impressive. Of course, a snorkeler sees only a fraction of those.

To tantalize my friends who are inspired by their Hawaiian experience, I often tell them they have only scratched the surface of piscine biodiversity. To experience a frisson of fishy ecstasy they should venture west to New Guinea or the Philippine Islands where they would share the water with about 2,800 unique fish species.[1] That island-rich part of the Pacific Ocean is at the heart of the diversity of marine fishes.

As one moves east from there, across the Pacific and back toward Hawaii and America, the island stepping-stones become ever fewer and further apart, and species drop off with distance. As a consequence, Hawaii is depauperate, that is, relatively poor in fish species, though not in their absolute abundance. If the friend shows a serious interest in fish diversity, I go on to describe the wonders of cichlid-fish speciation.

One large lake in Africa alone may have from 500 to 2,000 species of cichlids, though no one knows the real count.[2] That is about the same number of fish species as in all the Hawaiian Islands, and there they are spread over 101 families; that averages out to five or six species per family. In that single African lake, just the one family, the Cichlidae, has an incredible multitude of species, though some other kinds of fishes are also found there. In addition, two other African lakes have similar numbers of cichlid species, and most of those cichlids are confined just to their home lake.

One author, among several, has given a conservative estimate for the profusion of species in the Great Lakes of Africa: 700 to 1,000 cichlid species in Lake Malawi, 500 in Lake Victoria, and 250 in Lake Tanganyika.[3] That makes a total of 1,450 to 1,750 different kinds of cichlids in the Great Lakes of Africa, without taking into consideration the numerous other species of cichlids in several smaller African lakes and rivers. To put this in context, about 3,000 species of freshwater fishes have been recorded from the entire continent of South America, and Sven Kullander has estimated the final count will reach around 5,000 species.[4]

You might counter that the high number of species in the African lakes is just a consequence of their separation in enclosed bodies of freshwater, whereas the marine fishes are distributed freely through the open ocean. Fair enough. The marine fishes disperse by means of eggs and larvae that float great distances in the ocean. Consequently, one can travel considerable distances across the tropical Pacific and still recognize many of the same species seen on the reefs one left far behind. For a more balanced comparison, let us count the number of species of freshwater fishes, say, in all of Europe west of the Urals. The most recent published score gives a grand total of only 215.[5] Therefore the difference is not just a matter of the marine versus the freshwater environment.

Not fair, you say. The cichlids are in some of the biggest lakes in the world, the likes of which Europe lacks. Well, then, let's look at the Great Lakes of North America, which are even larger than their African counterparts. In all of our Great Lakes there exist a total of 235 fish species.[6] Clearly, the incredible evolutionary display of cichlid fishes in Africa is not a matter of being in large freshwater lakes.

Other kinds of fishes have radiated in isolated old lakes in different parts of the world. Perhaps the best known are three families of sculpin-like fishes in the oldest and perhaps deepest lake in the world, Lake Baikal, in Russia. These fishes have evolved about 50 species. Minnows have radiated in a Philippine lake and in one in Ethiopia, as have silversides in Mexico and killifishes in Lake Titicaca in South America. In none of these, however, have the flocks of species produced more than one or two dozen species, though these may be remarkably diverse in morphology and ecology.[7]

The overwhelming diversity of cichlids in each of the African Great Lakes truly stands apart among all vertebrate animals. Think of this: The famous Darwin finches of the Galapagos, the birds that inspired Charles

Darwin to develop the theory of organic evolution, consist of only 14 species inhabiting more separate islands than species.[8]

The plethora of cichlid species in single lakes was known to a small group of evolutionary biologists by the end of the nineteenth century. A brief review of the phenomenon attracted more general interest in 1950,[9] but it wasn't until 1972, with the publication of the landmark book by Geoffrey Fryer and T. D. Iles, *The Cichlid Fishes of the Great Lakes of Africa: Their Biology and Evolution*, that the cichlids there became to evolutionary biologists the quintessential example of speciation gone wild. Fryer and Iles concentrated on the cichlids in Lake Malawi while making comparisons with those in Lakes Tanganyika and Victoria. Nonetheless, further research of the cichlids in that area was relatively slow to develop, partly because of the inaccessibility of the lakes.

Before the book by Fryer and Iles, however, museum work had begun. In the late 1940s, the Belgian Hydrobiological Expedition brought back a large collection of cichlids from Lake Tanganyika. That led to the first comprehensive study of the taxonomy of the cichlids there. The resulting monograph has been an anchor for the study of relationships among the cichlids of Lake Tanganyika.[10]

Things began to move when a few teams of dedicated ichthyologists commenced observing cichlids underwater in Lakes Malawi and Tanganyika. I'll mention just a handful of the pioneers here. Tony Ribbink and his colleagues from South Africa did extensive underwater surveys in the clear Lake Malawi, recording details of cichlid behavior, ecology, and coloration.[11] Kenneth McKaye, an American, went to the same lake and initiated underwater studies of the huge congregations of male cichlids displaying over open sand bottom[12] and was soon joined by countrymen in a continuing project.[13] In the meantime, an ambitious program was underway along the inhospitable north shore of Lake Tanganyika, led by Hiroya Kawanabe from Kyoto University; the findings of that team have been summarized recently in an invaluable volume, *Fish Communities in Lake Tanganyika.*

The cichlids of Lake Victoria received less attention, though Humphrey Greenwood was actively investigating their taxonomy at the British Museum of Natural History.[14] When Greenwood did his studies, he worked largely with specimens collected from the trawls of fishing boats and from seine nets drawn along the shore. Victoria seemed too murky for observations underwater. Fortuitously, Greenwood recommended to a Dutch ichthyologist, Kees Barel, a lecturer at Leiden University, that the morphology of the cichlids of Lake Victoria would be ideal material for the study of adaptations. Eventually, a crew of Dutch biologists produced a splendid body of literature on the cichlids in Lake Victoria.[15] One of them,

Tijs Goldschmidt, wrote *Darwin's Dreampond,* a charming novel-like book that deftly combined an account of the cichlids and their evolution with his personal experiences there. In that crew was a young German biologist, Ole Seehausen, who discovered that although the waters of Lake Victoria are not ideally clear, underwater observations are possible, and that spurred research on the behavior of the cichlids there and also resulted in a highly useful book, *Lake Victoria Rock Cichlids.*

Now research on the African cichlids is coming in rapidly, but the attention has shifted toward studies of relationships among them, as garnered through molecular genetics.[16] Coupled with these discoveries and adding fuel to the excitement over the evolution of cichlids are the recent geological findings that the entire cichlid fauna of Lake Victoria may have evolved within the past 12,400 years;[17] prior to that, the basin where the present lake exists was totally dry. Geological research also suggests that a few species in Lake Malawi have originated in just the past 200 to 300 years. Thus they had not yet begun to evolve when in 1632 the great astronomer Galileo Galilei published his revolutionary book *Dialogue Concerning the Two Chief World Systems—Ptolemaic and Copernican.*

Such quick formation of species overthrows much of what is thought about the time it takes to evolve new species. Among vertebrate animals (fishes, amphibians, reptiles, birds, and mammals), evolving into a new species has typically been estimated to require on the order of several thousand years. Because of their rapid "explosive" radiation into multiple species, the cichlids have become celebrities in the realm of evolutionary biology. In fact, they are regularly cited in textbooks as the most impressive evolutionary radiation among vertebrate animals.

The spectacular speciation of cichlid species in the Great Lakes of Africa justifies an entire book on cichlid evolution, so you will be surprised, perhaps shocked, even disappointed when I tell you that most of this book will not be about the radiations per se but rather about what cichlids do and what makes them so special. That may make it easier to understand how they speciate so quickly when I finally address that issue and its significance toward the end of the book.

Cichlids have attracted the attention of serious biologists through much of the twentieth century, even before their proliferation of species was well known. Biologists originally tended to focus their attention either on the river-dwelling African species or on cichlids from the New World tropics, not on cichlids from the Great Lakes of Africa. That was due mainly to the lack of access to those remote regions of Africa. But all cichlids remain fascinating because of their remarkable, almost unfish-like family behavior.

The usual perception of fishes is that after they launch their eggs into the sea, or into the stream, the offspring are on their own. The notion of caring, protective parental fish is foreign. However, where known, and we know this for a myriad of species, cichlids are without exception parental. The ways they care for their issue, moreover, are varied. Some protect schools of baby fish for weeks, whereas others carry the developing babies in their mouths for comparable periods. Some practice a combination of both methods. And various lines of cichlids have evolved the same adaptations independently a number of times, a phenomenon called parallel evolution. Not only that, some species have evolved what amount to communes, extended families of several breeding adults of both sexes and their offspring. Also remarkable, the juveniles in some species remain with their parents to help rear their younger siblings. The evolution of family life among cichlids is a major theme of this book.

Cichlids are inventive, too, when it comes to mating systems. They show all the mating systems known to other vertebrate animals. No other family of fishes has evolved such a diversity of mating systems, all starting from the basal condition of prolonged monogamy. That any cichlids are still monogamous is remarkable; it is a rare mating system among not only fishes but other animals as well, birds excepted.

Related to mating systems is the way cichlids choose mates. Their behavior differs depending on whether the species is monogamous or polygamous, and how it differs fits nicely into recent theory on the mating games animals play. Mating leads to fertilization of eggs, and here the cichlids have again been creative. I'll tell you about the unique, some might say bizarre, way the mouth brooders get their eggs fertilized, which is an evolved adaptation unlike that known for any other vertebrate animal.

Engaging in such complex behavior requires a finely tuned system of communication. In the process of refining and integrating their social behavior, cichlids have evolved an intricate set of signals to convey their intentions. These include crude sounds, but most of the signals are visual. Thus when two cichlids get together, whether to fight or to reproduce, they perform several displays in which they change their head and body shapes, erect their fins or fold them, propel pulses of water currents at one another with their tails, and all the while change their color patterns. In this latter regard, they vastly outstrip the chameleon. I will describe this highly evolved communicatory network.

We do not understand how these adaptations have led to such a wealth of species in the Great Lakes of Africa. Some scientists have argued that the key lies in the cichlids' ability to exploit the different kinds of food in their environment. To make the point, consider two coral-reef families of fishes to which cichlids are closely related, the wrasses (Labridae) and the parrot fishes (Scaridae). Both families are major contributors to the assemblage of fishes on coral reefs and have evolved many species. The wrasses, however, are limited to carnivory, though they take many kinds of prey. The parrot fishes are also constrained in their feeding behavior, specializing in food they bite from the substrate, meaning mostly algae. In some of the larger parrot fishes that includes pieces of dead coral that they grind up to get at the embedded filamentous algae. Cichlids embrace all those foodstuffs (no, not coral, but algae) and more.

Cichlids can handle varied foodstuff because they have evolved a distinctive arrangement of the muscles and jaws in their throats for processing a wide range of edibles. Their guts have also reacted to the various diets, though that response is not unique to cichlids. The evolved specialization of their jaws is thought to be fundamental to the ability of cichlids to generate new species as they adapt to different foodstuffs. Because of this, I take up feeding and jaws early in the telling of the cichlid story.

In closing the book I will just touch upon the economic role of cichlids. Most cichlids are too small to enter into the catches of commercial fisheries. Some do, however, and they provide most of the protein for native populations in Africa and a significant part of the meat eaten in some parts of Central and South America. One species occurs in Israel, where it is eaten; it is the fish associated with Peter in the Bible and is called the St. Peter's fish.

A few species are even large by conventional standards and are prized fish for the table. What may surprise many readers, however, is that one type of cichlid is now the most widely cultivated food fish in the tropics around the world. These are the tilapias. They are even marketed in North America and Europe.

Tilapia aquaculture, however, creates problems for the environment, as so often happens with the culture of other fishes, such as salmon.[18] Cultivated fish can be a source of disease for native fishes, and their excrement can contaminate the areas where they are farmed, much as intensive pig farming does.

Attempts to "improve" fisheries by releasing alien species typically damage the native species, often in unforeseen ways. I'll discuss one spectacular example in closing, an introduction that has led to the extinction of at least one hundred species of cichlids in Lake Victoria and has destroyed the traditional trawl fisheries for cichlids. But we have a long way to go to reach that depressing situation. To begin, I want to tell you what cichlids are and how they are arrayed in an evolutionary framework.

≈ *one* ≈

SO, WHAT IS A CICHLID?

"You study what?" said the woman seated next to me on the airplane. She had a half smile on her face, partly amused, partly curious.

"Cichlid fishes," I said again.

"So, what's a cichlid fish?"

No matter how many times people respond like that, I am always at a loss when I try to explain what a cichlid is. When people first see the word written, they often mispronounce it, apparently inspired by Chiclets gum. Cichlid is pronounced *sick-lid.*

The name does sound odd, probably because it doesn't connect with anything familiar. If you are a bird watcher, however, you might have been struck by the shared Greek root in the scientific name of some birds, the thrushes.

The ancient Greeks referred both to thrushes and some Mediterranean sea fishes as *Kichle*. Why the names were shared is uncertain and has provoked creative suggestions. The marine fishes are the wrasses, close relatives of cichlids. Several hundred years later, when scientific names were first applied to animals and plants, the groupings were much larger than those we use today. The first cichlids were lumped with the wrasses as *Labrus,* which became the family Labroidae. At the beginning of the nineteenth century, the few known cichlids were split off by the eminent German ichthyologists Bloch and Schneider under a new name, *Cichla*. About half a century later the cichlids were elevated to family status, the Cichlidae.

Sometimes when explaining what a cichlid is I start by searching for common ground, the shared experience, such as some other kind of fish that looks like a cichlid and that the individual might have seen. A green sunfish is a case in point. People who fish in North America may have caught

FIGURE 1.1 The green sunfish from North America bears a striking though superficial resemblance to a generalized cichlid.

a green sunfish, and that fish has the same general look to it as many cichlids. If they have never seen a green sunfish, I often mention other sunfish familiar to anglers, such as the bluegill, or even the related largemouth bass.

As a last resort I ask the person if they have ever seen a freshwater angelfish. Lots of people have, because it is both a common and a memorable aquarium fish. Its exquisite shape, a compressed disk led by a small pointed mouth and outlined by long trailing fins, together with a body of silver slashed by bold black bars, leaves a lasting impression. A stylized rendition of the angelfish is often used as a logo for aquarium stores and books and in other places, too. Those who saw the movie *A Fish Called Wanda* might even know that Wanda was an angelfish. Unfortunately for the point I want to make, the angelfish is a highly specialized cichlid, and only a few other species of cichlids look like it. Besides, many people think of the coral-reef angelfishes. They also look like a pancake on edge, but they are only remotely related to the cichlid angelfish.

Another difficulty is that to answer the question of what a cichlid is presupposes that the person has some appreciation of how fishes are classified by ichthyologists. Where do the cichlids fit in among the familiar species such as sharks, minnows, trout, herring, tuna, and red snapper? Most biology students cannot adequately answer that.

FISHES RULE

Many are stunned to learn that the waters of our planet are populated by more species of fishes than all the other species of vertebrate animals, more than all the amphibians, reptiles, birds, and mammals put together. No one knows for sure how many species of fish exist. Each year, on average, 230 new species are described,[1] which makes them not only the most species-rich of vertebrate animals but the fastest growing as well. Bill Eschmeyer supervises the cataloging of named fishes, and he estimates the number of species ultimately will reach between 27,500 and 30,000. Of these, cichlids contribute around 2,000 species, perhaps more; just numerically, they are one of the most important families of fishes.

Collectively, fishes are also more diverse than all the other vertebrate animals. Some ichthyologists divide fishes up into groups as trenchantly different from one another as birds and mammals. The jawless fishes, which are the hagfishes and lampreys, are put into their separate group, as are the sharks and rays.

In the course of evolution, fishes have many times produced spectacular bursts of speciation called radiations, like the bursts of rockets on the Fourth of July, which have faded from the scene, leaving only fossils behind.[2] The first vertebrate animals were fishes, such as the placoderms and ostracoderms.[3] These strange armored, jawless fishes were plentiful at the time when the first fossils became abundant, and thus existed long before the dinosaurs. They were replaced by successive radiations of more advanced fishes, such as the sharks and their relatives.[4]

Well before the time of the dinosaurs, fishes gave rise to the amphibians, and hence to the land-dwelling vertebrates. Thus fishes were key species in the evolution of four-legged animals, the tetrapods. These early lobe-finned fishes branched off into a few lines of evolution that diverged from those that led to our advanced fishes. Therefore, to think of modern fishes as hav-

FIGURE 1.2 The shape of the South American angelfish, *Pterophyllum scalare,* is one of the most extreme in the family Cichlidae.

ing been the precursors of four-legged animals is a mistake. They were not. They have had their own distinctive radiations, and if anything, they evolved away from being like tetrapods. I mention this because so many people know that fishes are in our ancestral lineage, but they extend this to the belief that those ancestors were just like modern bony fishes, which include the cichlids. You are most emphatically not descended from a cichlid.

SORTING OUT THE KINDS OF FISHES

The modern bony fishes, called teleosts (which means true bone) arose about the time the dinosaurs were becoming extinct. The teleosts started displacing their ancestral fishes, which were much like our existing bowfin, alligator gar, and sturgeon. About 95 percent of all living species of fishes, including cichlids, are teleosts. These remarkable and diverse animals, considered the modern fishes, can be arrayed starting with those representative of primitive stages in the evolution of teleosts to the most advanced. (Some evolutionary biologists prefer to say plesiomorphic as a more objective term than the familiar ancestral, basal, or primitive, which I continue to use here. I will also apply interchangeably the vernacular terms advanced or derived.) The true bony fishes are often divided, almost for convenience's

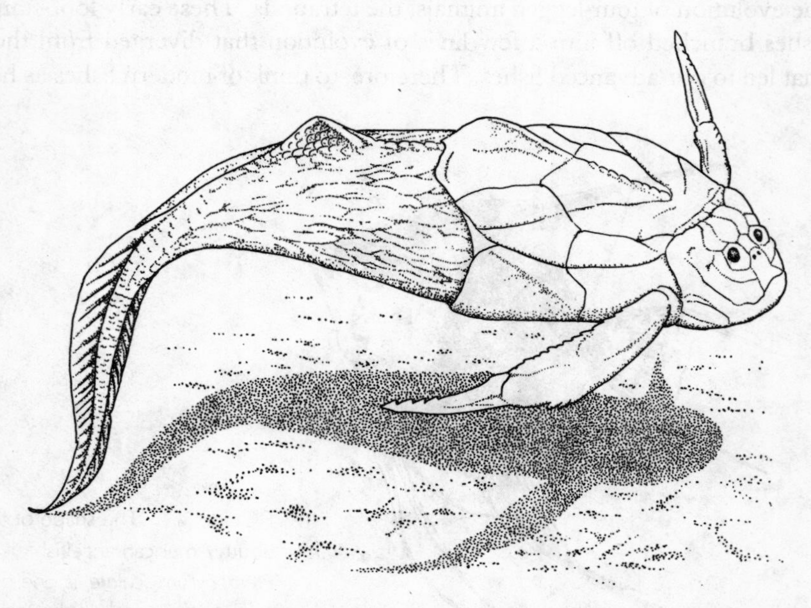

FIGURE 1.3 Found today only as fossils from 400 million years ago, this placoderm is representative of the first fishes, was heavily armored, and was common.

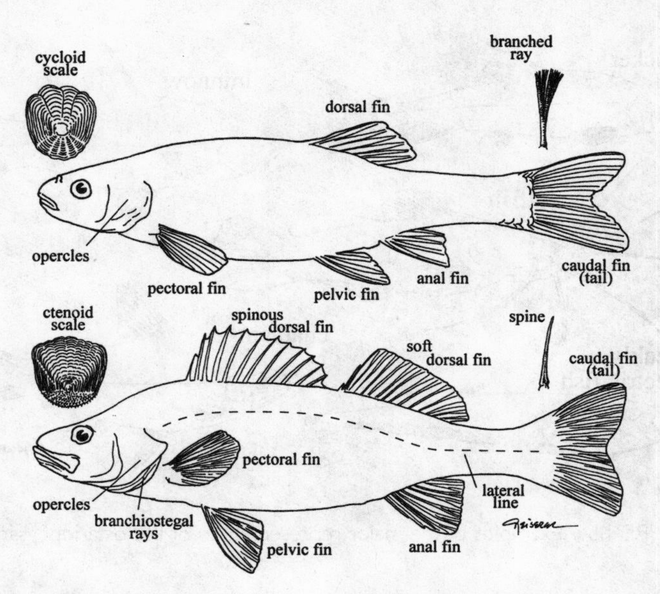

FIGURE 1.4 The upper fish is typical of the soft-rayed early fishes; compared with the fish below, the most obvious differences are the position of its pectoral and pelvic fins, the lack of spiny rays, and the presence of cycloid scale (both enlarged). The lower fish exemplifies more modern spiny-rayed fishes; in this case, the dorsal fin is present as two separate fins, the more forward spiny fin and the fin that follows it, which is supported mostly by branched rays; the scales are typically ctenoid.

sake, into the more primitive soft-rayed and more advanced spiny-rayed fishes, though this is only a rough division.

The Soft Kinds

The most primitive teleosts have relatively few species and are represented by the likes of herring and anchovies, eels, tarpon, and arapaimas (a group of freshwater fishes called bonytongues), and some deep-sea fishes. They are much like modern teleosts in their basic body structure and in details of their anatomy, such as their heart and gills. But at the same time they differ in many ways. Their mouth structure is less advanced, as are many other features of their general morphology.

That leaves the more advanced teleost fishes, consisting of about 20,000 described species in almost 400 families. The ancestral fishes here have the look and feel of being nonprickly. Representative familiar kinds are pike, trout and salmon, and true smelt. They lack spines in their fins, so they do not stick you when you handle them. Their scales, called cycloid, have

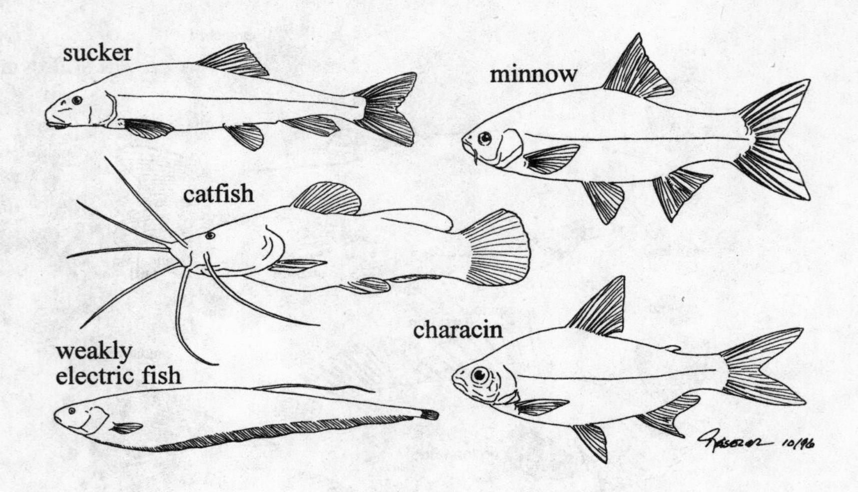

FIGURE 1.5 Examples of the major representatives of the ostariophysan fishes.

smooth rounded edges. And their jaws are simpler than those of more derived bony fishes.

The characteristic that makes fishes of this level of organization so readily recognized is the placement of their paired fins, the counterparts of mammalian legs and arms: The pelvic fins are located on the underside toward the rear of the fish, and the pectoral fins are positioned almost on the bottom side of the fish just behind its head.

Spiny Fishes

In contrast, the typical advanced bony fishes, such as cichlids, look and feel prickly. They have spines in their fins, and their scales are rough to the touch due to the tiny teeth on them. The sharp, piercing spines that lead the dorsal and anal fins, and the strong one that supports each pelvic fin, make it difficult for predators to swallow them.[5] So when I read the heading to an article in our local newspaper, the *San Francisco Chronicle* (1 April 1997), I knew both it was not an April Fool's Day joke and that the fish was most likely a cichlid, and it was.

> Man Chokes on Tropical Fish
>
> Bayou Vista, La. Stephen Hill Epperson, 36, popped a friend's 6-inch tropical fish into his mouth as a joke Sunday and died when it got stuck in his throat. The Jack Dempsey fish became wedged in Epperson's airway, said Dr. F. H. Metz, coroner for St. Mary Parish.

The Jack Dempsey is indeed a cichlid and performed the typical response of a spiny-rayed fish when engulfed by a predator: It locked its fin spines in the upright position to thwart being swallowed. Had the fish been a soft-rayed fish, such as the goldfish that have so often been safely gulped down by college students, the unfortunate Mr. Epperson would still be alive. The tragic ending of this story brings home a forceful message about one adaptive function of spiny fins in fishes in general and cichlids in particular.

But to continue with the characterization of the advanced fishes, the positions of their fins also make them easy to separate from the soft-finned teleost fishes. The once-rear pelvic fins have moved forward to lie on the underside of the fish just behind the head. As if to make room for them, the pectoral fins have migrated up onto the side of the fish. The list of such advanced modern fishes is endless and includes the snappers and grunts, bluefish, butterflyfish, bass, groupers, drums, and jacks, and on and on.

Most of the advanced species are in a group named the Percomorpha. That category is called a series in some classifications. The Percomorpha now rule the marine environment and the fresh waters of Australia but are otherwise not usually dominant in fresh waters. This huge category of fishes has slightly more than 200 families, including the Cichlidae, and about 12,000 species.

The percomorphs, however, frustrate most facile attempts to characterize them. The exact limits and composition of this cluster of species is still disputed by systematists. Many of the families have evolved bizarre specializations and body shapes that make them seem almost not fishes at all. The boxfishes and seahorses, to cite two instances, simply do not conform to one's concept of a fish.

Some advanced fishes seem to have reversed evolution to become primitive-appearing, such as the emergence of "eel impersonators" that are not related to eels at all. Even some of the relatively typical modern fishes have changed their scales from ctenoid to cycloid or none at all, as among many gobies. Others have reduced or lost spines in the fins and have moved the pelvic fins yet further forward to attach to the throat.

A ladder-like arrangement from primitive to advanced, as I have been intimating, is misleading. For instance, one of the most challenging groups to classify is the superorder Ostariophysi because they do not fit easily into a linear classification. These are the minnows, suckers, catfishes, gymnotids, characins, and a primitive group that has a few marine representatives.

The Ostariophysi are characterized by a baffling mixture of primitive and advanced features. For instance, their scales and the arrangement of their fins could be said to be primitive, yet they have the most highly specialized ears of any fishes. Further, some of the Ostariophysi have evolved secondary spines in their fins by coalescing and reinforcing fin rays. Many different kinds of catfishes have such spines in their pectoral fins, and they are often equipped with poison glands as well.

These are the foremost fishes of the fresh waters of the world.[6] For this reason, they bear remembering because their relationship to the cichlids for dominance of freshwater habitats comes up near the end of the book when I consider the amazing evolutionary radiations of cichlids.

The other difficulty in classifying families of perch-like fishes is that seldom does one or a few characteristics serve uniquely to distinguish one family from the other. Instead, the focus is on a singular mixture of traits, like the code to a combination lock. These traits often involve details of the complex skull, which are not immediately apparent to the observer. We can expect to see molecular biology help clarify relationships in the coming years.[7] Within the Percomorpha is the order Perciformes, which are the more typically perch-like fishes; the group is also hard to define because of the combination of characteristics that sets them apart.

PUTTING THINGS IN ORDER

Let's pause to consider how animals are classified. I will not be calling on this vocabulary much in what follows, but understanding the hierarchical nature of classification helps keep the confusion down. Think of living things arranged in several libraries, each building corresponding to a Kingdom, and each containing up to millions of books. The library we are interested in houses the animals. It has multiple floors, and each floor corresponds to a Phylum of living organisms. On each floor, rooms full of related books are called Classes. And within each room are many sets of book shelves, each set of which holds all the books belonging to one Order. Each shelf in the Order represents a family, and each book on the shelf is a Genus (plural Genera). A book, a Genus, can have from one to several pages, and these are our species.

In this odd library, the books are loose-leaf so that species can be moved from genus to genus as the page is better understood. The same applies to all the books: The genera may be moved about among families. Generally, the higher the level of classification, the less likely are any moves, though a few still occur.

A final quirk is that the librarian sometimes has difficulty deciding whether a book and its pages should be moved up or down the hierarchy. Taking the easy way out, subcategories are used between most of the categories. We know these, for instance, as Subgenera or Supragenera, Suborder or Superorder. Even these semi-steps don't suffice, so terms such as Series, Grade, and Division have sprung up, or at the lower end, subfamilies and tribes. Table 1.1 presents a bare-bones hierarchical classification of

TABLE 1.1 Hierarchical Classification of Cichlids

Classification
Phylum Chordata: the vertebrates and their early forms
Subphylum Vertebrata: animals with vertebral column
Superclass Gnathostomata: jawed fishes
Grade Teleostomi: bony fishes
Class Acanthopterygii: ray-finned fishes
Subclass Neopterygii
Division Teleostei: modern bony fishes
Superorder Acanthopterygii: spiny-rayed fishes
Order Perciformes
Suborder Labroidei
Family Cichlidae

cichlid fishes, which you may want to refer to at times. Note that the specific terms and the form in which they are given, as well as where to divide the steps, varies among authorities. I have taken as my guide the recent treatise of ichthyology by Gene Helfman, Bruce Collette, and Douglas Facey.[8]

The classification of animals and plants that we use today has a long and colorful history. Although others preceded him, the great eighteenth-century biologist Carolus Linnaeus was the one who laid out the formal scheme of classification that we still use today. (He was born Carl von Linné but as happened to so many of the species he described, he changed his name to the Latinate form Carolus Linnaeus.)

The Linnaean system is called binomial because each species has two names, and they are arranged like a Chinese person's name. I say that because the second name in this binomial nomenclature is the unique name of the species, and the first name designates the genus to which the species belongs. Hence several species may share the generic name if they are closely related. By convention, binomial names are italicized. When a person's name follows the genus and species, that tells you who first described the species. But the name is often given in a form that contains yet more information.

Take the species I have studied for several years, which has the common name of the Midas cichlid. Its original name in its full glory was *Heros citrinellus* Günther 1864. The generic name comes from the Greek *Heros* for a female with heroic qualities. The species name finds its root in the citrus tree (Gr. *kitrea*, L. *citrea*) because the first specimen was yellow. The two-part scientific name is followed by that of the person who formally described the species, Günther, who was an ichthyologist at the Museum of Natural History in London. The year, 1864, informs the reader when Günther's description of the species was published in order to make it easier to search the literature for the original description.

Lay people often think that one of the great discoveries of biology is a new species. Actually, that is a relatively humdrum though important milestone. New species are described regularly. A further misconception is that when a scientist discovers a new species she can, or will, name it after herself. Not so. That is considered bad form in the scientific community. The describer typically names the species for some distinctive feature, such as the yellow color of *Heros citrinellus*. Sometimes the place where the species was found is used for the species name, simply adding *-ensis* or *-ense,* depending on gender, to the end of the name, as in *nicaraguense*. When the species is named after a person, the usual ending is *-i,* as in the colorful and popular cichlid fish from Lake Victoria called *Haplochromis neyererei* in honor of the great African leader and first president of Tanganyika, Julius Nyerere.

In practice, the name of the describer and the year are often dropped in nontaxonomic articles. If the name is used frequently in an article, and the context is clear, the genus is indicated by just the first letter, for example, *H. citrinellus*.

The science of how organisms are named is called taxonomy, and the people who do that are taxonomists, in addition to being, say, ornithologists or entomologists. Some distinguish between taxonomy and systematics, though the two areas overlap broadly and individuals who do the one commonly do the other. Taxonomy addresses the work of describing diversity and how to identify species as well as arranging them into systems of classification. Systematics tends more toward relationships at higher levels of classification. I use the terms interchangeably in this book.

Taxonomy has a number of practitioners and remains a dynamic field of research. One consequence of that dynamism is that animal names, and how they are classified, often change, much to the consternation of others who prefer a fixed taxonomy that does not challenge one's memory.

Returning to the Midas cichlid, a later taxonomist reclassified it by putting it into a different genus. Thus he created the new combination *Cichlasoma citrinellum* (Günther). The parentheses around Günther's name indicates that he described the species, but not in the genus *Cichlasoma.* The ending of the species name changed to agree with the gender of the new genus.

The binomen (two-name combination) does not inform you, however, about the times when subsequent ichthyologists collected specimens of *C. citrinellum* and erroneously described them as new. When such mistakes are discovered, the latecomer is said to be a junior synonym of *C. citrinellum* and disappears into its synonymy.[9]

Fish hobbyists hate it when ichthyologists revise the names of species and genera, especially when a well-accepted species name vanishes into the synonymy of some obscure species name. Understandably, that produces howls of complaints about the metamorphosing nomenclature. Scientists are more conservative about the names of plants and animals than the amateurs; and taxonomists have specified rules of nomenclature. Common names, in contrast, vary regionally and change whimsically through time, not to mention the reality that the common name is different in almost every language, whereas the scientific name remains unchanged. Despite that, I try to use widely accepted common names when possible. Unfortunately, the Cichlidae is a large family and most species do not have common names.

But, you may well ask, why do scientists change a fish's name? Generic names get switched around as the understanding of relationships improves. As more species are described, and as we gain a more fundamental understanding of relationships, species are regrouped to reflect the new knowledge. Name changes simply reflect the lively state of the field. The generic status of *C. citrinellum,* for instance, is unclear as I write.

The formerly large genus *Cichlasoma* embraced species from southern Texas to the Amazon basin in South America, but recently it was restricted to a small number of species in South America. As a result, the generic status of the numerous other species were left in limbo, though many aquarists have elevated the subgenera suggested by Robert Miller to the status of full genera. *C. citrinellum* has been placed in other genera in the past, but which of those names should be used remains controversial. The most likely generic name in the future is *Amphilophus.*[10]

Until the question is resolved, systematists have decided to keep the same generic name; but to let the reader know the genus is tentative it is by convention marked in single quotes, that is, *'Cichlasoma' citrinellum.*[11] I am less conservative with the names of African species because revisions of their nomenclature are further along.

Classification depends on degree of relationship. Thus fishes in the same genus are judged to share more traits than fishes in other genera in that family. Likewise, all the genera in a family share some unique traits that unite them.

Unfortunately, no one can provide a well-defined convention for reaching the same classification, even within a specified group of animals such as fishes. Consequently, the reality of any classificatory scheme is often debated and is subject to continuous revision.[12] A few taxonomists even want to do away with the binomial nomenclature, effectively getting rid of genera.[13] A more radical proposition that is gaining followers is to do away with species names altogether. But such controversies about the nitty-gritty of classification need not concern us here. My objective in this exercise on classification is to give you an appreciation of the way fishes are arrayed and where cichlids fit in the scheme of things, and why things are not as neat and simple as all of us would wish.

COMBINING THE TRADITIONAL AND THE PROGRESSIVE

Given the spectacular radiations of the cichlids, one can only be impressed by how little research has been done on their classification, that is, until you appreciate how few systematic ichthyologists are available to pursue such studies. The allure of molecular biology contributed to the withering of more traditional studies of the classification of animals. That is changing now, fortunately, because of the growing appreciation of biodiversity, spurred by concern over the worldwide ecological crisis. Also, molecular biology, which contributed indirectly to the demise of traditional systematics, is now becoming a valuable tool for understanding relationships among animals and plants and is thus aiding current efforts along those lines.

The marriage of traditional museum taxonomy and molecular biology creates an odd couple. Their environments seem as different as their practitioners. A fish systematist is well grounded in evolutionary biology and comparative studies of animals and their natural history. He or she is surrounded by glass jars of fishes, taken from vast shelves of specimens that reach to the ceiling. He measures many features of the fish, such as the diameter of the eye, the distance from the point of attachment of the pelvic fins to the tip of the snout, and so on; and the scales and fin elements are counted.

Deeper research entails chemically clearing and staining dead specimens. The preserved fish is made transparent in glycerin. The bones and cartilage are stained red and blue, respectively. That reveals the details of shapes and connections among the many components of the skeleton. Larger fish are literally cooked to remove the flesh for the same purpose, to examine the osteology; this is a traditional class exercise in many ichthyology courses, such as the one I taught for many years.

Systematists then ponder the many attributes to see how the different species match up. Certain traits are deemed key to understanding relationships and thus how the species might have evolved. Such traits are given more emphasis than others.

The molecular biologist, in contrast, is typically trained in biochemistry and genetics and moves about in a scrubbed laboratory with high lab benches and lots of technical equipment. The object of analysis is some aspect of DNA (deoxyribonucleic acid). DNA is the stuff of genes, so one could fairly say that the genes themselves are examined. Several techniques are now available for such analyses, and they often yield slightly dissimilar results when applied to the same species.[14]

Different techniques are proving useful for different purposes. Take, for example, the DNA lying in tiny organelles in the cellular cytoplasm, called mitochondria. They control the flow of energy in the cell by regulating the exchange of electrons. Each mitochondrion contains a small ring of DNA. Barring the occasional mutation, this mtDNA is passed unchanged exclusively from mother to offspring. It does not combine with the mtDNA of the father because it is almost always excluded during fertilization, and that simplifies analyses, compared with recombined nuclear DNA.

The genes in mtDNA also evolve up to five to ten times faster than those in nuclear DNA, and for that reason mtDNA was at first deemed not useful when seeking relationships among species. However, with the advent of improved technology, mtDNA has almost become the DNA of choice for constructing phylogenies of animals.[15]

At the other extreme is the study of microsatellites in nuclear DNA. They consist of what are called tandem repeats of remarkably short pieces of nucleotides. These bits of DNA are abundant but difficult to locate, and they vary greatly from individual to individual. Their high variability provides a rich opportunity for revealing differences between closely related species or between populations within a species. They are also helpful in estimating the effective size of a population, the degree of inbreeding within it, and the rate of migration into the population, as well as kinship and parentage.[16]

One advantage of molecular analyses in general is the possibility of getting an estimate of the time span separating different species or lineages. This can be done because some of the genetic material is selectively neutral. Natural selection does not alter such genes, so their rate of change (mutation) is relatively constant through time.[17] Each mutation is like a step, and the number of successive steps required to reach a determined difference allows one to calculate how much time was needed.

When the molecular approach was first applied to classifying animals and plants, it was met with resistance. Some scientists were hostile. And results from molecular and traditional biology were pitted against one another. Now, however, data from the two approaches are increasingly viewed as complementary.[18] Both methods rest, ultimately, on the degree of similarity between two or more species. The more traits they share, the closer the relationship.

That elementary assertion has its pitfalls though, such as when unrelated species evolve the same body plan as a result of similar ecology, a phenomenon called convergence. The familiar and extreme example is the wings of bats and birds. Details of their morphology unambiguously demonstrate that the wings evolved independently, and for reasons of function they have converged on the same solution to becoming airborne.

Convergences often are not so obvious. Then careful study is needed. Molecular genetics is especially useful when morphology of different species has converged on the same adaptation, especially when the species are relatively closely related. The fundamental truth, nonetheless, is that both morphologists and molecular biologists are looking for degrees of similarity.

How does one assess which species are most like others when many of them possess a dazzling array of traits, some of which are similar and others different? Before the widespread use of computers, systematists relied on a respected logic that involved the weighting of certain traits. Unfortunately, different systematists sometimes reached uncomfortably different schemes of classification using this procedure.

But this is the computer age. Now a knowledgeable biologist with a desktop computer can enter a given set of data and ask the computer to come up with the best arrangement of species, said to be the most parsimonious. One has to be careful here. The "best" arrangement depends not only on the specific set of data supplied to the computer but on the algorithms of the different software applications as well.

The classification can be done using just morphological traits, or just molecular ones. Increasingly, however, both kinds of data are fed into the program. Also increasingly, traditional systematists are learning molecular techniques.

The computer can be programmed to produce what is called a cladogram, in the vernacular, a tree of relationships. In fact, it can crank out a number of different potential trees depending on the instructions it receives.

THE CICHLIDS AND THEIR KIN

The family Cichlidae lies within the order Perciformes, class Acanthopterygii, and its species are fairly representative of a general perciform in their basic body plan. For many years ichthyologists were uncertain about the relationship of cichlids to other perciforms, and an element of uncertainty remains even though our understanding of their affinities has improved in recent years. Cichlids are now placed in a suborder called the Labroidei. That is a reversion to an arrangement used in the middle of the nineteenth century by J. Müller, a classification later abandoned by other ichthyologists.[19]

The Labroidei accounts for about 15 percent of all the species of teleost fishes.[20] The families in the suborder differ in many ways, some of which would seem problematical. Nonetheless, they are united by a similar arrangement of the elements in their complex throat. Müller called the

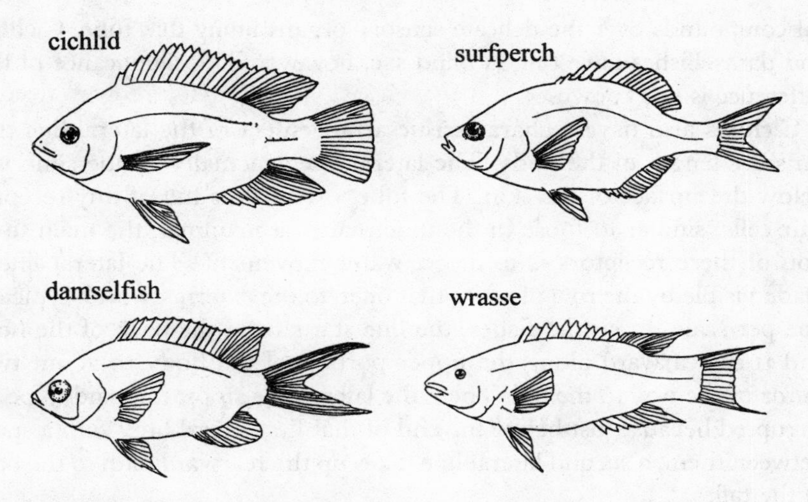

FIGURE 1.6 Representative sketches of the four major groups of labroid fishes. The wrasse group also contains the parrot fishes.

group the Pharyngognathi to draw attention to the advanced development of jaws in their throats; the name literally means throat-jaws. I will treat the significance of this structure in the next chapter.

As conventionally constituted, the Labroidei includes the parrot fishes (family Scaridae), wrasses (Labridae), surfperches (Embiotocidae), damselfishes (Pomacentridae), the poorly known small family Odacidae, and the cichlids (Cichlidae). Prophetically, one ichthyologist predicted the suborder might well change in the light of more research.[21]

Molecular genetics now provides an additional means for testing this arrangement. Based on a one-gene character, a radical overhaul of our classification of perch-like fishes was recently proposed. But as the researchers themselves cautioned, their arrangement should be regarded as provisional,[22] and that is why I don't present it here. Of most interest to us is whether the cichlids fall out as a coherent family. Are they descended from one ancestral stock, that is, monophyletic? The latest study concludes that they are.[23]

DEFINING A CICHLID

A combination of several features sets the cichlids apart from other fishes. Most often noted is that they have only one nostril on each side of their snout. Except for damselfishes, teleosts typically have two nostrils on each side of the snout. If a nostril of a typical fish were sliced in two, you would see that it is a U-shaped tube with two openings close together. Apparently water flows in through one hole and out through the other, passing chemical compounds over the delicate sensory organs lining that tube. Cichlids and damselfishes have only a blind sac, however. The significance of this difference is not known.

Cichlids also have a characteristic arrangement of the lateral line that runs the length of the body. The lateral line is actually a tube sunk just below the surface of the skin. The tube, or canal, is full of tiny receptor hair cells, similar to those in the inner ear of a mammal; the main function of these receptors is to detect water movement. The lateral line is made visible by the row of pores that open to the external world, typically one per scale. In cichlid fishes, the line starts from the back of the head and runs rearward along the upper portion of the body to about two-thirds of the way to the tail. There the lateral line stops; it is said to be interrupted because just below the end of that first lateral line, with a space between them, a second lateral line takes up the rearward path to the base of the tail.

As though to make life difficult for systematists, some cichlids have but a single, uninterrupted lateral line. *Teleogramma* and *Gobiocichla* are African genera dwelling in rapidly flowing water in the Zaire River.[24] They have adapted to life in the fast lane by evolving elongate, tubular bodies. Perhaps as a consequence of the lesser space, the lateral line has become continuous.

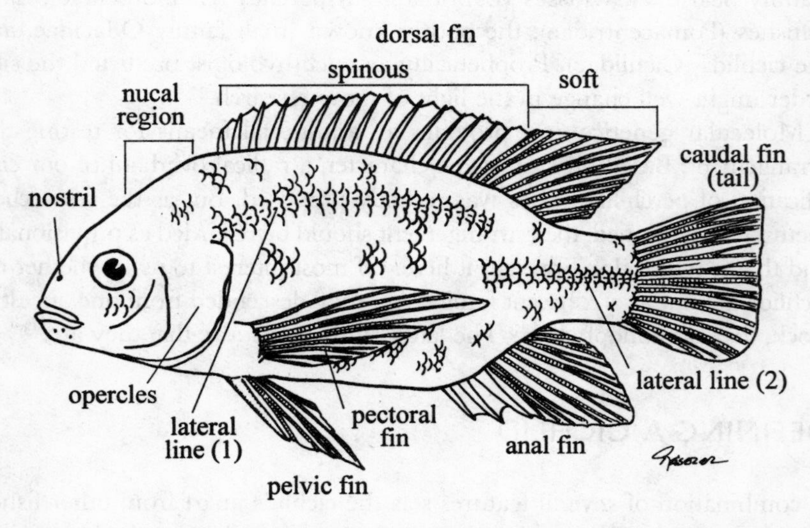

FIGURE 1.7 The major external features of a generalized cichlid. The dorsal fin is joined as one, with a leading spiny and trailing soft-rayed portion. Only one nostril is present on each side of the snout; the lateral line has two branches and is said therefore to be interrupted.

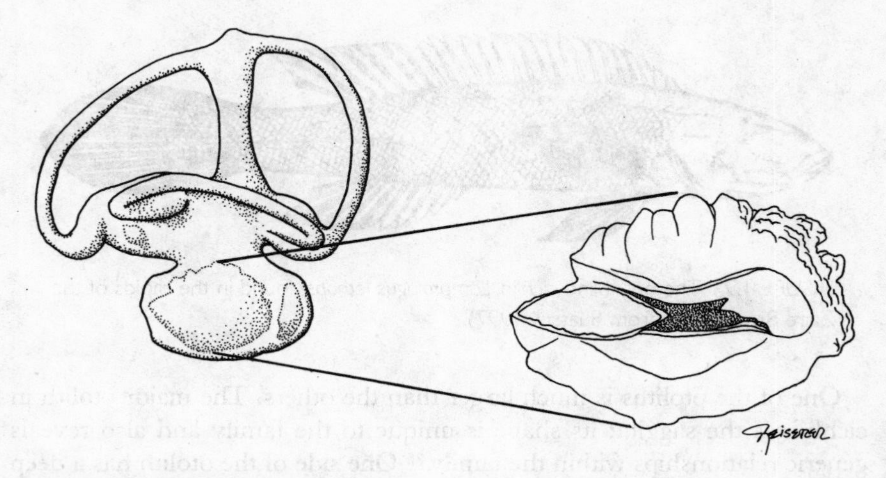

FIGURE 1.8 To the left are the semicircular canals that provide sensory information for balance. At the base of each of the three canals is a tiny bony element, the otolith. The largest of the three otoliths, the saggita, is shown, enlarged, to the right. The deep furrow in the saggita, recognizable as a dark central area, is characteristic of the Cichlidae (after Figure 1, Gaemers 1984).

Although cichlids share with other families of fishes the single opening of the nostrils and the interrupted lateral line, they are the only family of freshwater fishes that has both of these traits. Therefore, if you see a freshwater fish with that combination, you can be sure it is a cichlid. Cichlids do have some features not shared with any other family, but they are seen only by dissection. One such distinctive feature is found in their ears. Fishes do have ears, but they do not open to the exterior like ours, and so they are not visible. The ears lie right where you would expect them, toward the back of the head, a bit higher than the eyes.

The ear proper of teleost fishes consists of three looping membranous canals, each loop lying in a different plane. At the base of each canal is a tiny bony disk, and these disks are called *otoliths*. Only a few millimeters long in adult fish, each of the three otoliths consists of a dense mix of argonite crystals in a mesh of a protein called otoline. Because the otoliths are much heavier than water, and the tissues of the fish, they are the source of a mismatch with the fish itself and lag behind the movement of the surrounding tissue. The otoliths rest on a bed of sensory hair cells, much like the hair cells in the lateral line. When the otolith moves relative to the rest of the ear, it deforms the hair cells, causing them to send neural impulses to the ear proper. The parallels with our ears are impressive.

FIGURE 1.9 The only blind cichlid, *Lamprologus lethops,* found in the rapids of the Zaire River, Africa (from Stiassny 1997).

One of the otoliths is much larger than the others. The major otolith in cichlids is the saggita; its shape is unique to the family and also reveals generic relationships within the family.[25] One side of the otolith has a deep furrow, the function of which is unknown.

Another odd feature found exclusively in cichlids is the way the stomach connects to the small intestine.[26] In most fishes, the intestine departs from the stomach from its side, not its far, or rear, end, and the intestine exits on the right side of the stomach. Cichlids differ in that the intestine departs from the left side of the stomach.

The cichlids are something of a morphological paradox. On the one hand they are incredibly diverse. Their body shape ranges from tubular, to that of a typical perch, to that of a disk. An extreme modification is found in *Lamprologus lethops*, a tiny cichlid confined to the rapids of the lower Zaire River in Africa. It has a cylindrical body and a flattened head, small scales, and totally lacks pigment, rendering it pink. In addition, it is the only blind cichlid.[21]

Nonetheless, in the face of this versatility their basic body plan is conservative. By conservative I mean that the fundamental morphology–the position of the fins, the nature of the scales, the way the parts of the jaws are arranged–has been little altered. Thus what seem to be major changes in appearance have evolved with little alteration of the basic plan.[11]

The most profound differences from other kinds of fishes are associated with the external jaws and the throat, which I will explain in the next chapter. This relative uniformity of the basic body plan makes it hard to work out relationships within the family and explains why genetic studies are proving so helpful here. Sufficient morphological differences exist among them, nonetheless, to arrange them into genera and to group the genera within the family, though much remains to be resolved.

One other feature characterizes the family, though it has nothing to do with structure: All take care of their young. The variation on this theme is impressive, but in one form or another they all tend their offspring. I'll explain the evolution and diversity of this behavior further on.

WHERE AND WHENCE THE CICHLIDS

Unlike the other labroids, cichlids are predominantly freshwater fishes. They are one of the few perciform families that inhabits primarily freshwater rivers and lakes. Many cichlids, however, retain their ancestral ability to survive in marine water; some do this only temporarily, but others remain in marine water for long periods.[27] Some species regularly enter estuaries,[28] even where the water periodically becomes saltier than the ocean. And a few species have adapted to extremely harsh water conditions, such as to soda lakes in Africa.[29] The ability of some species to penetrate and survive in marine water when necessary has probably helped them colonize new freshwater environments.

The Family History

Whether a scientist or a hobbyist, people like to know when and where a group of animals originated and how they reached their current state. This is done typically by examining the present geography and relating it to the distribution of landmasses millions of years ago.

Early in the history of the earth, land coalesced into one huge continent, Pangea, which later gradually fragmented into separate constituent continents. The northern one, Laurasia, subsequently further fissioned into North America, Europe, and Asia. The family Cichlidae apparently arose

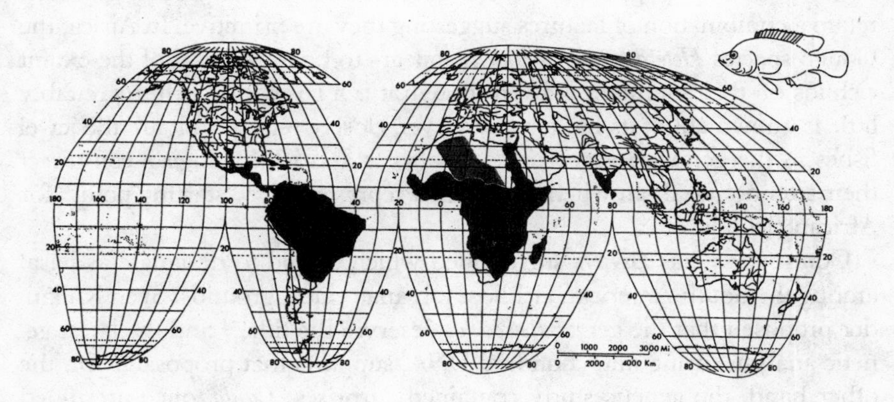

FIGURE 1.10 A world map revealing, in dark, the distribution of cichlid fishes (from Berra 1981).

in the southern landmass, Gondwanaland, after the separation from Laurasia. This suggests an age for the family of up to 160 million years before the present.[30] If so, they shared the earth with the dinosaurs during the Jurassic or early Cretaceous.

Around 150 million years before the present, Gondwanaland commenced fractionating. First, the subcontinent that was to become India and Madagascar slipped away from what is now southeastern Africa, taking with it primitive cichlids. Eventually those two landmasses drifted apart, leaving the genus *Etroplus* in India and several species of related cichlids on Madagascar. The species in those two areas remain more closely related to one another than to other cichlids, even though Madagascar is now physically so close to Africa. These species are thought to be the most basal of all existing cichlids in the family tree.

Later, around 65 million years in the past, a rift developed as Africa and South America slowly edged away from one another with the newly created Atlantic Ocean filling the void. Although clearly in the same family, the cichlids in each of those areas now constitute two separate lineages.[31]

That model for the origin of cichlids is controversial.[32] The problem lies with the fossil record, or more correctly, the lack of one. The oldest known fossils for cichlids are all in the neighborhood of 30 million years old. Further, the entire Percomorpha are thought to have originated after the Cretaceous, and that is because no older fossils have been found. If the cichlids originated in the Cretaceous, as claimed, then they would be older than the entire Percomorpha. Obviously, something is wrong here. No one knows the answer, but I am inclined toward the model of continental drift and suspect that older fossil cichlids, and those of percomorphs, remain to be found.

However they originated, cichlids continued to evolve. Some, however, retain a combination of features suggesting they are primitive. In Africa, the Congo species *Heterochromis multidens* appears to be most basal of the extant cichlids on that continent.[33] Even though it is a large food fish, regrettably little is known about its biology. However, closely related to it are the jewel fishes in the genus *Hemichromis*, so hobbyists and ichthyologists can direct their attention to them when they seek a representative "starting point" for African cichlids.

Controversy has always surrounded which cichlid to designate as basal among the South American cichlids. On anatomical grounds, Sven Kullander proposed that the genus *Retroculus* deserves the title,[34] and the latest genetic analysis, using mitochondrial DNA, supports that proposal.[35] On the other hand, the genetic study contained surprises. *Cichla,* once presumed the most primitive of the Neotropical cichlids, has traditionally been considered closely related to the elongate piscivores in the genus *Crenicichla*. But now *Crenicichla* is placed within the large, diverse assemblage of earth eaters, the geophagines, whereas *Cichla* is seen as allied with the popular Oscar, *Astronotus ocellatus.*

The Character of Cichlids

You may be wondering how much this really tells you about what a cichlid is. If you are a hobbyist, probably quite a bit. If you are the general reader, you probably prefer to learn about what they do. Indeed, what sets cichlids apart in my mind is their complex behavior. Many seem to have personalities, and those who keep them as pets come to view them as intelligent. But they are not for everyone.

Cichlids have a reputation for being aggressive and destructive, and many are. The more typical species of moderate to large size can be downright rambunctious. Hobbyists either detest them or love them. Those who love them, not surprisingly, tend to specialize in keeping cichlids because they require tolerance on the part of their keeper and the ability to adapt to the cichlid.

Not all cichlids are excessively aggressive, however. Some are relatively docile, such as the angelfish, and that partly explains their popularity. But even an angelfish is apt to grow too large for the home aquarium and to devour tidbits like guppies and neon tetras, to the dismay of the aquarist. But I don't want to anticipate too much of what is to come. This book is about cichlids and what they do, and that takes a lot more telling.

≈ *two* ≈

JAWS TWO

I admire creative horror movies. I am bored, however, by such movies when the monster is little more than an exaggerated version of some creature that already exists, though it worked for me when I was a boy. I was frightened out of my wits by a cine spider the size of a truck. The makers of the movie *Aliens* dreamt up a monstrosity that is creative: It differs from anything ever seen on earth. And it is scary, to say the least.

The unnerving property of the monster is the deployment of its jaws: First one pair of shark-like dentures bursts from its mouth, and then another pair springs from inside the first. In the same way more sets of jaws successively pop out of the preceding pair, like multiple cuckoos from a clock. The thought that instantly grabbed me, however, was less horror than comparative biology. The jaws of the *Alien,* in principle, had already been invented by fishes.

Fishes have at least two sets of jaws. One is the obvious outer pair that is plain to see. The other is an invisible set of jaws in the throat. The reason for this arrangement is not immediately obvious, but recent studies of cichlid fishes have been of help here.

In trying to explain why cichlids have been so successful, in an evolutionary sense, investigators were drawn to the function of their jaws and teeth. Cichlids have evolved important advances in the way their jaws work, especially the inner pair, and that sets them apart from most other kinds of fishes. Comprehending how their jaws operate has led to a better overall realization of how the jaws of fishes function. It has also helped us understand how fish jaws and their teeth relate to feeding behavior. And getting food is fundamental in the biology of all animals.

Competition for food is intense among most animals. People commonly think of such competition as among species but actually members of the same species are typically the most important competitors. The reason is simple: All members of the species are adapted to feed in the same way on the same things. The message here is that the process of natural selection focuses on the individual, not on the species. Further, in that context, all evolved adaptations have but one goal, to reproduce. To do that, they have to avoid being eaten, but because the antipredatory behavior of cichlids is, for the most part, unexceptional I will deal with it only briefly, further on. (Incidentally, predators on cichlids are often other cichlids.)

In this chapter, I describe initially the wide range of ways that cichlids feed, going into more detail over some of the unusual and remarkable feeding habits. After that, I take up the tools that cichlids use in feeding because that is one of the most important clues to understanding their remarkable propensity to evolve new species. I proceed in this order because I want you to see first what they have accomplished trophically. That should help you appreciate the importance of their trophic apparatus in obtaining this diversity of food.

If a cichlid is to grow large enough to reproduce it has to capture food. Thus it must obtain enough nourishment not only to reach adult size but also in sufficient excess to manufacture eggs and sperm and to carry out the behavior of reproducing, both of which require enormous energy. The structure of a cichlid fish, especially its mouth and jaws, is adapted to what it eats and how it obtains and processes it. So let's take a look at the various diets of cichlids and then at the morphology that makes such feeding behavior possible.

THE MENU

Cichlid species are numerous, and they feed on a remarkable range of edibles. Almost all of them are diurnal; that is, they feed during the day. A few species, however, find their food at night or during dawn or dusk. In Lake Tanganyika, which has a great diversity of cichlids that have evolved there over millions of years, some species are nocturnal feeders. For instance, the small predators *Neolamprologus toae* and *Lamprologus lemairii* go on the prowl when it turns dark,[1] and deepwater species of the genus *Trematocara* move into the shallows to feed at night.[2] A popular species among aquarists, *Cyphotilapia frontosa*, feeds actively during crepuscular periods. Some of the plankton-feeding species have become specialized to foraging either at dawn or dusk, for example, *Paracyprichromis*. For the most part, nonetheless, cichlids are up and about, finding their food, during the day.

I categorize the prevailing modes of feeding for the purpose of exposition, but unfortunately the fish don't always cooperate. Individuals often feed in more than one way, and some of the categories are inherently overlapping.

Food That Floats

Virtually everything that is edible and small enough to be eaten in the aquatic realm is consumed by one species of cichlid or another. Some feed on tiny planktonic algae, the phytoplankton. Those cichlids swim through the water, often in a school, filtering out the algae. Other species specialize on larger, but still tiny, planktonic animals, the zooplankters; those species typically form loose aggregations with each individual feeding on its own; they bob up and down as they snap up individual plankters borne toward them by the moving water. Zooplankton must be an especially desirable type of food because all kinds of cichlids that specialize on other types of food abandon their usual diet and take plankton when it is richly available.[3]

Cichlid Vegetarians

I often encounter people, occasionally even biologists, who believe no fish eats plant matter. Several kinds of fishes do, however, such as surgeon fishes, parrot fishes, and blennies. Most, but not all, such species are confined to the tropics. Understand that herbivory in this instance refers almost entirely to lower plants, the algae. At higher latitudes, algae is leafy and luxurious in its growth, as exemplified by giant kelp and numerous smaller plants.

In the tropics algae typically grows on rocks as a thick fuzz that is tough and filamentous, like a month-old set of whiskers on a man's face. Unlike a short beard, one hopes, it is also a habitat for all kinds of other organisms, including tiny crustaceous and worm-like animals. Diatoms also prosper there; these tiny unicellular organisms are neither plants, though they contain chlorophyll, nor animals, though they can move.[4] Because of this mix of organisms in an algal carpet, the growth is known by its descriptive German name, *Aufwuchs,* though most of the time I'll just refer to it as algae or algal growth.

To harvest algae and diatoms, the many species of reef-dwelling cichlids press their mouths against the *Aufwuchs* and rasp it off. For this, they have specialized jaws and teeth. The shapes of their teeth, the ways their jaws are positioned, and how the jaws themselves are shaped, varies among species in accordance with their diets. Even within the algivores, different styles of feeding have evolved. These have been broken down into grazers, who comb unicellular algae and diatoms from threads of attached filamentous algae; browsers, who rip off bits of algae from the rocks; scrapers, who rapidly rasp off tough algae in bouts of rapid scrapes with their mouths; and tappers, who entangle filaments of algae in their sharp teeth and then jerk the algae loose.[5] In the process of scraping algae, the herbivores ingest a multitude of tiny animals on the algae such as diatoms, bryozoans, and tiny crustaceans. Thus they are not purely herbivores.

This diversification of feeding modes reflects the intense competition for the *Aufwuchs* coating the rocks. Many species of cichlids literally defend their turf, their feeding territories. Other cichlids, less aggressive and more opportunistic, roam about in schools seeking feeding opportunities. Such schools commonly consist of several species. They invade feeding territories en masse, overwhelming the territory holder and plundering his larder.[6] I observed precisely the same behavior among algal-feeding surgeon fishes on the coral reef.[7]

Still other cichlids consume leafy algae or higher aquatic plants. To do that, they tear off chunks of vegetation. In some situations, certain river-dwelling cichlids in South America even devour fruit or seeds that fall from the forest.[8]

Because herbivores accidently eat small animals, and sometimes purposely prey on them, they might be thought of as omnivores. However, feeding on plant material requires special modifications of the teeth and of the gut, so they are best treated as herbivores.

Not a Gourmet Item

Sponges are not plants, but I bring them up here because they present a situation superficially similar to that of algae. In the marine environment, sponges often carpet the bottom, particularly in the tropics. Only a few fishes consume these animals, however,[9] because to ingest sponge is like eating cardboard laced with indigestible fibers, ground glass in the form of siliceous spicules, not to mention noxious chemicals.

Just a few cichlids devour freshwater sponges, but as time passes, more examples are being found. Two sponge-eating species occur in Cameroon, West Africa. One of these, *Pungu maclareni,* lives in the crater lake Barombi Mbo.[10] (This cichlid's specific name, one of my favorites, honors "that good naturalist, Mr. P.I.R. Maclaren, who used his opportunities as fisheries officer to add to the collections of the British Museum. In 1957 he met his untimely death as the result of an encounter with a crocodile."[11])

Two other cichlids, *Tilapia spongotroktis,* from the Cameroon crater lake Bermin,[12] and the popular *Julidochromis marlieri* of Lake Tanganyika,[13] eat mostly sponge. Sponge spicules turn up in the guts of a few omnivorous species of cichlids, most likely ingested incidentally from bottom debris or from browsing microorganisms or algae on the surface of sponges. I suspect more examples of sponge-eating cichlids will turn up.

Meat on the Table

Carnivorous species ingest a variety of prey, but to most the term connotes eating other fishes, that is, being piscivorous. Such behavior prevails in the larger cichlids such as the peacock cichlid, *Cichla ocellaris,* in South America and *Boulengerochromis microlepis* in Africa. Some are adapted to running down their prey, whereas others, more typically, have evolved ambushing; they lurk quietly where they are difficult to detect. Two species in Lake Tanganyika accompany a larger herbivore, using it as a mobile ambush site to get closer to their prey,[14] behavior paralleled in marine reef fishes.[15]

Species that pick from the bottom, however, commonly do not dine on fish. Some of them specialize on snails, others on shrimp, yet others on soft-bodied aquatic insects, or insects that drop into the water. Still other species have even more pointed snouts and small mouths, an adaptation for picking tiny crustaceans such as amphipods out of crevices.

Leftovers

When I first learned that some fishes are specialized to feed on detritus, and that was many years ago, I somehow found that implausible. It was just one more time when I was out of step with nature. Among the detrital-feeding fishes are several species of cichlids who sustain themselves from detritus that they sift from the bottom. The typical detritivore plunges its pointed snout into soft bottom, pulling up a mouth full of mud, sand, or debris. Or they may suck up debris that falls into cracks on the reef;[16] then a frequent morphological adaptation is large puffy lips, which apparently act as a gasket to the mouth around the cavity as the fish sucks out its contents,[17] though others have given different interpretations of the function of fat lips.

The edible particles, such as organic detritus, tiny clams, snails, and insect larvae, are sifted out and the inedible material is rejected. Thus the so-called detritivores devour much more than detritus. The geophagines of South America are outstanding examples of this mode of feeding. Geophagus literally means earth eaters. Detritivores are also common among African cichlids.[18]

Dependence, Thievery, and Collaboration

Fish species that pick prey from the bottom have to be versatile and opportunistic. On coral reefs, one often sees them following larger fishes that stir up the bottom, incidentally flushing out small quarry for the small predators,[19] much like cattle egrets following cows to grab the insects the cow scares up. Several species of cichlids in the Great Lakes of Africa similarly depend on a second, usually larger, species of cichlid to stir up prey for them.[20]

Six of the eleven cichlids native to Lake Barombi Mbo, also in Africa, carry this behavior a step further. They are opportunistic kleptoparasites, stealing prey from other animals as do mammals and birds, such as the parasitic jaeger. Because fishes generally engulf their victims whole, they rarely engage in such thievery. But these cichlids lurk near feeding crabs, who rip up the bottom with their claws. When a small insect larva is exposed, a cichlid darts in and tries to pirate it.[21]

Yet other cichlids, for instance, *Limnocaridina latipes* in Lake Tanganyika, feed on shrimp in a way that suggests collaborative hunting.[22] Still other species have what seem to us to be bizarre specializations.

Give Me Thy Parasites

Symbiosis, wherein two species interact for their mutual benefit, has drawn considerable attention among coral-reef fishes, especially in the case of cleaning symbiosis. It is a strange and, to humans, repulsive form of feeding. The cleaner gets nourishment from the parasites it removes and devours, and the serviced fish benefits from the removal. By convention, the fish that is cleaned is called the host. In many instances, the cleaner also eats necrotic flesh from wounds on the hosts.

Some species of coral-reef fishes are highly specialized to clean host fish that seek them out at what are called cleaner stations. Apparently to advertise their occupation, coral-reef cleaner fishes are typically brightly colored–even unrelated species are often decorated in the same way. Favored color patterns are contrasting stripes, especially black on yellow or blue.

The host fish characteristically pose before the cleaner. In this behavior, they spread their fins and tilt out of their normal plane, such as head up or head down. The hosts' colors also often change. Apparently, ectoparasites match the usual color of the skin of the host; by changing color, the host makes the parasite easier for the cleaner to see. These cleaner fishes, typically gobies (family Gobiidae) or wrasses (Labridae), may even enter the mouths of large hosts such as groupers, looking for a meal without becoming one for the grouper. Commonly, however, species are not highly specialized as cleaners and engage in that behavior only when young.

The host-cleaner relationship is seldom seen among freshwater fishes, though it has been observed sporadically in a variety of species such as the American sunfishes and the threespine stickleback. But more and more species of cichlids have been discovered to engage in cleaner-host relationships. This relationship is not as highly developed as among coral-reef fishes, however.

The first report of cleaning in cichlids was of one angelfish, *Pterophyllum scalare,* cleaning another, in an aquarium setting, and it was brief.[23] Most of the cleaning was confined to juveniles cleaning one another. Typically, an angelfish stimulated to be cleaned approached another and presented its side to the other's face. It raised its head slightly, bent its body toward the would-be cleaner, and trembled its median fins. The angelfish in the cleaner role then nipped at the side of the fish posing before it. What was eaten was not clear, but one fish that posed had a white infection, probably fungus, on its side, and that was picked at by the groomer. As the angelfish grew older, the behavior largely disappeared. Thereafter, it occurred only between mated fish.

Rick Wyman and Jack Ward were the first to report in detail interspecific cleaning in a cichlid.[24] Juveniles of the Asian cichlid *Etroplus maculatus,* the orange chromide, have specialized in cleaning their larger cousin, *Etroplus suratensis,* the green chromide. Wyman and Ward saw that when infected with patches of a skin fungus, individuals of the larger species, acting as hosts, approached the smaller species, darkened their coloration, and posed. Obviously, they were inviting the small species to clean them. The posing green chromides stood almost vertical in the water, head up. They spread the median fins and flickered their black pelvic fins. At the same time, the body quivered slightly.

Typically, a small individual of the orange chromide recognized this signal right away and swam to the larger posing green chromide. There it inspected the body of the host fish and bit at it. The green chromide positioned itself to offer the infected part of the body to the orange chromide. Its dark color also contrasted with the whitish fungal threads and helped the orange chromide find them. Within ten days, all the green chromides in the large aquarium had been rid of their infection.

Wyman and Ward theorized that this cooperative behavior had evolved from the signals directed by parents to their young. When guarding their young, known as fry, the parental green chromides assemble them by flickering their pelvic fins and quivering slightly (orange chromides do the same, but without the quivering). The fry ordinarily eat from their parents' skin, so all the elements are there to produce the cleaning behavior.

Cleaning by cichlids in African lakes has been reported but not studied in such detail. One early example from Lake Malawi is the cichlid *Melanochromis crabro.* It apparently plucks a branchiuran ectoparasite, *Argulus africanus,* from the skin of a catfish *Bagrus meridionalis.* Cleaning behavior, however, was never directly observed.[25] The conclusion was inferred from finding the parasites in the stomach of *M. crabro,* along with algal matter that this cichlid regularly eats.

The coloration of the cichlid also suggests some element of specialization for the cleaner role: The name *crabro* comes from Latin for hornet, referring to starkly banded yellow and brown coloration, suggesting some degree of convergence with cleaner coloration among marine fishes. In addition, the dentition of *M. crabro* is well adapted for removing ectoparasites.

More recently, Jay Stauffer discovered another cichlid cleaner in Lake Malawi.[26] Like *M. crabro*, *Pseudotropheus pursus* feeds mainly on *Aufwuchs,* but unlike *M. crabro* the supplementary ectoparasites it gets come from the surface of other cichlids. It also manages to eat some scales from its hosts, and cleaning and eating bits of the host have much in common.

Scale Rippers

The line between cleaning a host and attacking it is a thin one, even among marine fishes. Another cichlid in Lake Malawi, *Docimodus evelynae,* cleans fungal infections on the skin of other species of cichlids in shallow water. But as *D. evelynae* grows, it moves to ever deeper water, where instead of cleaning other fishes it attacks them. It bites out chunks of flesh from the large scaleless catfish *Bagrus meridionalis,* as does the only other cichlid in this genus, *D. johnstonii.*[27] *D. evelynae* also rasps scales off of other cichlids, digesting the attached skin.

Ripping scales and fin membranes off other fishes is a widespread behavior called lepidophagy (Gr. *lepis, -idos* = scale; Gr. *phagein* = to eat).[28] Certain coral-reef blennies are well known for feeding this way.[29] While diving on coral reefs, I have often felt one of these tiny blennies bouncing off my neoprene diving suit as it tried in vain to bite out a piece of my hide.

In 1947, the perceptive pioneer in the study of African cichlids, Ethylwin Trewavas, observed some unusual scale-eating behavior among some cichlids in Lake Malawi.[30] The scale eater has a broad file-like band of teeth in each jaw that is well adapted to rasping scales off the tail of its quarry; the guts of these fish were full of cichlid scales. She also noticed that the color pattern of the scale eater *Corematodus shiranus* "is exactly similar to that of females and non-breeding males" of *Oreochromis squamippinis.* Trewavas hypothesized that the infrequent scale eater engages in a sort of mimicry, joining schools of its victim; through its resemblance, it is able to get close enough to rip off scales. Another Lake Malawi cichlid, *Genyochromis mento,* is also a lepidophage.[31]

As expected, similar scale-rasping was discovered in cichlids in Lake Tanganyika.[32] Most studies have concentrated on *Perissodus microlepis,* though several species have the same feeding habit. *Corematodus shiranus,* for instance, lurks over the bottom and sometimes ambushes its victims, darting out of a hiding place to snatch a mouthful of scales. The fishes it attacks are frequently those that graze on the bottom and typically are flat-sided fishes.

Muderhwa Nshombo observed a remarkable bit of behavior by a female lepidophage, *Plecodus straeleni,* in Lake Tanganyika, Africa.[33] She entered the volcano-like nest of another species of cichlid, *Cyathopharynx furcifer,* while its male owner was hovering well above the nest. She began circling in the nest, as though she were a female of his own species in the act of spawning. Apparently her behavior and appearance were sufficiently similar to the spawning behavior of the female of the male's species that he descended and joined her. At that moment, the female attacked the male, tore out a meal of scales, and sped off.

The most highly developed lepidophages are asymmetrical to improve on this mode of attacking.[34] Among the several different families in which this mode of feeding has evolved, only the seven Lake Tanganyika cichlids in the genus *Perissodus* are so specialized that their members have their snouts bent right or left to varying degrees. The bias is genetically set (one gene, two alleles), and the crooked snouts can be seen in the fry. Remarkably, the attacker can launch a strike only toward the side to which its jaw is bent.[35]

An algae-grazing cichlid in that lake, *Telmatochromis temporalis,* has also been shown to have asymmetrical jaws. When eating, it opens its mouth to the right or left, though the direction of the skewed mouth and the side from which the fish feeds do not necessarily correlate, so the significance of this morphology remains unknown.[36]

Sometimes a fish crosses the line between lepidophagy and cleaning, both on the coral reef and among African cichlids. Jay Stauffer found that the occasional cichlid cleaner *Pseudotropheus pursus,* in Lake Malawi, had not only parasites in its gut but also scales of other cichlids.[37] And Tony Rib-

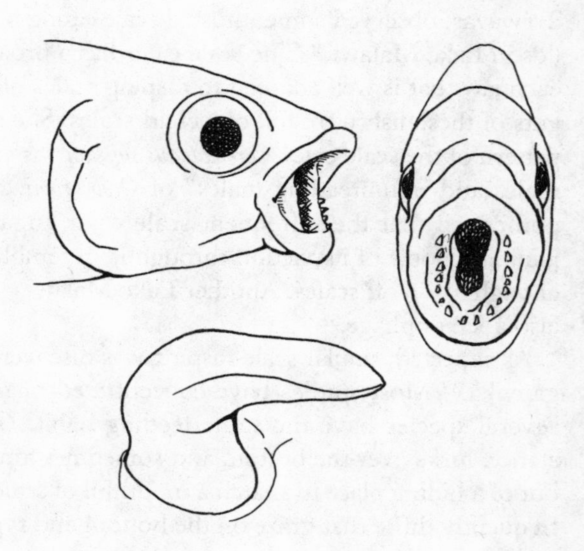

FIGURE 2.1 Two aspects of one of the scale-eating cichlids, *Perissodus eccentricus,* of Lake Tanganyika. The upper two images show the way the snout and mouth are bent slightly to the right. Below them is one of the strange teeth in the outer jaw, an adaptation to rasping scales.

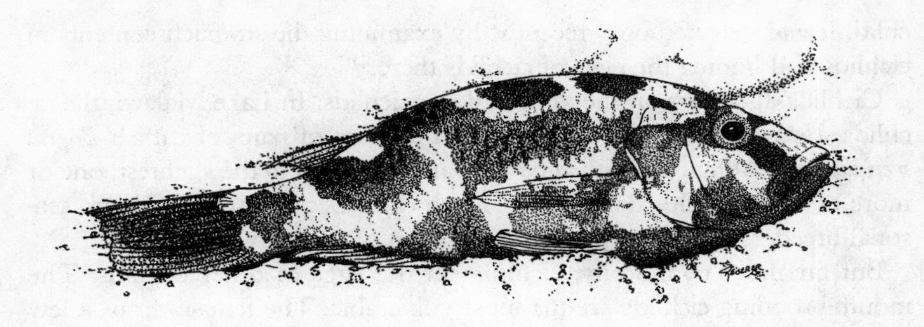

FIGURE 2.2 A fish predator, *Nimbochromis livingstonii,* of Lake Malawi, lying motionless on the bottom. The mottled coloration resembles that of a dead fish and thereby entices small fish scavengers to approach so they can be captured.

bink reported that another cichlid species in that lake, *Docimodus evelynae,* changed from one mode to the other as it grew.[38] When it was a young fish it cleaned fungi from the skin of other cichlids. Later in its life history it ate plankton, insects, and periphyton. As an adult, however, it fed on the scales and fins of other fishes.

Not Quite Dead

One piscivorous cichlid in Lake Malawi, *Nimbochromis livingstonii,* employs a macabre style of feeding that could be considered the reverse of cleaning behavior. This species feigns death by lying on its side on the bottom. To increase its resemblance to a dead fish, it changes color, becoming blotched with dark and light areas, as though beginning to decompose.[39] Small fish move in to eat from the surface of the "corpse," much as a cleaner eats from the surface of the host. However, here the small fish becomes a meal for the corpse-mimicking predator. This mimicry of a corpse by a predator to bait its prey was unique among vertebrate animals when first reported, but now another and unrelated cichlid, *Lamprologus lemairii* in Lake Tanganyika, has been found to do the same thing, a remarkable example of convergent evolution.[40]

Cradle Robbers

Perhaps even more intriguing is the eating of fish eggs, called oophagy (Gr. *oon* = egg); it also goes by the more inclusive term paedophagy (Gr. *Pais, paidos* = child), which is meant to include preying on eggs and larvae. Eating the young of other fishes is commonplace among fishes. The term has a more specific connotation for cichlids, however, in that it refers to the exploitation by one species of cichlid of the parenting behavior of another cichlid. It was detected only recently, by examining the stomach contents of cichlids and finding the eggs of cichlids there.[41]

Cichlids also prey on the young of noncichlids. In Lake Malawi, the cichlids *Melanochromis crabro* lurk around the nests of pairs of catfish *Bagrus meridionalis.*[42] When the catfish parents are distracted in the slightest, one or more of these cichlids dart in and snatch eggs. Because the catfish are seasonal breeders, this is an opportunistic form of oophagy.

But far more commonly, cichlids eat the eggs of other cichlids. The mouth-brooding cichlids are the most vulnerable. The female drops a few eggs in the male's exposed nest, and they lie momentarily on the bottom where the male fertilizes them. After spawning, the female swims off with the eggs in her mouth. A female carrying eggs stands out because her cheeks are swollen as though she has the mumps. The developing eggs remain in her mouth for up to weeks and would seem to be safe from predators. But the puffy-cheeked female is obvious to human observers and so probably to the predators as well.

In Lake Malawi, Ken McKaye and Eric van den Berghe have observed predation on the eggs of some cichlids by a number of other species of cichlids, and they studied one case in detail. The spawning cichlid was *Otopharynx argyrosoma;* it nests in large colonies over a sandy bottom. The egg-eating cichlids were *Otopharynx ovatus, Protomelas insignis,* and *Trematocranus labifer.* Two of the oophagous species defended their raiding sites over certain nests, but the other, *T. labifer,* roved about in the breeding colony.

Each species took eggs in much the same way. The predator watched as a pair started to spawn, then slowly descended, aiming at the nest. The moment the female released eggs, and before the female could pick them up in her mouth, the egg eater darted in and snapped up some of the eggs. The nesting male tried to drive the predator away, but the predator was successful about half the time.

The persistence of the egg predators, and their payoff, is impressive. Approaching was seen about every three to four minutes, and about 5 percent of the approaches progressed to darting into the nest. The eggs are large, by fish standards, being about 3.4 mm in diameter. On average, a predator consumed one to three eggs per hour, which comes to about 0.5 grams per hour. For a small fish, that is a substantial meal of highly nutritious eggs. Seen from the nesting pairs' perspective, they lose about one out of thirty-five to sixty eggs spawned. That amounts to a fairly strong selective pressure to prevent the eggs from becoming another cichlid's omelette.

At least one cichlid in Lake Malawi, *Champsochromis spilorhyncus,* has apparently evolved another tactic for paedophagy. I say apparently because this has been seen only once, though it involved two predators. This paedophage takes advantage of the very behavior of the mother that would most seem to protect her young, the act of calling the young to her and taking them back into her mouth. Two *C. spilorhyncus* approached a mother, *Mylochromis sphaerodon,* with a school of young who were feeding on plankton up off the bottom. The paedophagous fish repeatedly approached the mother, and she became increasingly agitated. As she did so, the young sank to the bottom and formed a compact group. The female then hovered over them and performed calling movements, signalling them to enter her mouth. As they quickly queued up, one of the predators zipped in beside the mother and gobbled up a large portion of the school. This scenario was repeated three times.[43]

The egg predators in Lake Malawi are committed to paedophagy. They can do this because eggs and young are available year-round. In the New World cichlids, comparable egg predation was unknown until just recently. Observing underwater in Lake Xiloá, Nicaragua, McKaye and van den Berghe were struck by the similar but less-developed oophagy in the Midas cichlid, *'Cichlasoma' citrinellum.* In a typical scenario, a single Midas cichlid defended an area over the nest of the large piscivorous cichlid *'Cichlasoma' dovii*, driving away other Midas cichlids.

'Cichlasoma' dovii is not a mouth brooder but is, rather, what is called a substrate brooder. The adhesive eggs are deposited on the bottom in a tight patch. When the eggs hatch into helpless larvae, called wrigglers, the mother *'C.' dovii* guards them in a pit. As a consequence, both eggs and wrigglers are continuously exposed to possible predation.

A Midas cichlid over a nest attacked at about twice the rate seen in Lake Malawi, and it ate both eggs and wrigglers. McKaye and van den Berghe estimated that each successful attack resulted in a meal of about twenty eggs, each about 2 mm in diameter. With roughly ten successful bites per hour, and approximately 0.062 grams of eggs per bite, that comes to about 0.62 grams of eggs per hour. That figure is impressively close to the 0.5 grams of eggs per hour estimated for the oophagous cichlids in Lake Malawi.

The eggs of *'C.' dovii* need two days to hatch. During that time, the egg predators may remove from 40 percent to 100 percent of the eggs. However, oophagous Midas cichlids are relatively rare in the population. McKaye and van den Berghe found only about one per 1,000 square meters. They reasoned that the rareness of oophagy might have two causes. First, the eggs are available only during a breeding season, so the Midas cichlids cannot afford to specialize on a food that is available only part of the year. Second, the rareness might be associated with the risks involved. The mother *'C.' dovii* is a large, fierce protector. She can kill the much smaller Midas cichlid with a single slam. McKaye and van den Berghe found six dying or freshly dead Midas cichlids near nests of *'C.' dovii.* That level of mortality indicates stringent selection against oophagy in the Midas cichlid.

Because eggs and larvae are so nutritious and vulnerable, we should expect oophagy to be widespread, and indeed it is. More than twenty species of paedophagous cichlids are known from Lake Victoria.[44] It is also found among the cichlids in Lake Tanganyika. Nshombo has reported in detail how females of one scale-eating cichlid in Lake Tanganyika also prey on eggs in the nest.[45]

In this case, the victim is a mouth brooder, *Cyathopharynx furcifer,* that aggregates in colonies to spawn, and the predator is a lepidophage, *Plecodus straeleni.* The females lurk near the nest of a male *C. furcifer.* When a female *C. furcifer* lays her eggs there, the *P. straeleni* dashes in and smashes into the side of the spawning female; the female flees, and the *P. straeleni* engulfs the eggs. The male *C. furcifer* does his share. He attacks the egg eater, who often fights back, pulling off and eating scales from the male before being driven away. In fact, the male often chases away the other *P. straeleni* lurking nearby, some of whom suddenly wheel and rip off a scale or two from the male.

Most of us had assumed that once the eggs were in the mouth of the mother, now well away from the place where she spawned, they were safe. Wrong. Some cichlids in Lake Malawi recognize that the egg-laden female contains a sumptuous meal for them. These predators maneuver below the mother. Then, with a lightning-quick strike, they ram the mother in the throat. Embryos are spewed out and, of course, eaten at once. Sometimes the mother is killed in the process. Nature is not kind.

When Opportunity Knocks

I do not want to leave you with the impression that the various species of cichlids are all specialized or that the specialists are locked into a single way of feeding. Actually, even the specialists show considerable plasticity in their feeding. Species that regularly browse the *Aufwuchs* feed opportunistically on other food when it becomes easily available. For instance, at times zooplankton drifts in over the reef in dense clouds.[46] Then even fish that specialize on scraping algae rise from the reef into open water to harvest the rich bounty. This has been observed in both African and New World cichlids. As we will see later on, the jaw morphology of some cichlids has been shown to be capable of adjusting to many different types of food.

Feeding behavior within a species changes as the individual grows. As young cichlids become larger, they pass through several phases of feeding. While still guarded by their parents, the tiny fry are all predators on microorganisms such as zooplankton and small copepods on the bottom. Michael Gottfried started looking at the morphology of the Midas cichlid when he took my course in ichthyology.[47] That led, eventually, to a study of the developmental transition in the structure of the feeding apparatus of that species. The juveniles have relatively long, pointed heads with large but delicate jaws; they feed on mobile prey, including their siblings. As they grow, the head becomes deeper and blunt, and the mouth becomes small relative to the head. Consequently, the adults feed on a variety of poorly mobile prey and plants. To do so, they need a strong bite.

Often, the gut also changes during the development of cichlids and many other kinds of fishes. If the diet shifts toward herbivory, the intestine becomes much longer and consequently has to coil in the body cavity.[48] Clearly, the feeding behavior of the fish is tied to its morphology.

TOOLS FOR FEEDING

In our terrestrial world we are accustomed to thinking about feeding adaptations of birds and mammals, including ourselves. Sometimes extreme specializations are obvious, such as the tongue of the anteater and the filtering baleen of whales or, among birds, the needle beak of a nectivorous hummingbird and the ripping hook-like beak of an eagle. However, unlike birds and mammals, fishes do not have the luxury of limbs.

To appreciate the problem confronting a fish when feeding, imagine how you would feed in the wild if you were bound in a straitjacket, effectively removing your arms and hands. The straitjacket also binds your legs so much that you can cannot use them to manipulate food. You can move, but the only way to grab and process food is with your mouth. How well would you capture mobile quarry? How would you dismember prey that are too large to swallow? If you tried to feed on *Aufwuchs,* could your teeth scrape off the nutritious growth?

As we've just seen, a fish has to find, capture, engulf, and process a vast variety of foods. As you might expect, different species have solved this problem in different ways depending on what they typically eat. And different types of food can present radically different problems to solve. No group of fishes illustrates this better than the cichlids. Although the major features of their feeding morphology are shared with most teleost fishes, certain fundamental changes in their structure are radical advances compared with those same fishes.

Cichlids have made unrivaled advances in how they obtain and process their food. As a starting example, consider this anecdote described to me by a postdoctoral investigator in my lab, Ron Coleman. He was feeding live goldfish to a pair of highly piscivorous cichlids of the genus *Crenicichla.* Aquarists call them pike cichlids for their resemblance to pike. The female grabbed the first hapless goldfish head first, and it disappeared into her mouth. Wanting to feed the next goldfish to the male, Ron quickly released a second goldfish. But the female engulfed that goldfish, too. Ron could detect that her throat jaws were chewing up the first goldfish while holding the second in her mouth. A third goldfish was dropped into the aquarium in a final attempt to get one to the male. But that goldfish was also snared by the female, who held it in her protruded jaws. The female continued to masticate one prey, stored another in her mouth and held on to the last one in her distended jaws. How is that possible?

To explain, I will take up the various components of the jaws and associated structures. An appreciation of the feeding apparatus is central to understanding the versatility of feeding adaptations and the production of multiple species that are treated later.

Lips

The first part of the feeding apparatus is the lips. Just watch a goat or a chimpanzee feeding and you will see how the lips reach for food and embrace it. If you had no lips, the coffee running down your chin would certainly discourage you from frequenting a coffeehouse. Fishes have lips, too, but they are not equipped with special muscles the way mammals' are. They are little more than a strip of tissue in some kinds of fishes, notably the pursuit predators who just grab their prey out of the water. But lips are commonly fleshy and useful as gaskets around food objects, serving much the same function as your lips on a coffee cup.

Huge puffy lips crop up in a wide variety of bony fishes. Within the cichlids, such lips are seen in *'Cichlasoma' labiatum,* the so-called red devil, in Nicaragua, in *Lobochilotes labiatus* in Lake Tanganyika, and in several other cichlids of the Great Lakes of Africa. Such lips have also evolved in *Tilapia bakossiorum,* a new species found in tiny Lake Bermin of Cameroon, Africa.[12] Several explanations have been suggested for their form, such as sealing the mouth against an irregular surface, the better to suck up food there.

Teeth

Behind the lips lie rows of teeth lining the upper and lower jaws. A characteristic number of rows in New World cichlids is from four to seven, with the outer row having the largest teeth. The ancestral tooth shape is a modestly elongate cone culminating in a single point. Cichlids have produced many variations on this theme. The most frequently seen is teeth that are flattened like ours but that have cusps in various patterns. Unlike ours, we hope, cichlid teeth often seem stained, as if the fish were a habitual smoker: They are yellowish-brown with darker tips that sometimes seem almost red. As with sharks, teeth that are lost can be replaced.

Tooth shape reflects feeding habits. Conical teeth characterize predators and generalists. They are useful for grasping prey. Flattened teeth with cusps are associated with feeding on vegetative food, including *Aufwuchs.*

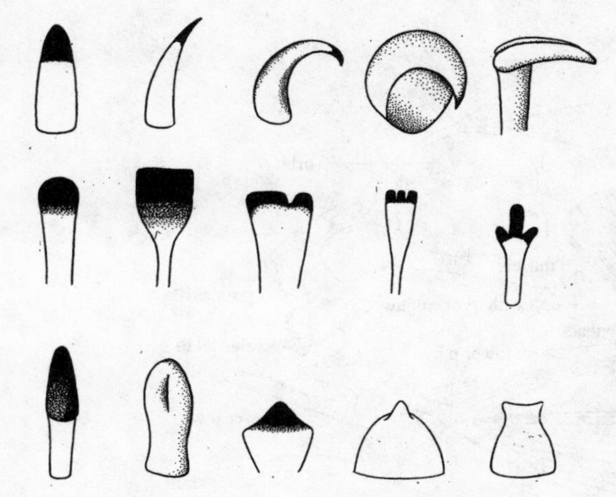

FIGURE 2.3 Teeth from the outer jaws of several different species of African cichlids, demonstrating the vast array of shapes that have evolved in cichlids with different feeding habits.

Some teeth in those kinds of fishes form a virtual brush on the jaw and are apparently adapted for combing diatoms off filamentous algae. The teeth of herbivores may also be inclined outward, as in buck teeth, and may have flexible connections to the jaws to allow them to spring back at the end of a rasping bite at the bottom.[49]

Rickety Jaws

From the example of the pike cichlid capturing three goldfish, you can see that the outer jaws and teeth of piscivores are for capturing and holding prey. You might think of them as the hands, or multiclawed paws, of the fish. And they do show dexterity and mobility crudely reminiscent of the paws of rats and raccoons. Consider that our upper jaw is fused to our solid head–all of a single piece. The only mobility is at the hinge of the lower jaw where it connects with the skull. Some side-to-side movement is possible for grinding, but the main action is a simple up movement to crush the food against the top jaw. Fish jaws are so different that any relationship to mammals seems impossible, especially when you consider the supporting structures.

The evolution of fish skulls, and their jaws, illustrates a trend in the evolution of the heads of higher vertebrates that may seem counterintuitive to most people, who tend to think of evolution pursuing a course of increasing complexity. Just the opposite has taken place with the skull and jaws. Enormous complexity has been replaced by ultimate simplicity.

The jaws of mammals and birds, for instance, are paragons of simplicity. As just mentioned, the skull is of a piece, including the upper jaw. Only the lower jaw is a separate unit. In comparison, the skull and jaws of bony fishes, as so well illustrated by a cichlid, look like a classic Rube Goldberg

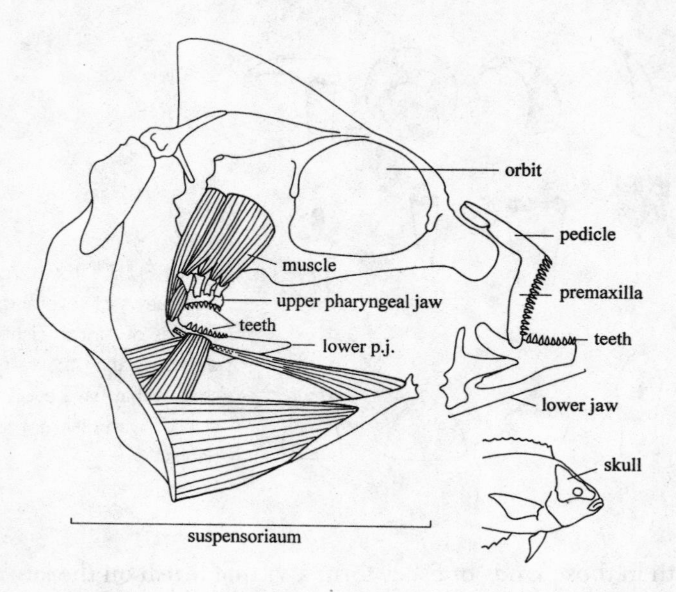

FIGURE 2.4 A cut-away drawing of the head of a cichlid. It reveals the small cranium, the elongate pedicle on the premaxilla (upper jaw), the sling of muscles supporting the pharyngeal jaws, and the area that contains the suspensorium (the scaffolding of the complex jaw apparatus) (courtesy of Melanie Stiassny).

contraption. The first thing that strikes the eye is that the solid part, the skull proper, is small and looks as if it is surrounded by an intricate cage of numerous, loosely to tightly connected delicate bones.

The elaborateness is needed because in the absence of prehensile limbs the cage serves two primary functions, capturing and handling food and pumping water across the gills. Thus the opercles, the cheeks of the fish, open as they draw oxygenated water into the area around the gills and close as they expel the water out the opercular slits. That coordinated action, the breathing of the fish, is a relatively elementary undertaking compared with grasping food, which is much more complicated.

Grab the Food

Only in recent years have we begun to understand how the complex arrangement of the jaws functions. A first step was sorting the ways various fish feed into three categories. Bear in mind that this is an oversimplification and that the different methods may grade one into another. In the first method, the fish just bites into the bottom or into a leaf; this is called

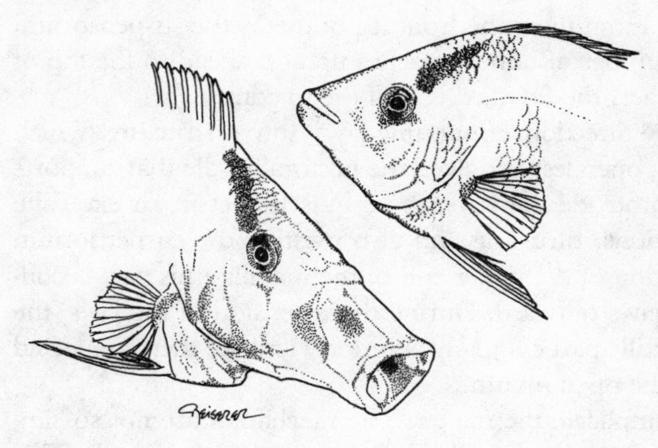

FIGURE 2.5 The South American cichlid *Caquetaia myersi,* showing at the upper right the normal position of the jaws and, diagonally at the lower left, the jaws when extended into a tube.

manipulation because the food object may be worked around, handled, by the jaws. In the second method, termed suction feeding, the fish, especially an ambush predator, is stationary and throws its outer jaws at the prey while simultaneously expanding its mouth cavity; that creates an inward jet of water, carrying the prey into the mouth cavity. The third method is known as ram feeding, in which the pursuit predator rushes at the quarry and opens its mouth to engulf it.

Some cichlids are pursuit predators, but most use a combination of suction feeding and manipulation.[50] This can be read from the morphology of their outer jaws, especially the upper one, which is the jaw that best reveals the refined function of the jaws. The upper jaw consists of four bones. In the center are two premaxillas that are joined in the middle, and they bear all the teeth. The jaw extends to the rear on each side as the maxillas, which are toothless. The maxillas broadly overlap the premaxillas, lying above them when at rest.

The special feature of the premaxillas is the pedicle, the long vertical protrusion that extends up from the jaw proper and is key to the ability to protrude the jaws. The pedicle, sometimes referred to as the ascending arm, lies in a groove running up the front of the head. The pedicle is usually longer than the horizontal, tooth-bearing part of the premaxilla in New World cichlids and is often so in the African species.

But how does a cichlid literally throw its jaws at prey? Frankly, I am not sure I fully understand how that is done, and the experts in the field are not totally in agreement. All agree that the cage of bones around the skull is the responsible structure. The upper jaw is connected in an almost devious way to the rear of the head through a V-shaped complex of bones. The back arm of the V, the hyomandibular (itself a complex of bones) extends down and slightly forward from the top, rear end of the skull to the corner of the upper and lower jaws. From there the front leg of the V, the suspensorium (also consisting of a number of bones) reaches up and ahead to the top of the skull where it contacts the forward end of the maxilla.

Muscles run in every direction, connecting jaws, throat structure (which I've not yet described), opercles, and even the pectoral girdle that supports the pectoral fins. To protrude the jaws, these muscles act in an exquisite symphony of contractions, thrusting and also twisting the suspensorium which, through the action of the upper end of the maxilla (this part is contentious), propels the jaws outward. During the extension of the jaws, the premaxilla and the maxilla part company anteriorly and are then connected only at the corner of the open mouth.

But then cichlids complicate the matter. The mechanics are not so simple. Cichlids can protrude the premaxilla without opening the mouth. This passive protrusion is done with little action of the suspensorium and connected members. Apparently the premaxilla can be protruded passively when the body muscles lift the head, though this action may be assisted by other muscles associated with the jaws. In the most extreme evolution of this adaptation, the jaws extend forward a distance about equal to the length of the head of the fish.

As the jaws start their outward journey, the mouth is still closed. At the same time, the mouth cavity expands, creating negative hydraulic pressure. When the mouth finally opens, the prey is sucked in. The jaws retract, throwing the prey into the mouth cavity, where processing begins. Actually, the mouth can be divided into two cavities separated by gill arches. The first one, the mouth proper, is called the buccal cavity; the second one, the opercular cavity.

As negative pressure builds up during the process of striking the prey, the pressure first drops in the buccal cavity. Then, in sequence, the pressure rises in the buccal cavity with the prey, but drops in the opercular cavity, to further transport the prey. This cycle is much simpler in those species that take small prey by picking at the bottom than in piscivores that use ram feeding.

As the example of the pike cichlid illustrates, prey do not always pass directly into the mouth but rather may require some handling before being engulfed. Karel Liem has studied this aspect of feeding in cichlids in elegant detail. Using cine X-ray film, he watched the movements of the array of bony elements inside the head of the fish as it manipulated prey. From that he could make inferences about how the muscles work.[51]

Liem found what may be the secret of the plasticity of feeding behavior of cichlids. They can move the outer jaws in a multitude of ways. For instance, the two sides of the jaws can move independently to accommodate the food item. And cichlids can hold back the lower jaw while extending the upper. That turns the opening of the mouth downward to grasp prey below them, and so on. The outer jaws, therefore, truly take on the properties of a simple hand.

The adaptability of the jaws may have something to do with their enclosed tongues. The tongue typically is attached for much of its length to the floor of the buccal cavity, though the tip is usually free and variously shaped. How the tongue is used does not seem to have attracted much attention from functional morphologists so far. However, Karel Liem described to me the feeding behavior of a cichlid, as seen on X-ray film, that made him suspect that the tongue and other parts of the mouth must play a role in situating food.

The fish he studied had positioned a snail between its pharyngeal jaws. But it could not crush the snail because the strongest axis of the shell was lined up with the direction of the bite. The snail was then returned to the mouth where the fish rotated it into a different position and then sent it back to the pharyngeal jaws. Individuals repeated this process until the snail had its weakest axis in the path of the bite. Rotating the snail in the mouth must be done in some part by the tongue, though the fish probably mostly uses water flow created by the mobile buccal and opercular cavities.

This use of currents in the mouth has given rise to the term aquatic (hydraulic) tongue; it is employed by many kinds of fishes and in different ways.[52] Puffer fish, for instance, blow a jet of water at the bottom to uncover food items. I have seen a cichlid, *'Cichlasoma' maculacauda,* feeding the same way.

The Gill Basket

Once past the jaws and on its path to the gut, the prey, whatever that might be, first meets the gill arches. Anyone who has cleaned a fish, or watched the process, has probably seen the gill arches that lie inside the head. A fish typically has three such arches. On the rear side of each arch, out of harm's way, are red structures resembling stacks of leaves; they are the soft delicate gills proper, and that is where the blood coursing through them exchanges carbon dioxide for oxygen.

On the inner side of each arch, cartilaginous tooth-like structures called gill rakers protrude into the chamber that receives the food. The gill rakers vary enormously in shape, depending on the diet of the fish. For instance, they are generously spaced, short, sturdy, and sharp in piscivorous species. On the other hand, in planktivores the gill rakers are numerous, long, thin, and closely spaced to filter out tiny food items. Sometimes the gill rakers also have true teeth on their inner surfaces, another adaptation to processing prey.

The function of the gill rakers is to start the processing of the prey, including separating the wheat from the chaff, so to speak. Inedible parts of the ingested substance can sometimes be seen dropping out of the gill slits of the fish, though some edible parts are also often lost there. This is particularly apparent in those species that take in mouthfuls of mud or sand, the gill rakers filtering out the desired morsels. Once in the opercular chamber, the second pair of jaws, the pharyngeal jaws, take over.

Jaws in the Throat

The pharyngeal jaws of cichlids and their relatives in the Labroidea set them apart from other perch-like fishes. Fishes in the Labroidea were originally called the pharyngognaths to draw attention to the distinctive nature of the pharyngeal jaws. Liem believes the pharyngeal jaws of cichlids are the critical feature that enabled them to evolve into so many species in the rift lakes of Africa.[53] For this reason, it behooves us to take a close look at pharyngeal jaws of cichlid fishes and see how they differ from the way such jaws work among other perch-like fishes.

The upper pharyngeal jaw lies at the top of the gill basket, where the gill arches meet. Similarly, the lower pharyngeal jaw is situated at the bottom of the gill basket at that convergence of the arches, and where it faces the upper jaw. In many perch-like fishes, the upper jaw is made up of separate, symmetrical left and right plates, each consisting of four connected subplates, and each of the subplates bears teeth. Because their upper pharyngeal jaw does not interface with the base of the skull, the upper jaw floats freely.

In perch-like fishes, the lower pharyngeal jaw consists of two large independent plates, symmetrically right and left, and bears teeth. These pharyngeal jaws are able to rotate and to retract; the basic movement is a simple rocking action.[54] The upper and lower jaws do not, however, come together, that is, occlude, as we normally expect jaws to do. Thus the pharyngeal jaws in most fishes mainly transport food back into the esophagus, though some chewing also occurs.

With only a few adjustments, cichlids have improved on this design in a major way. The upper pharyngeal jaw is only slightly divided, and the four elements of each have coalesced to function as a unitary tooth plate. More important, the upper jaw has developed a specialized protuberance that links with the base of the skull. The coming together of the jaw and the skull has produced the basipharyngeal joint, which is unique to the family Cichlidae. Consequently, the upper jaw is no longer free or floating as it is in other teleosts.

The lower pharyngeal jaw has also become exceptional by means of minor change: The two plates are now fused into a single tooth-bearing plate, a feature they share with other labroid fishes. On top of each rear cor-

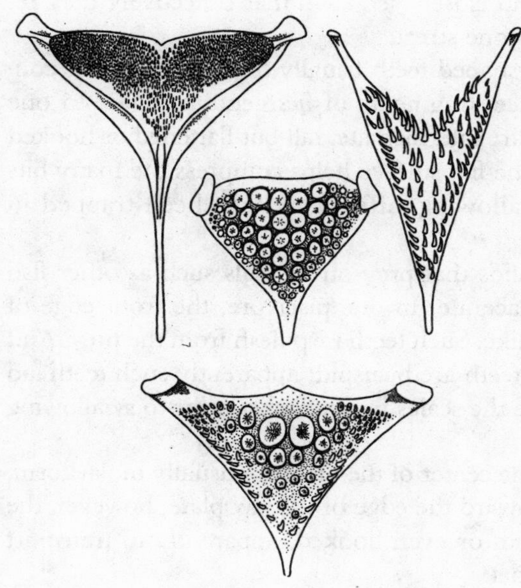

FIGURE 2.6 The lower pharyngeal jaws of four different species of African cichlids. Note the extreme differences in shape and in type of teeth. In the upper left is the jaw of a tilapia that eats planktonic algae; the papilliform teeth are small. At the far right is the jaw of a piscivore; the papilliform teeth are elongate and lance-like. In the middle is the jaw from an extreme mollusc eater; the jaw is extra sturdy and has few but large molariform teeth for crushing. The bottom jaw is that of another molluscivore, but one that also eats other organisms; the jaw has a mixture of papilliform and molariform teeth (after Fryer and Iles 1972).

ner of this unified plate, they have also evolved a bony process for the attachment of more muscles, effectively doubling the number of muscles that can be used to raise the lower jaw, producing much greater force. In essence, the lower jaw has become suspended in a sling of powerful muscles.

As with other teleost fishes, the pharyngeal jaws can rotate, and they can move food to the rear. But in addition, the upper jaw of cichlids can swivel on the basipharyngeal joint. And the sturdy lower jaw, being a single plate and having more muscles, can work with the upper jaw to masticate prey.[45] The jaws, moreover, can handle a variety of prey types.

The very same jaw can now crush tough prey and also perform delicate movements as needed, such as rolling and cutting objects. Liem argues that this structure, unique to cichlids, has therefore given individual cichlids the capability of exploiting a wide range of edibles and, at the same time, the potential to evolve further feeding specializations as the opportunity arises to exploit new food resources.[55] Some of this specialization is evident in the teeth on the pharyngeal jaws, especially the lower jaw.

The pharyngeal teeth can be broken down roughly into three types. The first type of tooth is delicate, long, and hook-shaped. The second type, called papilliform, resembles the teeth in the outer jaws, being basically slender, pointed, and knife-like; they can take on the appearance of a lawn of teeth. The last type, called molariform, looks like an eraser on the end of a pencil, squat, rounded, and closely set, such that collectively they resemble the paving of a cobblestone street.

Cichlids with delicate hook-shaped teeth usually feed on food that consists of tiny pieces, for instance, fragments of filamentous algae. In one algal-feeding cichlid, the teeth are intermediate, tall but flattened or hooked on top to form a pavement. The flat surface helps compress the many bits of food into a packet for swallowing, after they have been trapped in mucus.

Papilliform teeth typify cichlids that prey on animals such as other fish or insect larvae, which they lacerate. In one piscivore, the front edge of each tooth is pointed and saw-like. Such teeth rasp flesh from the prey. And in one of our scale eaters, the teeth are bicuspid; apparently such teeth aid the pharyngeal jaws to arrange the scales into a packet prior to swallowing them.

In snail eaters, the teeth in the center of the plate are usually molariform, to crush the hard molluscs. Toward the edge of the jaw plate, however, the teeth may change to papilliform or even hooked, apparently to transport the masticated food to the throat.

This is but a sampling of the many different configurations of the dental equipment of the pharyngeal jaws of cichlids, especially the African species. I have not mentioned that the pharyngeal jaws themselves vary in shape and sturdiness in relation to diet. For instance, the mollusc eaters have more massive pharyngeal jaws, for crushing, and the plankton feeders have comparatively delicate jaws.

Understanding the feeding biology of cichlids, and the morphology that has evolved along with the behavior, is important for appreciating the multiple flowering of cichlid species, which will be treated later. With regard to the trophic morphology, these are the important points: The outer jaws may be the most adaptively prehensile of any known family of fishes. And the pharyngeal jaws of cichlids are better able to process foods of different types than those of any other known family of freshwater fishes. I will come back to this theme at the end of the book, after telling you about other aspects of cichlid biology, such as the remarkable flexibility of sexual identity and how it develops.

≈ *three* ≈

PLASTIC SEX

Imagine the world has become fiercely overpopulated and polluted. New killer viruses are rampant. Resources are severely depleted. Biologists forecast that all humans on Earth will soon perish. By pooling all the resources left, society can send one and only one spaceship to a distant planet.

One hundred healthy young women and men, fifty of each, are selected to colonize the planet. They blast off with the glowing vision of starting a new society that, with the failed experiment of Earth in mind, will avoid the serious errors of their forebears. The journey goes well, but not the landing. Half the people on board are lost. The survivors are all women. What to do? To some kinds of fish this would not be a problem because they can actually change sex from female to male, or from male to female–they are said by biologists to be sexually plastic.

HOW IS SEX DETERMINED?

Let's start with a consideration of sex determination in mammals such as humans to appreciate how radically fishes differ. I'll then treat fishes in general and discuss where the cichlids fit in.

It is common knowledge that the father determines the sex of the offspring among humans. This is a consequence of heterogamety (Gr. *heteros* = different, *gamete* = wife, *gametes* = husband) in which two different types of gametes are produced. A gamete is a mature germ cell, such as an egg or sperm. The product of union between an ovum and a sperm is a zygote (*zygos* = pair). If the fertilizing sperm has a Y chromosome, the child is a male, and if it has an X chromosome the child is a female. Except for the sex cells, all cells are diploid (*di* = two, *ploos* = fold), in that their chromosomes exist as pairs. These cells, called the somatic cells, divide into two daughter cells to multiply or to replace dead cells. In this process of mitosis (*mitos* = thread, so named for the appearance of the dividing chromosomes), each daughter cell receives the complete set of pairs of chromosomes.

Sex cells differ from somatic cells. When they divide, in a process called meiosis (*meiosis* = to make smaller), each daughter cell, now a gamete, receives only one of each pair of chromosomes and is thus haploid (*haploeides* = single); one can thus think of the gamete as half of the potential individual it will produce. The two gametes produced by females, the ova, each have one X chromosome. But of the two gametes made by males, the sperm, one has an X chromosome and the other a Y. When ova and sperm unite at fertilization, one half of the zygotes will be male and the other half will be female because the two types of gametes are produced by the male in equal numbers. And that accounts for the number of male and female offspring being equal.

SEX RATIO

But this is not the only reason equal numbers of males and females are created. Mammals sometimes produce lightly skewed sex ratios. In humans, the sex ratio is regularly biased slightly in favor of males at conception. Males die more readily than females, however, so by early childhood the sex ratio is about equal. Thus differences in mortality between the sexes is one way the mother could control the sex ratio. In some cultures, male offspring are much more desirable. So why, then, is an approximately 1:1 sex ratio the general rule?

Ronald Fisher, a great geneticist and theorist in evolutionary biology, outlined what has come to be called Fisher's fundamental theorem to explain the equal sex ratio. (Now it turns out Fisher was elaborating on a conclusion deduced by Darwin in 1871.[1]) The theorem goes as follows: A mother might reason that a son can fertilize many women, but a daughter can have only a finite and relatively small number of children during her life. Therefore, produce sons to increase your genetic representation in successive generations. In evolutionary parlance, increasing the number of your genes in the next generation, relative to everyone else, is called increasing one's Darwinian or evolutionary fitness, or just fitness for short. (The word fitness in this context sometimes confuses those not familiar with this usage. Most of us think first of physical fitness, as seen in a trained athlete.)

Many families in undeveloped nations are now selectively aborting female fetuses, producing an excess of males. Fisher's fundamental theorem addresses the problem in doing that. When the sons grow up they will discover a shortage of potential mates. That will result in competition for wives. Necessarily, many men will have no offspring at all. They will have zero evolutionary fitness, unless they help close relatives raise related offspring.

When that happens in animals, the value of female offspring will rise. In later generations, consequently, females who produce more daughters will be favored by natural selection. If an excess of female offspring results, then male offspring will regain the reproductive advantage and will be of more value. The "system" might shunt back and forth a bit, but the ultimate solution will be somewhere close to an equal number of sons and daughters.

Consideration of the process of sex determination, the X and Y chromosomes, and of the disadvantage to the excess sex, has stimulated a field of study called sex allocation. Spurred by discoveries of different forms of sex determination, biologists have been actively investigating departures from the 1:1 sex ratio and the evolutionary forces driving them. Fishes have been central to that enterprise.

NOW ONE SEX, NOW THE OTHER

Only a few decades ago fishes were thought to fit the general scheme of sex determination through meiosis. Researchers, however, were hard-pressed to find clearly identifiable sex chromosomes in the smattering of species carefully examined. A few species were also known to change sex or to have both ovaries and testes in the same individual, such as some of the flatfishes and sea basses. Gradually, more and more kinds of fishes have been found to be hermaphroditic in one form or another.

The most frequently encountered kind of hermaphroditism is protogyny (*protos* = first, *gyne* = female). Individuals of a protogynous species start life as females. When they become large, they turn into males.

The first clear demonstration of sex change remains one of the best. For his doctoral dissertation in Australia, Ross Robertson spent long hours recording the behavior of the cleaner wrasse, *Labroides dimidiatus,* underwater on the Great Barrier Reef. In this pretty blue fish, one male dominates a small group of females and juveniles at a cleaner station such as a coral head. The females have their own dominance hierarchy. In the evening, the male spawns first with the dominant female, then with the next, and so on down the pecking order.

When Robertson removed the male cleaner wrasse, a remarkable change came over the dominant female. Within a few days she began behaving like a male, dominating and courting the other females. Within about a week, her ovaries had changed into testes, and she, sorry, he, was able to fertilize the eggs of the remaining females. Many other wrasses and parrot fishes, to name but a few kinds of marine fishes, are protogynous hermaphrodites.[2]

Already we see a departure from the fundamental sex ratio found in heterogametic animals such as birds, mammals, and many insects. The younger fish are all females and the older ones are males. Because the mortality rate is roughly constant from an early age, the older the fish, the fewer of them exist in the population. Consequently, young, small females greatly outnumber the older, larger males.

Michael Ghiselin proposed an evolutionary explanation for protogyny.[3] It is favored by natural selection if the female has a reproductive advantage when she is small and the male the advantage when he is large, as in the case of the cleaner wrasse.

Robert Warner refined and generalized this principle by graphing the size-advantage model.[4] Small fish are excluded by the large males from reproducing as males, so they do better by remaining female; when they reach a size such that they can compete with the large males, and increase their reproductive success, they switch sex and function as males. Thus in polygynous mating systems, the expectation among fishes is that small adults will be females, large ones males.

But this is not always the case. Sometimes females are the larger sex. Two examples are illustrative. The first is typical of mobile fishes that move about in groups, such as porgies or sea breams (the Sparidae; some sparids are also protogynous). Here, when a female is ready to release her eggs, several males swarm around her and synchronize the emission of their sperm with her release of eggs.[5]

Egg size is fairly constant for a given species. The larger the female, therefore, the more eggs she can pack in and release at one spawning. Hence a fish increases its reproductive success, its fitness, by being a female when it is large. Male reproductive success is limited by two factors: the number of eggs a female partner produces, and the considerable number of competing males in a spawning aggregation, which he cannot control. (I am leaving aside the interesting issue of whether sperm competition plays a role, with larger males producing more sperm.) So here it pays to be first a male, and then when large to change to female. This type of hermaphroditism is protandry (*andros* = male). It is less common than protogyny and occurs in other situations.

The best-known case of protandry among fishes is provided by the anemone fishes.[6] These are damselfishes that live on coral reefs in a symbiotic relationship with a large sea anemone. The reproducing unit is a pair with a few smaller anemone fish sharing the anemone with them. When they spawn, the eggs are laid under the umbrella of the anemone where they are guarded by the male. (In fishes that have parental care, the typical situation is that the male provides the care.)

During the time the male attends the eggs he seldom eats. The female, in contrast, continues to feed on plankton flowing past the anemone to collect energy to produce the next clutch. She is also substantially larger than the male and is the dominant member of the pair. If the female is removed, and no new female moves in, the male promptly changes into a female. The next anemone fish in the dominance hierarchy becomes a fully functioning male.

The reason protandry occurs in anemone fishes probably relates to two things. First, the female is larger and so can produce more eggs than if she were the smaller member of the pair. That benefits both the female and the male. Second, if the female had to care for the eggs, which would reduce her feeding behavior, the interval between spawnings would inevitably become longer. That would be to the detriment of the fitness of both the male and the female.

THE GRECIAN MODEL

The mythical offspring of Hermes and Aphrodite were the utmost example of simultaneous hermaphroditism. Some species of fish have achieved this ultimate state: An individual fish has both functional ovaries and testes. The extreme case is the tiny rainfish, *Rivulus marmoratus,* of Florida. It produces one egg at a time, fertilizing its own egg as it emerges.[7] Its reproduction is therefore sexual, but genetic recombination is limited to the reshuffling of its own chromosomes.

Sex is said to have two costs for an individual.[8] One is that each time an egg and sperm are joined the genetic contribution of each parent is reduced by one-half. The other cost is that of producing males. Half of the female's reproductive output consists of males. Whether these conceptions of the costs of reproduction are real or the result of faulty logic remains in the realm of evolutionary controversy. George Williams, however, argues that the only real cost is the first, that half of one's genes are lost with each fertilization.[9]

One way to avoid diluting one's genes is to fertilize one's own eggs. Some simultaneous hermaphrodites that can do that nonetheless have evolved behavior aimed at exchanging gametes with other individuals. That suggests the benefits of outcrossing outweigh the cost of genetic dilution. Consider the little sea basses of the Caribbean Sea called hamlets ("To be or not to be, that is the question"). Eric Fischer found that each hamlet has both ovaries and testes. The ovaries are much larger than the testes, which make up only about 10 percent of the germ tissue. Hamlets, therefore, produce almost as many eggs as would a completely asexual mother. At the same time, hamlets retain whatever advantages accrue to sexual recombination because each spawns with another individual.[10]

That seems such a wonderful reproductive scheme that one wonders why more animals do not reproduce that way. The likely answer is that the system is vulnerable to cheating. A rogue hamlet could greatly increase its evolutionary fitness by playing the following game: Take the role of the male and persuade another hamlet to spawn with you as a female. After fertilizing "her" eggs, immediately desert and find another hamlet who will play female.

To prevent cheating, the hamlets have evolved a reproductive strategy Fischer called egg-swapping. First, as I originally described, they spawn just before sunset, leaving little time to locate and persuade another hamlet to spawn.[11] And they drag out the spawning by engaging in protracted, elaborate courtship between each act of spawning. That leaves little time for either fish to find another mate that evening.

Second, they release only a few eggs during each spawning act. Before releasing another batch of eggs, the first one that played "female" insists its partner should now provide eggs for it to fertilize; the protracted courtship resembles an argument between the fish about who will play which role. Fischer verified that hamlets do reverse male and female roles repeatedly during an episode of spawning. That prevents one fish from fertilizing all the eggs of the other and then leaving to seek another dupe.

Other types of hermaphroditism have been uncovered in gobies in Japan.[12] They are neither clearly protogynous nor protandrous nor simultaneous hermaphrodites. Rather, they apparently can change back and forth between functional males and females, as can other gobies.[13]

Until recently, most of us believed that once a change in sex had happened, reverting to the previous sex would not be possible. Our ideas need modifying. In these gobies, and probably in other fishes that do this, the gonads are often not completely male or female. They retain some elements of testes and ovaries. The "decision" for the goby concerns how much of the gonad to allocate to male or female function. They can alternate between being male and being female.

SEXUALLY PLASTIC CICHLIDS

For years fish hermaphroditism was thought to be confined to marine fishes. The adaptive significance follows from our spaceship model. Most marine fishes, especially those in the tropics, release millions of eggs or larvae into the ocean by spawning huge numbers of gametes almost daily and nearly year-round. Their spawning behavior is adapted to trying to return their offspring to the area of their home reef. Most die at sea. But whether the few survivors return or are carried to distant locations, if only two individuals meet where they land, their chance of reproducing is assured if they can adjust their sex, one becoming male and the other female.[14]

But now we know that sexual plasticity also occurs among freshwater fishes, even though they don't face a severe "spaceship" situation. However, sex change has been demonstrated in relatively few freshwater species, suggesting that selection for hermaphroditism is weak. Examples include the rainfish, mentioned above, a number of minnows, some trout, labyrinth fishes and, yes, some cichlids.

As Richard Francis has shown in his review of the growing literature, when fishes are examined closely, many show signs of sexual plasticity.[15] The basic pattern seems to be protogyny: Start life as a female, then decide whether to remain female or change to male. Apparently, many fishes make the decision for maleness early in life and this unobtrusive protogyny goes unnoticed. The change to male is sometimes revealed by small remnants of nonfunctional ovarian tissue in the gonads of males.

Sex and Acidity, or Is It Alkalinity?

The physical environment sometimes plays a role in early sex determination. The effect of temperature on sex determination during the embryonic stage has recently been described for several reptiles such as turtles, alligators, and lizards.[16] Temperature determines sex in some fishes, too.[17] Low temperatures at higher latitudes favor males, and higher temperatures at lower latitudes favor females in the Atlantic silverside, *Menidia menidia,* along the eastern coast of North America.[18] Unfortunately, such correlational studies make it impossible to say when in development the environmental effects play a role.

The pH (the relative acidity/alkalinity) is known to have an effect on the sex ratio of a few fishes, including some cichlids. This was first discovered almost accidently. Batches of the African cichlid *Pelvicachromis pulcher* were raised in waters of differing pH. The most acidic bath, that with the lowest pH, produced almost all males, and neutral water resulted in mostly females.[19] Uwe Römer and W. Beisenherz recently explored this question in more detail, looking at thirty-three species of the South American dwarf cichlids.[20] They found a similar effect. In general, the higher the pH, that is, the more basic the water, the greater the proportion of females produced in a given batch of young fish.

Cooking Up Male Cichlids

For dwarf cichlids of the genus *Apistogramma,* temperature plays a strong and clear role. The higher the temperature, the higher the proportion of males. Römer and Beisenherz carefully identified when the temperature works its effect. It does so during a sensitive period that occurs about 30–40 days after hatching. They also studied one African species of cichlid, *Pseudocrenilabrus multicolor;* neither temperature nor pH influenced its sex ratio, so the phenomenon is not general.

Temperature also influences sex in another cichlid, however, the African mouth brooders *Oreochromis niloticus* [21] and *O. aureus*.[22] Progressively higher temperatures induced ever higher proportions of males in a family. The young fish were sensitive to the temperature effect over a period lasting about three weeks from when they were ready to emerge from their mother's mouth.

Could dwarf cichlids benefit from tuning the sex ratio to the environment? We simply do not know. The more immediate question is, do temperatures in nature vary enough to make this possible? Römer has visited the Rio Negro in South America and measured temperatures where some *Apistogramma* live. (Hobbyists use the nickname of apistos for cichlids in this genus.) He found the range of temperatures to be even greater than those he and Beisenherz employed in their laboratory experiments. So far, however, he has no observations of the apistos breeding at different temperatures, nor has he any indication of how they might benefit from varying the sex ratio. This remains a fascinating and open question.

Likewise, the fry of the mouth brooder *Oreochromis niloticus* are released by their mothers in water temperatures that can influence whether they will become males. But the situation in this species is more complex than I described above. Researchers have been able to demonstrate that individuals of this species also have a degree of genetic sex determination;[23] genetically, they behave as though females are XX and males are XY, as in mammals. Examining the chromosomes with a microscope, however, failed to reveal any indication of sex chromosomes. Demonstrating the genetic effect called for hormonal manipulation.

Well, Not Completely Plastic

The gender of a young *O. niloticus* can be determined by feeding it food laced with sex hormones during the same time period when temperature would control its sex. In that way, genetic females were diverted into males with functioning testes. Proving the genetic determination of gender here turns on the fact that these males were genetically XX, that is, would have become females without the hormonal treatment. When such males were mated with normal XX females, all the offspring were females because all were XX.[24]

Other lines of evidence indicate a genetic component in the determination of gender. Hybrids between two species of cichlids often result in offspring of all, or nearly all, one sex. Evidently, the more widely unrelated the two parent species, the truer this is.[25] Aquaculturists have long been aware of this effect and take advantage of it when they want to stock ponds with only male tilapias;[26] if both sexes are present, the fish quickly overpopulate and grow poorly.

You might well wonder if any fishes are "normal." In fact, in many kinds of fishes the adults seem to be gonochorists (*gonos* = offspring, *choris* = separated), clearly either male or female, as in mammals and birds. The size-advantage model predicts that in gonochoristic species, males and females have about the same Darwinian fitness regardless of size and age. In monogamous species the two sexes share about the same reproductive success. And many cichlids are monogamous.

Beating the Sex Ratio

With the evolutionary model of protogyny in mind, one would predict that it should occur in cichlids in which large males have a pronounced advantage in mating. One large male mates with many smaller females.

Protogyny was first sought among African cichlids in which one male mates with several females, though the choice of species was happenstance because the theory had not yet been developed. Protogyny was hinted at in a Lake Malawi cichlid in which some males had a bit of ovarian tissue in their testes, but the possibility was dismissed by the investigator.[27] The first well-documented example of protogyny in a cichlid was discovered in a dwarf South American species, *Crenicara punctulata,* whose social system features one large male mating with many females.[28]

I had two early experiences with cichlids demonstrating that at least some species have the capacity to change sex. In the first instance, I was studying a species of tilapia *(Sarotherodon melanotheron),* the blackchin mouth brooder, from Africa. This species is not monogamous, though they look much alike and males and females stay together during a prolonged courtship.

To prepare the fish for experiments on courtship behavior, I kept them separated by sex. Invariably, a few weeks after separating them, I discovered a male or two, recognizable by a yellow patch on the cheek, among the females. I thought at first I had made a mistake in sorting the fish. Slowly, I came to appreciate that these individuals had changed sex from female to male.

At that time I was also studying the orange chromide, *Etroplus maculatus,* and for similar reasons. This primitive Asian cichlid is monogamous, and the sexes are difficult to tell apart, though it can be done. When separated according to sex, females commonly paired with one another.[29] They spawned synchronously but the eggs were not fertilized. At that moment I was preparing to leave the University of Illinois for Berkeley, in 1966, and had little time for careful observation.

I noticed, however, two female orange chromides spawning together. They had both previously spawned with a male and had successfully raised young. I smiled inwardly at their futility. To my surprise, two days later, when eggs normally hatch, a small portion of the clutch did hatch. One of the females must have developed enough testicular tissue to produce some sperm.

Thus the orange chromide has the capacity to be a protogynous hermaphrodite, even though this potential is seldom evoked. I thought about this in the context of a primitive cichlid retaining the sexual plasticity of its ancestors, but subsequent experience further opened my eyes.

At Berkeley, I devoted myself to studying an "advanced" monogamous species from Nicaragua, the Midas cichlid *('Cichlasoma' citrinellum)*. They do not fit the model for protogyny. Consequently, I had no reason to expect it to be sexually plastic despite my earlier chance observations on other cichlids. But then I received a letter in 1977 from an aquarist, Michael Herzog of Göttingen, Germany, who knew of my interest in this species.

He had a few Midas cichlids in his aquarium, and so he knew them individually. He sent me photographs of one individual that had laid eggs and then much later fertilized the eggs of another female. Clearly, this advanced species of cichlid also has the capacity to change sex from female to male though it is rarely expressed.

Sex, Dominance, and Stress

Richard Francis and I stumbled on remarkable evidence of sexual plasticity of a different nature from that previously reported. And the finding suggests cichlids might in general be sexually plastic. We had noticed that in the offspring of a pair, the larger youngsters regularly became males and the smaller ones females. We assumed the obvious: Growth is slower in females and faster in males. We had the explanation backward, as it turns out. The fish that grow faster become males, and those that grow slower become females. We demonstrated that experimentally.[30]

The young of most kinds of fish show what is called growth depensation when raised in captivity.[31] No matter how well you feed them, if held in a group, they grow into a wide range of sizes. We took advantage of this in our experiment. When the sizes of juvenile Midas became clearly spread out, we divided them equally into two groups. We then had a group of Larges and a group of Smalls.

After some months, the group of Smalls had caught up to the group of Larges, and both groups then had about the same distribution of sizes. A few months later the fish would start to become sexually mature. Under our original conception, most of the fish in the group of Smalls should have been females and most of the fish in the group of Larges should have been males. Not at all. The little fish in the two groups, Larges and Smalls, were almost all females. The big fish in the groups of Larges and Smalls were almost all males. The obvious conclusion is the fish did not start life as males or as females, that is, as gonochorists. Instead, they became males or females depending on their relative size.

We have some reservations about what can be drawn from our experiment. The data are robust, but was the experiment realistic? First, we would like to know if this holds up in nature. With many adults producing thousands of offspring, how do the juveniles sort themselves out in a lake? Are they in situations where they can assess their status vis-à-vis others in their group? How does this produce a 1:1 sex ratio, which is the case in the field? Is it simply a matter that the first pairs to produce young during the breeding season have mostly male offspring and the late-breeding pairs have mostly female descendants? No one knows. And can this be a general model for cichlid fishes? I think not.

The laboratory finding from the Midas cichlid, nonetheless, says something important about sex determination in fishes. At least one species of cichlid has the capacity to become either male or female as it grows, and the "decision" depends on social relationships. In the preponderance of fishes, the most manifest consequence of size in a group is dominance and subordination. Regularly, the larger fish dominate the smaller. We assume that is the driving force of sex determination in the Midas cichlid, whether through hormonal effects of being the bully or of being the persecuted. This finding makes for a fascinating comparison with sex and gender in mammals.

Experiments on the determination of gender in mammals such as rats and guinea pigs have shown that the gonads are in command.[32] Around birth, just before or after, depending on the species, the gonads release sex hormones that go to the brain and are said to "organize" the nervous system for one gender or the other. Thus a female infant sends sex steroids such as estrogen to the brain, and those hormones mobilize the nervous system into female structure. Likewise, the testes of a male infant send male steroids to the brain to establish the "maleness" of the central nervous system. (This is more critical in males than in females because the default sex of mammals is female.) Compare this with our cichlid fishes.

Dominance is achieved through aggressive behavior. In another cichlid fish, *Haplochromis burtoni,* a dominant territory-holding male develops an enlarged group of neurons in the hypothalamus of its brain.[33] In nonterritorial, subordinate males, the group of neurons is much reduced, but if one of them acquires a territory the neurons grow. The enlargement is probably associated with the release of steroid hormones. Thus the cluster of neurons can wax and wane, befitting the status of the male. This puts the central nervous system in an entirely different light when contrasted with the traditional view of the brains of humans as morphologically immutable after early childhood.

Territoriality has but one function in the Midas cichlid and in tilapias: reproduction. Feeding is done elsewhere. When the neurons enlarge concomitant with breeding territoriality, they start producing large amounts of gonadotropin-releasing hormone (GnRH). That hormone carries the message from the hypothalamus to the pituitary gland. True to its name, GnRH stimulates the release of gonadotropins. Those hormones flow through the circulatory system to the testes, causing them to enlarge as they crank up production of sperm.

A side product of the enlargement of the testes is the release of their own hormones, androgens, into the blood stream. This higher level of androgen produces the changes associated with male behavior, such as increased aggressiveness. Thus when a male achieves the status of territory holder, that results in enlarged GnRH neurons, which in turn stimulate development of the testes.

Of course, other physiological mechanisms might be operating. For example, dominance relations, especially among cichlids confined to the small quarters of an aquarium, produce stress, and that may contribute critically to growth depensation seen among cichlids in that circumstance.[34] Stress is associated with the release of the hormone cortisol, which is related to the sex hormones. Cortisol profoundly influences the reproductive physiology of at least one kind of cichlid.[35] Thus hormones resulting from stress, rather than sex hormones, might be influencing the sex of these developing cichlids.

In fishes, then, sex is determined in the brain, and the sex of the brain controls which type of gonad is developed, testes or ovaries. (For the interested reader, Richard Francis has developed this thesis in fascinating detail, including how the critical cells move around in the brain.)[36] Francis's model fits with the observations on other fishes that are sex changers. In our cleaner wrasse or anemone fish, the dominant fish suppresses the subordinate ones. So, a similar mechanism could be at play.

THE SEX CHROMOSOMES

Let's turn back to the sex chromosomes. How can such sexual plasticity happen, you might ask, in the presence of sex chromosomes? Are fishes not heterogametic? Well, yes and no. When genetic studies of sex determination have been done, the outcome often is typical chromosomal determination of sex. The problem is that when examined microscopically, specific sex chromosomes often cannot be detected.

The property of distinctive sex chromosomes is called heteromorphism. The X and Y chromosomes, particularly the latter, are easy to distinguish in mammals. Only rarely can they be made out in fishes, and then the examples are scattered among different kinds of fishes, as though independently and probably recently evolved. The better-known instances come from the family Poeciliidae, the guppies, swordtails, mollies, and their relatives.[37] But even within the Poeciliidae, sex chromosomes are not always evident, and different populations of the same species differ in their sex chromosomes, or in the apparent lack of them.

Amphibians and reptiles are intermediate between fishes and mammals in the development of sex chromosomes.[38] They are to some degree sexually labile, though less so than fishes and more so than mammals. Thus fishes represent an early stage in the evolution of chromosomal sex determination. Their sex remains to varying degrees plastic even when they have sex chromosomes.

Cichlids seem to fit nicely into this scheme, though generalizations must be made with reservations pending wider and deeper study of them. Some have the capacity to alter their sex from female to male. But consequential questions remain. Do they take advantage of the physical environment, especially temperature and pH, to adjust the sex of their offspring? And in the cases where sex seems so labile, do the fish have any degree of genetic control? If so, individual broods might at times have highly biased sex ratios.

Much remains to be discovered about the control of sex in cichlid fishes, and the consequences for their mating systems. But despite all this flexibility, their mating systems are no more varied than those of animals with fixed sexes, such as birds and mammals, as we shall see in the next chapter.

≈ *four* ≈

MATING GAMES

Mating systems are ordinarily not an issue to people in the industrialized Western world. After all, the prevailing view is that humans are monogamous, or at least should be. The assumption that monogamy represents the natural order is inscribed in law. Deviations from monogamy are just that. Harems, a fabled example of polygyny, are viewed as curiosities, although one man married to more than one woman is commonplace.

Seen in the wider context of biology, we are exceptional. The vast majority of animals are not monogamous, and that includes mammals, though with a few notable exceptions.[1] Monogamy is something of an evolutionary specialization; it has been derived from polygyny, in which one male mates with multiple females. Songbirds are often mentioned as the only group of vertebrate animals in which monogamy is the predominant system of mating. But even there, ornithologists are now debating whether monogamy is the appropriate term for songbirds. That is because recent discoveries have revealed such a high incidence of cuckoldry among them that exceptions are newsworthy.[2]

Cichlids differ because monogamy is the basal condition, evolved from an ancestral polygynous fish. Cichlid monogamy usually goes hand in hand with skilled biparental care of young. This, too, is unusual. When parental care is seen among fishes, the male is almost always the single caregiver,[3] so why are cichlids so different? Once having perfected monogamy, what selective forces acted on them to produce mating systems that are not monogamous?

A mating or social system responds to two major forces. First, a given species must adapt to its surroundings–where it lives and what it eats. A biparental system of caretaking may not work well in all situations. For instance, a cichlid that feeds over open bottom is exposed to high levels of predation. The extra vulnerability of the fry requires the attention of both parents. But when a reef cichlid plucks its food from the interstices of rocks, it will most likely nest in a crevice where only one parent is needed to guard the offspring; that provides the opportunity for one parent to forsake its mate to seek another. The second force is the intrinsic difference between the sexes, which I'll explain in a moment.

Another, though secondary, element in the equation is the mode of reproduction, and it can have profound effects. Cichlids fall out into two main groups, the substrate brooders and the mouth brooders. The basic substrate brooder is monogamous, and the eggs stick to the bottom. The parents closely attend, care for, and protect their eggs and larvae and go on to shepherd their tiny offspring when they swim.

Mouth brooding has evolved from substrate brooding several times,[4] so naturally there are different varieties. The prevailing condition among mouth brooders is polygyny, often in an extreme form. A female visits a male to spawn and then gathers the eggs up in her mouth and departs. The eggs hatch in the mother's mouth and are held there until the young are developed enough to swim on their own.

In this chapter, we track the evolutionary transitions from monogamous substrate brooders, in which both parents care for the young, to polygynous mouth brooding, where females are largely responsible for the offspring. At the extreme, this has led to harems. Those cichlids that have retained substrate brooding, whether as monogamous pairs or extended families, have a social system that depends on knowledge of individual relationships in the family. But those substrate brooders that have become highly polygynous, living in harems, have moved away to varying degrees from a system based on personal knowledge. A comparable trend has emerged among the mouth brooders. The monogamous mouth brooders also have personal relationships. The trend, however, has been to evolve harems and, most often, lekking, an anonymous mating system.

This is only a general map of what is to come. As we proceed along the main highway we'll encounter a few side roads dotted with issues such as sneaky spawning and the powerful impact of empty snail shells, of all things. We'll also learn how to deduce a given fish's mating system simply by looking at the differences between males and females of that species.

CONFLICT OF THE SEXES

Evolutionary theory rests on the proposition that the reproductive interests of males and females are in conflict.[5] Humans understand this intuitively, as reflected in our folklore and its prevalence in e-mail jokes. But why have evolutionary theorists concluded that the sexes are reproductively antagonistic? Some basic biology is in order here.

Sexually reproducing organisms produce gametes, the germ or sex cells, and one gamete from a single individual combines with one gamete from another individual to start a new organism, the zygote. All the animals you are likely to know produce gametes of two exceedingly different sizes, the small ones being sperm and the large ones eggs, and this is called anisogamy (*anisos* = not alike). Anisogamy sets the stage for the conflict between the sexes.

Males manufacture lots of tiny so-called cheap sperm. A male invests little in each sperm and accordingly can easily afford to fertilize the ova of many females, so he need not be discriminating. Although the number of eggs potentially available to a male is huge, he cannot expect to have access to all the eggs out there. The reproductive success of a male is constrained by the reality of the sex ratio. Males and females are usually equally abundant, so if a male is to fertilize the ova of many females, some other male will of necessity be able to fertilize only a few of them or even none. Females are accordingly the limiting sex, males the sex that is limited.[6]

The result is a suite of traits called the masculine syndrome: Males compete among themselves for the eggs of females. Although not the only means of contesting, the most obvious way males compete is by fighting for females, and so selection tends to favor the larger males. Access is just the first step, however. A male must also persuade the female that he is the best male to father her offspring. To induce her to accept him he displays in various ways, so selection often favors the evolution of male ornamentation and bright colors. As a consequence of male-male competition and female choice, coupled with lack of selectivity by the male, male reproductive success typically varies greatly.

The contrasting female suite of traits is referred to as the feminine syndrome: She produces relatively few expensive eggs (note the high tab for human donor eggs compared with the price of sperm), and she invests much more per egg than the male does per sperm. An important corollary to the limited quantity of eggs she makes is the ceiling on the abundance of offspring she can generate during her lifetime. A female, therefore, should exercise care when deciding which male to allow to fertilize her eggs. That should not be a problem, however, because multiple males offer themselves up. All she need do is choose what looks to be the best one. Thus, in our ideal polygynous species, overt aggression between the male and female is unnecessary and hence nonexistent.

Another consequence is that the female is said to be coy. She takes time to assess which male is the most fit. This assessment is done typically by observing his displaying or his prowess in competition with other males and by appraising the limiting resource he commands, such as the best place to spawn. She does not have to compete aggressively with other females, nor does she need to be conspicuous to attract a male. The result: Females are the smaller and protectively colored sex. In addition, reproductive success does not vary between females nearly as much as it does between males.

Although overt aggression is avoided, the interests of the two sexes clash. This is referred to as sexual conflict. In the archetypal scenario, the large, ornamented male tries to unite his sperm quickly with the eggs of any and all females he meets. To maximize his success, he does not tarry with any one female after fertilizing her eggs; the success of males hence varies greatly. The small, drab female, for her part, seeks to choose the "best" male, or the male controlling the best resource, and determining that may take a little time.

WHY PARENTING MATTERS

When parental care evolves, this equation is profoundly altered. Most fishes don't care for their eggs, and even fewer care for their swimming young. The reasons are usually, but not always, ecological. Coral-reef species cannot both protect their young and have them disperse to other islands, and effective dispersal is crucial. To broadcast their progeny, they launch millions of fertilized eggs into currents that then carry the zygotes out to sea.[7] However, when natural selection favors parental care, that changes the dynamics of sexual conflict.

Assume the pair is faithful and nurtures and protects their young. Their reproductive interests are now equivalent but signs of sexual conflict persist. Either parent is tempted to run off and leave the other stuck with the care of the young. The selective force driving this is that the time consumed by parental care cancels any further immediate opportunity to reproduce, and it thus results in some loss of future reproduction.

Indeed, the parental fish can be so occupied by its custodial duties that it cannot even take time to feed. Without energy-yielding food, preparation for another episode of reproduction is impossible, and that reduces the number of surviving offspring the parent can hope to produce in its lifetime. But if that were the entire story, no animal would provide care of offspring. The payoff for parenting is that a higher proportion of the offspring may survive than if the parents abandoned them.

Whether it is better to care for young or not, however, depends on the paramount features of the life history of an animal such that neither method, taken alone, can be said to be inherently better than the other. For instance, in addition to its energetic costs, parenting may expose the parents

TABLE 4.1 Classification of Cichlid Mating Systems

General	*Subsets*
1. Monogamy Male and female form enduring exclusive pair.	A. Courtship Male and female pair and court reciprocally for prolonged period. Separate after eggs are laid.
	B. Parent Male and female form pair that persists until offspring reach independence.
2. Polygyny One male fertilizes the eggs of more than one female.	A. Male territory Male holds territory where he is visited by females to spawn. Female spawns with only one male each reproductive cycle.
	B. Bigamy Two females hold breeding territories within that of the male, who may help care for offspring.
	C. Harem As bigamy, but may have multiple females. Little or no male parental care.
3. Polygynandry Each male fertilizes eggs of more than one female. Each female has eggs fertilized by more than one male.	A. Male territory Males hold well-spaced territories continuously. Males receive more than one female, and females may spawn with more than one male. Some species polygynous.
	B. Lekking Females visit multiple males clustered on transient territories. Some species polygynous.
4. Polyandry One female reproduces with multiple males. Each male spawns with only that female. Family cycles overlap.	
5. Extended Family Two or more group members of both sexes reproduce. Some offspring remain in the family.	

themselves to predation. The reproductive benefits to the parent, therefore, must compensate for those costs. If only one parent is needed to protect the young, then one or the other parent will inevitably cut its costs by deserting. If the other parent continues nurturing the offspring, it is stuck with the costs of doing so, but that would have been true even if the mate had not left.

Natural selection favors deserting if two requirements are met: First, the abandoned, uniparental family survives. Second, the deserting individual will produce yet more offspring by sharing gametes with a new mate. The issue, then, is which parent deserts, and there we have the conflict between the sexes again. Following the anisogamy argument, the male, who has invested less in producing the zygote, is more apt to abandon the family than the female. Evolutionary theorists have had endless fun exploring the multiple facets of this contest between male and female.[8] The pipefishes have been instructive here.

Pipefishes, basically unbent seahorses, present a natural but elegant test of the hypothesis that the sex that limits the reproductive rate of the other shows the feminine syndrome, and the limited sex the masculine one. I mention this because of a possible parallel in one species of cichlid that we'll learn about below, and I want you to see how it fits the theory. In pipefishes and seahorses, males get pregnant when females pump their brood pouch full of fertilized eggs. In some species, the males take longer to hatch out the eggs than the female does to produce a new clutch, thus limiting her reproduction. The males show conventional female behavior, whereas females act like males, displaying at, and aggressively competing for the actual males.[9] That nicely validates the hypothesis of the role of which sex limits the reproduction of the other. Almost always, that is the female.

To help you keep the mating systems straight, I have provided a table for reference. It relies on placing the mating systems into discrete categories. And we humans tend to think in terms of categories, even when the boundaries in nature are fuzzy or nonexistent. The mating systems of cichlids are to varying degrees fuzzy and at times slide back and forth between categories, depending on local conditions. Despite the frailty of this classification, it aids our thinking. Just don't apply the classification too rigidly.

FAITHFUL PAIRS OF PARENTS

To many of us, monogamous cichlids embody the abstract ideal of the human family: The pair cares for their offspring just long enough to promote their survival but not so long that the cost to future reproduction is excessive. In certain situations, such as among some substrate-brooding

African cichlids, the pair continues to share a feeding territory after parental care has ceased. The pair bond is thus abiding but for a reason beyond care of young. One persistent belief among cichlidophiles is that pairing among biparental cichlids is in general a lifelong commitment. That, however, is most likely an artifact of a pair being kept together in an aquarium. In nature, most such cichlids separate after rearing a brood.

One example is the orange chromide, *Etroplus maculatus*. It is a primitive cichlid in the evolutionary scheme of things, and its mating system is representative of most biparental cichlids. The male and female differ in color pattern only in detail (they are monomorphic), and the male of a pair is characteristically a little larger than the female, which is usually the case among monogamous cichlids. They breed twice each year, during the two monsoonal rainy periods of India and Sri Lanka.[10] Each monsoon is apparently only long enough for a pair to raise one family. Therefore, when the male starts a family, his best bet is to stay with his mate to help raise the young rather than desert in the slim chance he would find an available female. There simply isn't enough time to start a second family. Thus the pair is tempted to extend the period of parental protection because that can be done at relatively little cost. This may account for the unusually long time orange chromides protect their school of offspring.

One aspect of parenting sets the orange chromide apart from more advanced cichlids: The male and female share about equally in care of both the clutch of sticky eggs and the fry. More representative are the evolutionarily advanced substrate brooders in South and Central America and in Africa. Among them, males and females play slightly to markedly different parenting roles.[11] In Nicaragua, female Midas cichlids, *'Cichlasoma' citrinellum,* do all the direct care of the eggs and then of the larvae. Males are more involved in protecting the territorial boundary: They repel potential usurpers of the nesting cave.[12] Given that the female is almost exclusively the one who nurtures the eggs, males should be tempted to desert and probably would if the mother alone could provide sufficient protection of the fry.

Male Midas cichlids do not abandon their mates, however, even when offered other females who are ready to spawn;[13] in this they represent a relatively early step in the evolution of cichlid mating systems. When the fry swim, the male helps the female protect them from predators, but of the two parents, she stays physically closer to the fry.

In color pattern and shape, the sexes in the Midas cichlid are indistinguishable. Males tend to run larger than females, but the sexes overlap in size. When they mate, however, males and females sort themselves out by size. The male is always larger than the female, which is typical.[14] At the time of pairing the male develops a swollen forehead that makes him clearly different from the female for a few days,[15] which helps in sex recognition at the time the pair forms.

Monogamous biparental cichlids are also common in the rivers of Africa. Although they are unrelated to the New World cichlids, as cichlid systematics go, their behavior is much alike. Aquarists who have watched the family life of the gorgeous African jewel fish, *Hemichromis guttatus,* will have been struck by the parallels.

WHEN FEMALES OUTSHINE MALES

One of the hallmarks of monogamous species is that the male and the female look pretty much alike, though one sex, most often the male, may be slightly the larger. This lack of a pronounced difference between the sexes is called monomorphy (*monos* = one, *morphe* = form) or, alternatively, isomorphy. To some scientists, dimorphism refers strictly to difference in body size or shape. But many others, including me, use it in the larger context of any difference. When stressing the similarity between males and females in coloration, the more restricted term monochromatic is used as opposed to dichromatic.

Monomorphy can be a serious problem for people who raise parrots, macaws, and cockatoos because among these birds males and females are indistinguishable. A single macaw can cost over $1,000, so if you have one and are looking for a mate, you need to be certain of the sex of each. (This ambiguity has spawned a high-tech company, Zoogen Inc.; as of this writing, they will determine the sex of your bird through DNA analysis for about $20.)

The sexes in monogamous cichlids are often colored so much alike that they cannot be told apart just by looking at them. But fascinating exceptions occur among substrate-brooding cichlids, and I want to consider some examples more closely before we travel too far along the path of mating systems. Keep in mind that some of these examples are of cichlids that tend toward polygyny. The specific sexual dichromatism I want to tell you about is so different and exceptional for any type of animal that it deserves detailed treatment, though it is a digression from mating systems per se.

In a few monogamous cichlids, the females are more strikingly colored than their male mates, as though females were playing the masculine role, competing for male mates. However, that is not the function of the coloration. Before presenting some of the hypotheses for this reversed dichromatism, let's explore the nature of the difference.

The distinctive pattern among these colorful females, and in a number of unrelated cichlids, is vibrant coloration around the abdomen. The convict cichlid is the best-known case (it is so named because it is marked with dark vertical bars on a pale background). The male is much the larger fish of the pair, and the trailing tips of his dorsal and anal fins are drawn out into long filaments. This is consistent with the tendency of some males to have two mates at the same time.[16]

When the female is seeking a mate she adopts a dark, almost black, color pattern. The lower half of the middle of her body becomes densely flecked with gold, a color the male lacks.[17] The gold coloration may become less intense when she is parental, though this varies among strains of convict cichlids. But why should the female be so conspicuously marked? To announce her gender? Maybe.

Often conspicuous markings are associated with competition for some limiting resource. If so, the color might be a threat display to other females competing for the best male.[18] With this in mind, Simon Beeching and his colleagues recently explored whether having gold flanks provokes attack by other convict cichlids. They found that males and females behave differently. Males attacked neither more nor less in relation to the gold color. Females, in contrast, responded more to live and dummy females that had golden flanks, which is consistent with the competition hypothesis.[19] In line with this, Gerald Meral observed unmated females in Nicaragua who competed for nest sites and then courted passing males.[20]

The female's color also emphasizes the part of her body that swells as she fills up with eggs. Because of that, I suggested many years ago that the female might be drawing the male's attention to her midsection, as though announcing she is full of eggs.[21] Consistent with this, male convict cichlids select the female having the greatest girth, indicating an abundance of eggs.[22] The gold flanks of the female could be an adaptation meant to deceive the male into "believing" she is wider than she actually is, but that has not been investigated. On the other hand, the bright coloration of the female might inhibit attacking by her larger mate.[23]

Brilliantly colored bellies on females have evolved in many other substrate-spawning cichlids. Several are close relatives of the convict cichlid, being in the same subgenus *Archocentrus*. Females in other subgenera in Central America are also occasionally more conspicuously colored than their males, for example, brilliantly yellow females of *'C'. (Nandopsis) salvini*. Strikingly orange to red abdomens are found in female cichlids in South America, as well, such as *Microgeophagus ramirez,* known to the aquarists as "rams."

Female dwarf pike cichlids, genus *Crenicichla,* of South America typically have a rosy red belly and so are more brilliantly colored than are their mates. A female *Crenicichla regani* initiates courtship with a male by arching her body sideways into a U-shape and presenting her rosy midsection to him.[24] Curiously, once a pair has formed, the male does the same U-shaped display to the female, but his belly is not ornamented.

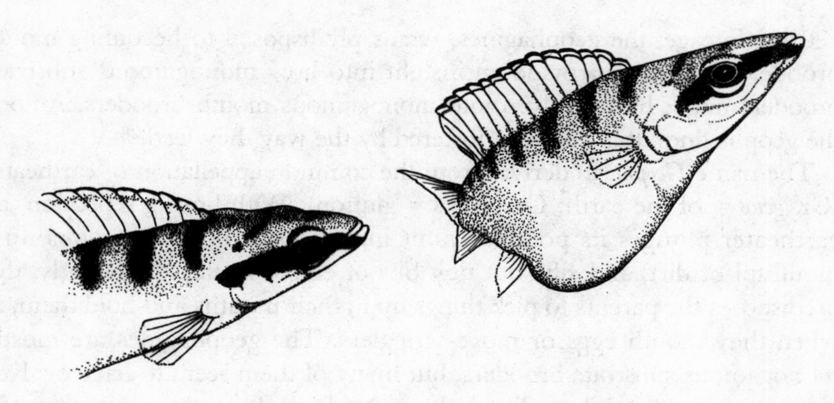

FIGURE 4.1 The female of the South American *Crenicichla regani,* right, bends her body to present her red abdomen to the male.

Reversed sexual dichromatism is also found in small African cichlids. My favorite is one I kept as a hobbyist, *Pelvicachromis pulcher,* often called the "krib" by aquarists because for so many years it went by the species name of *kribensis.* The belly of the female krib is cherry red, even purple. During courtship, she flexes her body sideways, presenting her red midriff to the male, and shakes.[25] Other examples from Africa include the appropriately named dwarf cichlid *Nanochromis transvestitus.*[26]

Once aware of the phenomenon, I kept discovering yet more examples, all of them substrate brooders. Reversed sexual dichromatism is probably more widespread among them than we have appreciated and may be restricted to them. The reason for this difference remains a mystery. We have no shortage of hypotheses, just a dearth of comparative studies. If we knew more about the correlates of reversed sexual dimorphism we would be in a better position to evaluate the hypotheses. Too often we do not really know whether these species are in truth monogamous. My impression is that the reversed sexual dimorphism becomes more pronounced as we look at the more polygynous species. When I discuss haremic substrate brooders, such as the dwarf cichlids of South America, keep this in mind.

Now let us return to our discussion of the progression of mating systems, switching to the few monogamous mouth brooders to see the parallels with and differences from monogamous substrate brooders.

MONOGAMOUS MOUTH BROODERS

In South America, less than 10 percent of the cichlids brood eggs or larvae in their mouths; the preponderance of species there are substrate brooders.[27]

Yet one lineage, the geophagines, seems predisposed to becoming mouth brooders, and they provide an insight into how monogamous substrate brooders might have evolved into monogamous mouth brooders. Among the geophagines, this might be fostered by the way they feed.[28]

The name *Geophagus* derives from the common appellation of eartheater (Gr. *geios* = of the earth; Gr. *phagos* = glutton). With boring repetition, an eartheater plunges its pointed snout into the soft bottom, scoops up a mouthful of dirt, and sifts out tiny bits of edible matter. Apparently, this predisposes the parents to pick things up in their mouths and hold them, as when they mouth eggs or move wrigglers. The geophagines are mostly monogamous substrate brooders, but many of them seem to have evolved monogamous mouth brooding independently. In fact, one population of a given species may be a substrate brooder while another population is a mouth brooder.

In the earliest stage of mouth brooding, the adhesive eggs stick to the substrate and are fanned and protected by a monogamous pair. In that respect, they are indistinguishable from substrate brooders. But when the eggs hatch, both the mother and father pick up the larval wrigglers and nurture them in their mouths. When the offspring are ready to swim, the parents remain as a pair and shepherd them, again as would ordinary substrate brooders. When danger threatens, the fry then take refuge in the mouths of their parents.

Moving to Africa, where mouth brooders predominate, the behavior of an unrelated monogamous mouth brooder in Lake Tanganyika must have originated in a different way. Mating in the scale-eating *Perissodus microlepis* seems at first like that of polygynous mouth brooders (which we will get to soon) in which the female lays nonadhesive eggs, picks them up in her mouth, and swims away. In *P. microlepis* the female takes the eggs into her mouth soon after spawning. But instead of swimming away, she remains in the breeding territory where she is tolerated by the male.

In most polygynous mouth brooders, the male would drive the mother away. Here, instead, the scale-eating male and female remain on territory as a monogamous pair. After a little more than one week, the mother spits out the fry. Either parent may then shelter the now mobile fry in their mouths, and both parents keep the youngsters in their mouths during the night.[29]

This kind of monogamy, featuring passing of the young from female to male, has subsequently been seen in several cichlids in Lake Tanganyika.[30, 31] Recently, a mouth-brooding cichlid has been discovered in Lake Victoria that does the same thing.[32]

These species are unusual among monogamous mouth brooders because their eggs, now usually nonadhesive, are picked up right after spawning. This is a pivotal step away from monogamy because either both parents may take them up and separate, or one or the other parent may grab all the eggs and leave. As an alternative evolutionary scenario, the species might have been polygynous, with the female departing with the eggs, but then selection favored the female remaining with the male, and male toleration of the female.[33] Among possible mating systems that could emerge from this is the harem.

HAREMS

In harems, or one-male groups, a single adult male and a group of females form a social unit in which the individuals appear to know one another. Characteristically, one large dominant male has an extensive territory that embraces two or more smaller female territories. He fertilizes the eggs of all the females.

For a harem to arise, the fish need to remain in one place. Their eggs are normally anchored to the bottom, so the mother cannot leave, and that is why the harem is most common among substrate brooders. Contrast this to a mother who packs her eggs in her mouth and is consequently free to take off.

Harems are common among tiny cichlids. Hobbyists speak of these and other bantam cichlids as dwarfs. They are less than three to four inches long (around 100 mm), and some are even smaller. Females and males in one species of the African genus *Nanochromis,* for instance, are sexually mature at a length of one inch (25 mm).

Haremic species of dwarf cichlids in South America occur in more than one genus, but the most heralded species are in *Apistogramma,* "apistos" for short. This genus of substrate brooders has seventy or possibly more species. Each dwarf female apisto sets up a small territory that has as its center a small cavity. When breeding, she becomes strikingly canary yellow with contrast-rich, inky black markings; this fundamental pattern is consistent across species, though the black markings vary considerably.

Male apistos are huge compared with their females, being as much as two to three times as long and approximately at least ten times as heavy. They have reached the climax of cichlidian sexual ornamentation. Males of many species have spectacular long filaments on the spines of their dorsal and anal fins, elongated filaments on their tails and pelvic fins, and gorgeous coloration. For each species, moreover, the males are typically radically differently colored. This stands in sharp contrast to the pervasive black-on-yellow coloration of females.

Harems are also common among the lamprologine cichlids in Lake Tanganyika, for instance, *Lamprologus savoryi.*[34] But a more informative example is provided by *Neolamprologus tetracanthus* because it illustrates how the environment can cause a substrate-brooding cichlid to be alternately monogamous and haremic.

Neolamprologus tetracanthus reproduces in shallow water in a barren environment, sandy bottom with scattered rocks and pebbles. In one study, predators were abundant at the north end of the lake, and the males there helped guard the fry. The mating system was monogamy.

At the south end of the lake, predators were relatively few and evidently because of that *N. tetracanthus* was more abundant. Each female had her own feeding territory, and the territories of individual males embraced up to fourteen female territories, the most extreme case of harem polygyny known. Males did not guard the fry.[35] Thus the combination of fewer predators and more densely settled *N. tetracanthus* diminished male parenting in favor of obtaining more mates, even in an exposed habitat.

Analogous flexibility of mating relationships exists among the mouth brooders, and they tell us a lot about evolution away from monogamy.

RUSH TO POLYGYNY

Transitional species are also found among South American mouth brooders. The progression from monogamy to polygyny moves with how long the eggs are stuck to the substrate.[36] The sticky period lasts from about two days in the biparental species to instant pick-up in the maternal forms. An example of the latter is *Geophagus steindachneri.* The mother engulfs the eggs right after they have been fertilized. The male chases her away and starts courting other females who are ready to spawn.[37] They are clearly polygynous.

Another intermediate species, though in a different way, is the St. Peter's fish, *Sarotherodon galilaeus*. It is a tilapiine fish distributed across the central part of Africa, ranging north into Israel and Jordan.[38] Through history, St. Peter's fish has supplied the fishery in the Jordan Valley, which led to its appearance in the Bible. Unlike the better-known tilapias in the genus *Oreochromis,* males and females of *S. galilaeus* are colored much alike, though males reach a slightly larger size than females. Right away, one suspects the species might be monogamous.

Whether *S. galilaeus* is monogamous or not, however, is a matter of definition. A pair forms, and courtship can be prolonged; in that sense they are monogamous. Unlike the Latin American eartheaters, the eggs are non-adhesive, and usually, but not always, they are picked up by both parents. The parents may hesitate a while before taking up the eggs, however, as though waiting for their partner to assume the burden of parenting.[39] Either the father or mother, or both, eventually pick up the eggs and then go their separate ways.[40] If either the male or female St. Peter's cichlid could find a way to manipulate the other into taking all the eggs, the species would become polygamous.

STICKING THE MALE WITH THE BILL

In one exceptional cichlid, however, the male winds up with the eggs instead of the female. This cichlid is an analog of the pregnant male pipefish. Evolutionarily, this species was probably recently monogamous. It behaves much like *Sarotherodon galilaeus:* A pair forms but dissolves after spawning.

This arrangement exists in just a few closely related species,[41] but the blackchin mouth brooder, *Sarotherodon melanotheron*, is the only one that has been studied.[42] The male nurtures the eggs and wrigglers in his mouth for about fifteen days.[43] Female blackchins can usually produce another clutch of eggs in around one week. Even if the female were to take a little longer than a week, the male would still need to recuperate from the fasting done while mouth brooding. Only then would he be up to accepting another clutch of eggs. Therefore, at any one time males available to brood eggs should be in short supply. As a result, receptive males, limit the reproduction of females.

This pattern led to the hypothesis some years ago that the sex roles would prove to be reversed, if investigated carefully.[44] Specifically, the male blackchin mouth brooders should show the feminine syndrome, and the females the masculine syndrome. This prediction has now been tested by scientists. Their data support the prediction that males are the less aggressive, more choosy sex, preferring larger females. Females are more aggressive and not as choosy.[45] In this respect, the blackchin tilapia departs from the main line that proceeds from biparental substrate brooders to female mouth brooders. Now let's return to that main highway.

HAREMIC MOUTH BROODERS

Haremic mouth brooders crop up just where you would predict: where the females have a reason for staying put. Feeding territories are rare among freshwater fishes,[46] but with ever more research, they are proving to be widespread among cichlids in the Great Lakes of Africa.[47] When a female mouth brooder holds a feeding territory, a large dominant male can then easily establish a territory that encompasses some number of females. Voilà, a harem. This can be seen in many *Aufwuchs*-scraping cichlids inhabiting the shallows of Lakes Malawi and Tanganyika.

Although each female *Tropheus moorii* has her own feeding territory in Lake Tanganyika, the territory is not fruitful enough for her to produce quickly a clutch of eggs. So, three weeks prior to spawning, the female moves into the male's territory. After they spawn, she departs and quietly broods her eggs in her feeding territory.[48]

Another ecological circumstance in Lake Tanganyika favoring a haremic mating system is seen in the mouth brooder *Ctenochromis horei.* The group studied by Haruki Ochi occupied a patch of underwater vegetation. Membership was stable, and the individuals appeared to know one another. Within the harem, the dominant male fertilized the eggs of most of the females, though some small males managed to sneak an occasional fertilization.[49]

In the African Great Lakes the haplochromine species vary in the degree to which they hold breeding territories, depending on feeding behavior. That makes it difficult to identify the form of their mating systems. For those species that move about searching in crannies for small animals, for instance, *Melanochromis labrosus* in Lake Malawi, territoriality is lacking or only weakly developed. Males of this and other species in Lake Malawi, such as in the genus *Labidochromis,* simply rove about and mate with females where they meet,[50] and their mating system has not been adequately studied. More typically, mobile species evolve a well-structured mating system in which males gather on an arena and females approach them to spawn. This arrangement is called lekking.

LEKKING

Lekking derives from a Swedish verb. It means to play, apparently because of the animated displays of male grouse who come together on what is called an arena or lek.[51] The males are tightly packed on the lek, each in its tiny display territory. The female observes the males at a distance. Then she approaches and mates with one of them. Mating is brief, the female leaves, and the males continue displaying. The most famous and early examples of lekking were provided by relatives of chickens, such as various species of grouse.[52] In those species, the females mate only once and hence with just one male.

No other mating system so fascinates evolutionary biologists as does lekking, for two reasons: First, lekking provides a golden opportunity to understand sexual selection. In many instances, sexual selection has led to extreme, even outlandish, differences between the sexes in lekking species, differences that would seem to handicap the male and so reduce his chances of surviving, such as the burdensome tail of the lekking peacock.[53]

In lekking, unlike other mating systems, the situation is reduced to its bare essentials because the female derives no obvious direct benefits from the male she chooses, such as parental care or sharing of resources. The one plausible benefit for the female is the quality of genes provided by the "best" male. This is the "good genes" hypothesis.[54]

The second reason lekking is so fascinating is because the males aggregate. A typical lek consists of a group of males, close together and interacting. Mating success among them varies hugely. So why do the unsuccessful males remain? One proposed benefit is that a cluster of males does indeed draw more females. But that is not enough. For a lek to evolve, the number of females attracted *per male* has to be greater than that lured to a solitary male. Direct observations usually show no such advantage. Something else is going on. The behavior of the animals in a lek offers some clues.

Lekking is remarkable for the way it functions.[55] If a female seeks a male with the best genes, no matter how she might determine that, she must compare one male with the next. Lekking males accommodate by gathering in a lek where they offer a market for female shoppers. That, however, does not explain what is in it for the males, and that remains an open question.

Today it seems difficult to imagine that when lekking was proposed for fishes, the idea was resisted by many ornithologists. Twenty years ago, Paul Loiselle and I presented a paper reviewing lekking in cichlids at a national scientific meeting where we were treated with hostility, although our published version was later well received.[56] The resistance is all the more noteworthy given the wide acceptance today of lekking among many kinds of animals, ranging from fruit flies to antelopes.[57,58]

Ken McKaye and his colleagues, observing planktivorous cichlids living in the open waters of Lake Malawi, have now described cichlid lekking in four species, one of which, *Copadichromis eucinostomus,* received detailed attention.[59] When these fish reproduce, the males move inshore to set up breeding territories in sand. They congregate in what are the largest leks among vertebrate animals in the world. The number of males in one lek varies from about 5,000 to 50,000 along 4,250 meters (roughly 2.5 miles) of shore. And they lek over a long breeding season. Each male builds a volcano-like nest of sand, called a bower by some. Females approach the lek in the morning, look the males over, and then descend among them to spawn.

Here is where lekking cichlids and birds such as grouse diverge significantly. Lekking birds are considered polygynous because certain males mate with many females, but not the other way round. Among many kinds of fishes, not just cichlids, a male fertilizes the eggs of several females, but the female distributes her eggs to more than one male. Thus they are polygynandrous (many-female-male).

In *C. eucinostomus,* females spread their favors among four to twelve males, and that seems to be a general pattern among lekking cichlids.[60] Each female spawns a few eggs with one male and then moves to a different male to repeat the process. Subsequently, she departs from the lek to nurture a mouthful of developing eggs elsewhere. Later in the day, the males forsake the lek and feed on plankton in the open water of Lake Malawi.

Some of the plankton-feeding cichlids in Lake Tanganyika lek in much the same fashion even though they have become emancipated from the need to spawn the eggs in a pit on the bottom. Haruki Ochi spent hours underwater observing male *Paracyprichromis brieni* lekking next to the face of an underwater cliff where females approached them individually.[61] A given female spawns two to three eggs with one male and then typically moves on to another male to spawn more eggs. Her eggs are large, so her total clutch has on average only eleven eggs. Spawning close to the face of a vertical wall rich in crevices provides places for small sneaker males to hide and dart out to attempt to steal fertilizations. *P. brieni* is thus also a polygynandrous lekking species.

The other species observed by Ochi, *Cyprichromis microlepidotus*, spawns at a greater distance from any reef structure. The mating system differs in some salient details from that seen in *P. brieni*. A female approaches a male and spawns just a few eggs per batch. However, she behaves like a female grouse, spawning with only one male, laying on average just nine eggs. Some females then simply swim off with their brood. That indicates polygyny. However, after finishing spawning some females approach a second or third male. No further eggs are spawned, but the females nuzzle the males around the vent, the act associated with getting sperm from the male. Consequently, this lekking species is variously polygynous or polygynandrous.

Ochi studied another lekking species, the silvery *Gnathochromis pfefferi,* that is notable because it is not planktivorous. Like planktivorous species, however, it lives out in the open, which favors mobility. This species stays close to the bottom where it feeds on small animals such as shrimp.

In the morning, males swim to their individual but clustered breeding territories, sometimes more than 100 meters distant, but often adjacent to their feeding territories. There they take up their positions in a lek around underwater vegetation, such as *Vallisneria,* a grass-like plant familiar to aquarists. Patches of vegetation are scarce and form the foci of the leks. Unlike many other lekking cichlids, the males construct no distinctive pit nest. Some males that Ochi observed kept their breeding territories for up to three months, and if they abandoned them, they were quickly seized by new males. Females visit the males to spawn. In the afternoon, the males return to their feeding territories, and the females retire to quiet areas around vegetation with eggs in their mouths.[62]

I have described in some detail the lekking of *G. pfefferi* because it might seem at first a contradiction to the usual pattern. If the males hold feeding territories, why don't they have a haremic mating system? The answer, I presume, is that females are not territorial. They range about in the same area, but they do not form a structured group attached to a given space.

Perhaps the males need to move to the patches of greenery to signal females where they are.

I have described, for the most part, just the major features of lekking. As more is learned, the richness and complexity of lekking becomes more apparent. As one example, the males of *Cyathopharynx furcifer,* a so-called bower-building species, can be sorted into three different types on the lek, each pursuing a slightly different strategy. The first is a large dominant male that builds a distinctive bower. The second is also a large male who holds a territory, but he does not construct a bower. The third kind of male is smaller and looks like a female, which suggests sneaking.[63]

Just how prevalent lekking is among cichlids is not known, though it is apparently widespread. Nor do we know whether lekking species are typically polygynandrous or polygynous, though the evidence at hand suggests the usual pattern is polygynandry. Lekking is probably most frequent among species that feed and spawn in open areas, such as midwater or along sandy stretches of shore or marshy areas with soft bottom.

One thing is clear: Lekking is the most extreme mating system among mouth brooders and is the culmination of the trend away from a system that depends on individual recognition to one that is anonymous. By its very nature, substrate brooding precludes lekking. Having reached this end point for the mouth brooders, we turn back onto the evolutionary highway traveled by the substrate brooders to explore the unique complexities of social life that they have evolved.

POLYANDRY—THE FEMALE WINS

Polyandry–where females have multiple male partners–may be the rarest mating system of all.[64] The best-known cases are among birds, such as jacanas, phalaropes, and the Galapagos hawk. Another famous case of polyandry is found among humans living in the Himalayas.[65] In such an austere ecological setting, two brothers sometimes share one wife when one man alone cannot provide the needs of a family. Note, however, that the husbands are brothers, and that the relatively more affluent people living in the same village are typically monogamous.

Too little is known about the newly reported possible case of polyandry in a cichlid fish to hazard a guess about its adaptiveness or, for that matter, to be completely confident of the finding. Two Japanese scientists, Satoshi Yamagishi and Masanori Kohda, spent two months studying *Julidochromis marlieri* along the shores of Lake Tanganyika.[66] They presented their findings cautiously.

This substrate brooder is routinely considered monogamous by aquarists. And indeed, most of the breeding territories observed in nature are held by a mated pair. Nothing out of the ordinary there, but then something strange popped up. The largest two females among the seven studied were found to have two breeding territories. That was unexpected. One of the females, the largest observed (91 mm total length, or about 3.5 inches), had a male mate in each of her two territories. That was unheard of. The other, and second largest female in the group, also had two territories, but just one contained a mate.

With only this evidence, this case of a female with two families might be dismissed as an oddity. Additional considerations, however, support the suggestion that the largest female was truly polyandrous and that the other females would be under the right conditions. First, the arrangement persisted throughout the two months of the study, and the two polyandrous males separately reared the young. Second, in each of the eight pairs they watched, the female was substantially larger than her mate. I know of no other pair-bonding cichlid in which the male is regularly smaller than his female. Third, females were the less devoted parents. When approached by divers, they fled before the males did. Among typical substrate-brooding monogamous cichlids, the males are regularly the ones who flee first. If this species is actually polyandrous, that mating system probably prevails in the other species of *Julidochromis* because the females in them are the larger sex. Cichlids again prove to be exceptional actors in the theater of organic evolution.

WHEN MALES COME IN TWO FLAVORS

When male members of the same species exist in two forms and reproduce differently, that gets the attention of evolutionary theorists. The little African cichlid, *Pelvicachromis pulcher,* the krib to aquarists, provides an arresting example of just that. Males come in two different color morphs, called reds and yellows. They are colored much alike except on their faces. In one morph, red extends from the belly up onto the sides of the head; the males look like they are wearing red lipstick. The head and mouth of the yellow form is, not surprisingly, yellow. Aquarists recognized the differences, which recur in other species in the genus, but they made little of it.

Two researchers, Elisabeth Martin and Michael Taborsky, decided to explore the significance of the color difference.[67] At the Konrad Lorenz Institute in Vienna they had available to them not one but two re-created sections of a West African stream, complete with the naturally occurring community of fishes, including competitors and predators. Without such a spacious and naturalistic setting, they might not have uncovered the rich and complex mating system of this species.

The two color morphs differ genetically and also in behavior. Red males can be either monogamous or haremic. Yellow males can also be monogamous, but never haremic. A third mating strategy emerged: Yellow males are often satellite males; red males never are. Up to three satellite males join a red male and his harem and take over defense of the territory against other males of their species as well as predators. In doing this, the yellow satellites incur considerable risk, as proved by frequent injury and a few deaths from assaults by the predators. The payoff for the satellite males is that they sometimes manage to "sneak" into the nest when the harem-master is fertilizing eggs and fertilize some themselves. This sneaking is often called parasitic fertilization (I'll describe sneaking in detail later).

But how can such a mating system be stable? To persist, the different morphs should have equal Darwinian fitness. Otherwise, one morph would eventually drive the other to extinction through the process of natural selection. Red harem males appear to have the highest fitness because they sired 3.3 times as many offspring as did monogamous red males, and seven times as many young as were fathered by the average yellow satellite male.

But now the interesting twist. The most dominant satellite males sired as many young as the monogamous red males. They did this by sneaking when their harem-master was distracted. This is the first time for any kind of animal that a sneaker has been proved to have reproductive success as high as a "conventional" mode of reproducing, here that of the monogamous males. Their data on longevity of males pursuing different strategies also suggested that lifetime Darwinian fitness is equal: The more offspring sired, the shorter the life span. I find this a slightly puzzling finding, given that satellite males are at higher risk when defending the territory.

SNAIL SHELLS SHAPE MATING

Lake Tanganyika is a cornucopia of cichlid mating systems, and this is nowhere better demonstrated than among the tiny species of the subfamily Lamprologinae. Some of these substrate brooders have moved out over the sandy bottom, which offers special opportunities for feeding. Unfortunately, it affords little protection from predators, except for empty snail shells. Several small lamprologines have adapted to hiding and reproducing in the shells; they are said to be ostracophilous, meaning shell loving. That shelter protects them from most predators, although the aquatic cobra can stick its narrow snout into the shell to capture them.

Many of the snail dwellers make bunkers out of their shells by interring them. Given an empty shell, the fish first tests it by entering. If the shell is suitable, the fish digs under it, eventually burying it so deeply that just the opening is exposed. In the process of submerging the shell, the fish pushes and pulls it to get it into just the right position. In *Neolamprologus brevis,* the opening faces down current, resulting in the flow slowing when it passes the raised lip, and that causes small planktonic animals to fall into the shell where they are eaten by the baby cichlids.[68] Just how the ostracophilous lamprologines find new abodes and set up house varies among the species.

Refuging in the snail shells limits the upper size of the fish. In some species, such as *N. brevis,* the male and female are so minuscule that a monogamous pair can cram into a single shell. Other populations of this species dwell in roomier crevices in rubble instead of in snail shells, and those individuals are larger than their shell-dwelling "cousins."[69]

The abundance of empty shells, notably those in the genus *Neothauma,* also varies from place to place, as does the prevailing species of snails, which differ in size. The pattern of distribution and size of snail shell influence the particular mating system of the cichlids.

Harems in Shells

In the haremic species *Neolamprologus ocellatus,* each male and female has its own shell.[70] But in *N. callipterus,* the ultimate haremic ostracophilous cichlid,[71] only the females dwell in the shells. And their shells are not interred but just lie in a cluster on the sand.

The male is much too large to enter any shell. On average, males are twelve times as heavy as their female mates, but the difference may reach nearly thirtyfold. This is the most extreme example of size dimorphism known for any vertebrate animal,[72] including elephant seals (if we ignore the parasitic males of the deep-sea angler fishes; females in other animals are sometimes much larger than males, but we are considering here only males larger than females). Males and females in *N. callipterus* are so different that when first discovered by scientists they were described as different species.

Ordinarily, evolutionary biologists would first assume that the extreme size dimorphism is driven by female choice of mate. On the contrary. The size dimorphism in *N. callipterus* is propelled by at least three other factors. First, the females must be small enough to enter the snail shell and spawn inside of it. Actually, females can use only the largest of the empty snail shells.

Second, the male must be so large that he can carry empty shells back to his territory, where he accumulates them; that sets a lower limit to the male's size. Thus the extreme size dimorphism appears driven largely by selection for males large and strong enough to transport bulky snail shells rather than by female choice.

The third factor is aggressive competition among males (called intrasexual selection). The male must be imposing enough to ward off other males who would usurp his territory or steal his shells, which are in short supply. Given a choice, females do not select the largest male offered.[73] Her concern is simply to have the roomiest shell she can get. That a male can acquire and hold a territory certifies him as a superior male. She would also benefit from mating with a male who can protect her from other males until she has finished her reproductive cycle; apparently, however, she does not choose on that basis. Consequently, males are not selected to be colorful, just big and strong.

Sato demonstrated that body size of a male is important in determining his reproductive success.[74] The biggest males have the most shells in their territories, those shells enjoy the highest occupancy rate by females, and such males maintain the longest control of their territories. Moreover, to maximize his Darwinian fitness each territorial male scurries about, gathering up loose snail shells. Eventually they carpet his territory like derelict autos in a junkyard. Males swim as far as 60 feet (about 20 m) to collect shells. It is all a male can do to struggle back to his territory lugging a large snail shell. Often he steals them from neighboring males, and vice versa, resulting in the abundance of shells fluctuating in territories. The "best" males accumulate up to one hundred shells, and with them a bevy of reproductive females.

Consequences of Kidnapping

Sometimes when the male steals a snail shell from another male, the shell already contains a female. How fortunate for the male, one might happily conclude–yet another bride in his harem. But that would be wrong.

Consider the now classic studies of lions [75] and some monkeys, as well.[76] When a new male lion takes over a pride of females he kills the infants to bring the females into heat as quickly as possible. Caring for a stranger's offspring would cost the male the opportunity to produce his own young. The tenure of the breeding male is brief, in the larger scheme of things, because other males will eventually evict him, so he must make the most of his opportunity.

Each male of our ostracophilous cichlid, *N. callipterus,* holds a territory for as few as 10 to more than 120 days. While reproducing, he gradually becomes weaker because he is so occupied he seldom eats. (And when he does eat, it is usually eggs of females in his harem.) The female abducted with the stolen shell is caring for eggs fertilized by a different male. Our kidnapping male cannot afford to waste time waiting for the abducted female to provide him with new eggs to fertilize. After all, she remains in the shell with her offspring for about seventeen days. And then she leaves because she is spent and needs to feed before she can spawn again. To the male, she is a total waste of his time and energy. What is he to do? He makes life miserable for his kidnapped female, driving her out of her shell and thus out of the colony.

A male cichlid that supplants a territorial male and takes over his harem behaves the same way. He picks up the shell, female and all, and shakes it. Any babies that fall out are gobbled up (so much for the benefit of the species). If he can grasp the tail of the female, he will. He tries to pull her out and eventually succeeds. She flees and her offspring are doomed. Within a day or two, a new female appears and spawns in the shell.

Shells large enough for females to enter are in short supply. Females full of eggs cruise around, constantly on the lookout for spawning opportunities. When a female finds a large, empty shell in the territory of a male, she moves right in. Because females full of eggs are readily available, males have evolved the tactic of eviction and infanticide to increase their fitness.

The bottom line is that the reproductive interest of the female may come into conflict with that of the male. The female needs only one male to fertilize her eggs. The male, however, is adapted to fertilize the eggs of as many females as possible, and as quickly as possible. Accordingly, if a female has offspring fathered by a different male, he tries to force her out of her large shell so he can attract a new female whose eggs he can fertilize. Because large shells are a limiting resource, one expects the females to compete aggressively among themselves. However, large size among them is selected against because the female must remain small enough to enter the shell.

Shells That Bind

The accumulation of patches of snail shells has made possible one further step in the evolution of cichlid mating systems. It has been observed in no other family of fishes. Indeed, it seems confined to insects, birds, and mammals, and even among them it is rare. The extended family, what we would call the macrofamily, is worthy of the special consideration that Steve Emlen of Cornell University has given it.[77] In what he terms a simple family, all the reproducing is done by only one member of each sex; wolves are a case in point. In the extended family, in contrast, two or more members of both sexes share in reproduction.

Thanks to concerted, but not yet published, research in the laboratory and in the field by Jürg Lamprecht of the Max-Plank-Institut in Seewiesen, and his doctoral student Uwe Kohler, we now have a good understanding of the extended family of another lamprologine cichlid, *Neolamprologus multifasciatus*.

Neolamprologus multifasciatus lives in those places in Lake Tanganyika where many snail shells accumulate in depressions 20 to 50 cm in diameter (roughly 10 to 20 inches). Although difficult to excavate from the hard, semi-mineralized sandy bottom, the fish dig the shells out in substantial numbers, forming a complex of apartments that would do modern public housing proud. The result is a relatively secure colony of *N. multifasciatus*.

A colony has up to nineteen members with one to three males, the largest of whom is dominant (by convention, the dominant group member is called alpha, the next dominant beta, and the third gamma). Up to five females also live in the group. The remainder of the fish are juveniles of various ages. The smaller juveniles share a snail shell with their mother.

Analysis of nineteen colonies revealed that only the alpha and beta males sired young. Seven of the ten beta males examined, using molecular genetics, were possibly sons of the alpha male as were the two gamma males. Thus those fish had probably been born into the colony and remained there until reproductively mature. In seven colonies, at least one of the females was a putative daughter of the alpha male, and six of those females had offspring. Thus groups consist of more than one generation of reproducing individuals.

Occasionally, individuals from other groups are accepted into the macrofamily. Joining a new group, however, is risky. In one experiment, the researchers transplanted individuals in their shells to new colonies, but the newcomers were quickly driven out. To start a new colony, a solitary fish must dig out shells that are deeply embedded in the hard-crusted sand, and that takes much time; meanwhile, they are exposed to predators.

If all the young remain in the colony, it becomes overpopulated, even though some of them are being picked off by predators. If a young fish leaves the colony and braves the open sand, it is a near certainty that it will end up in the gut of a predator. When the researchers set out unoccupied snail shells near colonies, they were occupied by fish within a few hours, proving some migrants nonetheless are roaming around on the open sand bottom. Finally, the researchers removed the alpha males from some colonies and within half an hour a large new male from a nearby colony invaded and took on the role of alpha male. The newcomer was probably a beta male that was larger than the resident beta male.

As an aside, the resident females feigned receptivity to new alpha males, as do female primates when a fresh male takes over a group. Lamprecht suspects infanticide by the new male may occur, but that question awaits further field studies.

The picture of the extended family that materializes is one of a relatively stable group of mostly related individuals. As Lamprecht and Kohler point out, the families meet three requirements for the development of extended families: They live long enough for overlapping generations; staying in the colony is a better option than dispersing; and the dominant individuals either do not benefit from thwarting the reproduction of others in the group, or they cannot.[78] Thus, at least one cichlid fish meets the criteria for an extended family. It is much like that of extended families among birds and mammals.

SNEAKY SPAWNERS

Through all this, I have only mentioned in passing a fascinating mating practice, one that is widespread among animals but poorly known among cichlids. I refer to males who parasitize the spawn of other males. This phenomenon is called variously cheating, sneaking, or parasitic spawning.[79] It goes back to the basic anisogamy argument. Recall the reasoning: Some males, through aggression, have access to a disproportionate number of mates. That means some, perhaps most, smaller or weaker males get few or no matings–the zero-sum game. Making the best of a bad situation, those subordinate males attempt to steal fertilizations through a number of tactics.

The absence of sneaking is also informative. Parasitic spawning has not yet been reported for monogamous substrate-brooding species, probably because the male and female stay in close attendance throughout the act of spawning. A haremic mating system, on the other hand, provides at least some opportunity for sneaking. As the alpha male devotes his attention to one of his females, other consorts are temporarily neglected, giving the sneaker his chance. Sneaking also occurs among lekking mouth brooders where the sneaker often mimics a female. In Chapter 10, I consider parasitic spawning in more detail.

THE MESSAGE

Cichlids illustrate plainly the way mating systems evolve. Starting with monogamous substrate brooders, sexual conflict is manifest, but muted, in the separation of male and female roles, even though the pair shares reproductive interests. Ecological factors, such as degree of exposure to predation, keeps the pair together. Any change in the environment that lessens the need for two parents to protect the young gives the male the opportunity to mate with other females, as seen in the harem. This is demonstrated nicely by the ostracophilous cichlids in Lake Tanganyika. Depending on the distribution of empty snail shells, the species are variously monogamous, haremic, or live in extended families.

Some might wonder why empty snail shells have not shaped the mating systems of mouth brooders. Many species in Lake Tanganyika and all the hundreds of species of cichlids in Lake Malawi are mouth brooders, and shells are readily available in both lakes. Moreover, small species, such as

FIGURE 4.2
Pseudotropheus livingstoni, a Lake Malawi dweller, is one of the few mouth-brooding species that occupies a snail shell.

Pseudotropheus livingstoni, take refuge in shells, but they never breed there. Evidently the way mouth brooders spawn works against remaining in an empty shell. This illustrates beautifully how mode of spawning rules out some mating systems.

For substrate brooders in general, individual recognition of mate is essential in monogamous systems to keep aggression under control. Individuality decreases in haremic species as mate recognition becomes less a matter of knowing one another and more a matter of recognizing the opposite sex quickly. The extended family probably evolved from a haremic mating system, but the importance of individual recognition again became foremost.

With the appearance of mouth brooding in monogamous species, the evolutionary rush was on to revert to classic male-female conflict. Males and females play radically different roles in lekking species, for instance, and individual recognition of mates may be nonexistent. Social contact has become just a matter of males fertilizing the eggs of the females–the anonymous society.

The comparative study of mating systems is done currently almost entirely at the level of the final outcome of several behavioral mechanisms. Feeding behavior, use of space, exposure to predation, sequestering a mate, and so on, are viewed as shaping the mating system. The mechanisms underlying choices and interactions were central issues in animal behavior until recently, however. Communication was seen as a central issue. For cichlid fishes, the start of an interaction features the performance of threat displays by one individual, to which the other may respond in kind, and this is clearly communication. Even if the interaction is between a male and female and may lead to pair bonding, the encounter looks at first as though the cichlids are starting a fight. To understand cichlid behavior, therefore, one must understand how they engage one another. That is the theme of the next chapter.

≈ *five* ≈

OH YEAH? PUT UP YOUR FINS!

So great was the reputation for fighting among cichlid fishes that one of the early favorites, *'Cichlasoma' octofasciatum,* was named after the pugilist Jack Dempsey who was renowned for his overpowering style of boxing. True to form, most cichlids the beginning aquarist is apt to buy are anything but amiable, unfortunately, often killing one another when put together. I'll bet that no expert cichlid aquarist has ever gone without having some, usually several, fatalities from fighting among his or her cichlids.

Most of the belligerence is inflicted on members of the same species. But cichlids are not chauvinists. Sometimes they go after other species, including noncichlids that resemble them. The good news is that cichlid hobbyists increasingly understand how to maintain their pets in ways that eliminate the peril of injury from fighting. Some cichlids, moreover, are relatively mild mannered and can be kept together with less chance of mayhem. But even then caution is called for.

One of the joys of having an aquarium at home is the tranquil beauty it provides. Landscaping an aquarium with delicate underwater plants, artfully arranged rocks, and driftwood is a source of pleasure and pride to the owner. But the typical cichlid that feels the urge to reproduce, which most soon do when well cared for, has other ideas. After becoming comfortable in their new aquarium, they start setting up house. To the aquarist, they are the tenants from hell. Like a bulldozer, the fish starts digging in the bottom, piling up gravel in all the wrong places, shoving rocks around and ripping out plants.

Understanding such aggressive behavior between cichlids, and its flip side, fear, is key to comprehending how these intelligent animals communicate with one another. Almost all interactions between adults start out with signals that are the prelude to combat, and this includes pair formation in the monogamous species.

WHEN TO FIGHT

Why do animals fight at all, and why are some cichlids so highly aggressive while others are relatively peaceable? Aggressive cichlids may seem to fight mindlessly when put into an aquarium where the subordinate fish cannot escape and are thrashed by the dominant one. Seeing such behavior, most people think the fish are territorial in the traditional sense, as exemplified by songbirds. Natural selection, however, works in a powerful way to limit fighting to only those encounters where benefits on average outweigh costs. And fighting is indeed costly. The most obvious price is injury, sometimes death. Even when a fish wins, the victory may be pyrrhic.

The fish could become so badly damaged as to impair its ability to escape a predator, to feed, to mate, or to retain its territory. Even when no obvious damage is sustained, fighting still burns up energy. Combat also consumes time that would be better spent in activities with direct payoff, such as feeding and mating. And because a fight is such an intense, focused enterprise, both fighters become oblivious to the approach of predators. Thus an animal should confine its fighting to a ritualistic phase of displaying and thereby avoid overt, injurious combat and reduce the risk of predation.[1] In an evolutionary sense, the animal should calculate when it pays to fight, how intensely to fight, and when not to fight. So, when does it pay to fight?

Fighting occurs only when some resource is in limited supply and is economically defensible. And by limited, or limiting, I mean that without it, the Darwinian fitness of the individual is reduced. A resource of low quality, such as scattered food or food of minor nutritional value, is not worth defending. On the other hand, if the resource is of high quality, say a concentration of nourishment rich in calories, then so many competitors might be drawn to the food that the price paid for defending it would negate the gain. Other resources can be limiting, too, such as scarcity of available mates, breeding sites, and refuges.

Most cichlids are intimately acquainted with the bottom where they live. Even those that feed well up from the bottom commonly take refuge in the reef when predators approach, again with exceptions. For many cichlids, a place of refuge is a sanctuary that can be, and often is, defended. A breeding site for cichlids has much in common with a refuge, and it is often in short supply.[2]

Cichlids also scrap over food, but in nature they seldom hold feeding territories. In this respect they differ from songbirds and instead behave like most other freshwater fishes. Let's look at why freshwater fishes rarely defend feeding territories.

WHY TERRITORIES OF MARINE AND FRESHWATER FISHES DIFFER

Territoriality requires two behavioral mechanisms. First, the animal must remain in one place. Second, it has to exclude intruders. That seems simple enough, but the necessity of both behavioral components is often overlooked.

Fish kept in aquaria often guard a territory where food is introduced. In nature, the same fish may not hold a feeding territory because the food moves around. Many kinds of freshwater fishes establish breeding territories but not feeding territories. Moreover, ichthyologists who study marine fishes are seldom knowledgeable about freshwater fishes, and vice versa. For these reasons, territoriality among fishes has been poorly understood.

When marine and freshwater fishes were finally compared, a sharp difference emerged.[3] Marine reef fishes, especially those inhabiting coral reefs, often control feeding territories, whereas freshwater fishes rarely do. A common situation on the coral reef is an individual or a pair defending a territory that supplies their food.[4] The territory also serves as a spawning site for males of many species, though the females leave their territories to spawn within those of the males.[5] Herbivorous surgeon fish, in contrast, guard feeding territories[6] but migrate away from them to spawn in groups.[7] Thus many species of reef fishes have feeding territories, and many of them even protect multipurpose territories.

Few freshwater fishes have feeding territories. The notable exceptions are trout and juvenile salmon. Many of those individuals, however, are not territorial at all or alternate between territoriality and roving.[8]

Cichlids mostly follow the pattern typical of freshwater fishes: no feeding territories. Individuals, however, sometimes defend a feeding site, but only briefly.[9] Thus they possess the behavioral mechanisms necessary for feeding territories, but the temporal scale must also be reckoned with. These so-called territories are fleeting, enduring perhaps a minute or two. Then the fish moves on to find a new patch of food. Such flexibility of behavior suggests that territoriality lies at one end of a continuum of degrees of competition for limiting resources.

Such flexibility offers insight into why many marine fishes are territorial over food and freshwater species only rarely are. Within the lifetime of a territorial coral-reef fish, barring the odd cyclonic storm or tsunami, the environment is stable. A damselfish farming its garden of algae can count on its algal food persisting long enough during its life to be worth defending. For a freshwater fish, that is not the case.

The primary environment of a freshwater fish is the river or stream. Lakes come and go. And flowing waters are inherently unstable. Freshets frequently rearrange the channel, scour soft bottoms, and shove rocks and boulders into new positions. Trees fall, opening patches of stream to more sun and hence increased productivity. The consequence is that food is episodic in time and place.[10]

Freshwater fishes clearly have the behavioral mechanisms to become territorial. Staying fixed to a site seldom pays, however, at least long enough to talk comfortably about a feeding territory. But if some freshwater fish have evolved feeding territories, where would you expect to find them?

SOME CICHLIDS DEFEND THEIR TURF

Following the marine model, one should look at herbivorous fishes in the few truly ancient lakes that have long periods of stability. I reasoned that at least some of the cichlids in the African rift lakes should have feeding territories because their feeding biology is so similar to that of damselfishes.[11] Additionally, cichlid fishes are marine derivatives. As such, they should be preadapted to life in lakes and perhaps more disposed to forming feeding territories than are the archetypal freshwater fishes, the minnows, catfishes, and their relatives.

I started researching this hypothesis by combing through the pioneering, careful monograph by Tony Ribbink and his colleagues on the mbuna type of cichlids in Lake Malawi.[12] These mouth-brooding cichlids graze on the *Aufwuchs* in shallow water and thus in feeding behavior parallel their algal-eating marine relatives, the damselfishes. They noted, almost in passing, that some mbuna do indeed have feeding territories.

These feeding territories may have evolved from breeding sites.[13] The widespread, presumably ancestral, pattern of breeding in the mbuna is that the male holds a territory to which the female comes to have her eggs fertilized. She then departs with the eggs in her mouth. Potentially, the male can breed almost continuously in the benign tropical climate of the ancient stable lake. That is, he can do so if he can continue to feed right where he receives the females.

From the monograph of Ribbink and colleagues we can see a representative continuum of territorial states, from herbivorous males that maintain territories exclusively for breeding to species in which the males, the females, and even the juveniles have feeding territories. The pinnacle is achieved in the form of *Pseudotropheus elongatus,* which Ribbink called "aggressive" in guarding algal gardens that are maintained by males, females, and large juveniles.

That the territory is valuable to the female is shown by her behavior. She leaves her territory to spawn with a nearby male. Then she returns to her territory and there gestates her eggs orally, even though the females of most species cannot feed while doing so. She probably benefits directly from refuging in the territory and from having the food it harbors available to her when her offspring finally emerge from her mouth and soon depart. More recent research in Africa has documented feeding territories in several species.[14] The females of some species of haremic cichlids remain on territory for successive rounds of reproduction. That should be possible only if these substrate-brooding females are able to feed on the territory as well as to care for their fry.

The Lake Tanganyika cichlid *Gnathochromis pfefferi* displays yet another level of complexity. They have feeding territories that they defend only in the afternoons and where they also sleep. Each morning they migrate to their separate breeding territories where they receive females for spawning. The females are not territorial; instead they have consistent, overlapping home ranges.

That *G. pfefferi* has a feeding territory separate from the one used for breeding suggests the possibility of feeding territories evolving independently from breeding ones. Therefore, we need to keep an open mind on how feeding territories have arisen among the cichlids in the Great Lakes of Africa. Feeding territories in some of the species may well have evolved without regard to reproductive behavior, but then later become spawning sites.

Rituals Among Dear Enemies

No matter how aggressive, breeding cichlids must not fight needlessly. Breeding pairs often have neighbor pairs close by. The neighbors already have the territory they need, so they are not a serious threat, though they might try to encroach a bit. The pair learns who their neighbors are, and they expend only the energy needed to keep them in their place. The result is that their encounters consist of demonstrative aggression. Serious fighting is reserved for potential territorial usurpers. This understanding between territorial neighbors is sometimes called the dear-enemy phenomenon.[15]

One of the more entertaining illustrations of ritualized boundary encounters is done by the orange chromide, *Etroplus maculatus*, of India and Sri Lanka. When a territorial pair confronts a neighbor pair at the boundary, the male faces the neighbor male and the female faces the other female. In each pair of rivals, one swims forward with its tail end while backing up with its front end, using the pectoral fins. When the back end prevails (= attack) the orange chromide surges forward a bit, but when the front end dominates (= fear) the fish moves backward.

As one fish of a contesting pair moves forward its rival backs up. The advancing fish starts penetrating its neighbor's territory, and as a result fear starts to overwhelm its aggression. Then the front end takes control and the fish retreats while its rival advances toward it. The two fish thus seem to do a coordinated dance back and forth across the boundary. This lock-step, back-and-forth behavior has been called pendling after the action of a clock's pendulum.

This shift away from damaging fights to more demonstrative display seems to form a gradient that runs from the more isolated monogamous substrate brooders to the colonial, lekking mouth brooders. In lekking tilapias, the territorial males display frontally with open mouths, but also exchange tail beats. They seldom have damaging fights, probably because full-blown combat would become almost continuous and hence too energetically costly.

In the lekking tilapias, the territories of males are sometimes so compressed that the diameter of a territory is only two to three body lengths of the male. Each male digs out a large pit that serves as a territory in which females lay their eggs. When the male takes up a mouthful of sand, it often spits it out at its neighbors, raising parapets that become the territorial boundaries. When the territories are maximally packed, they converge on a hexagonal shape, just as in the cells of a honeycomb in a beehive.[16] In such species, boundary displays are highly ritualized.[17]

Beyond the known neighbors, the territory owner must contend with a variety of intruders, especially when the territory is held for feeding. Driving away all potential thieves would be a Herculean task and an enormous waste of energy. The clever fish learns who else would like to eat what he guards. Members of its own species are the obvious competitors, but other species can also be trophic rivals, and they should be recognized as such and driven away. Indeed they are, among the mbuna cichlids of Lake Malawi. Mbuna are least tolerant of their own species and of those species that are similarly specialized.[18]

Ganging Up on Owners

On the coral reef, the best patches of algae are often controlled by highly aggressive herbivorous fishes. The less aggressive species are denied access to the best areas. What are the excluded species to do? This is what I discovered.

While diving on the Kona Coast of Hawaii, I noticed large schools of the manini, a classic T-shirt surgeon fish *(Acanthurus triostegus)*. This is the "Casper Milktoast" of the herbivorous surgeon fishes found there, and it is not territorial.[19] As I watched them, they moved slowly across the reef in a long column. They were a roving gang whose goal was to overcome the defenses of the territory holders.

From time to time the lead fish in the school headed to the bottom where they entered the territory of another herbivorous surgeon fish, the purple tang *(Acanthurus nigrofuscus)*. The poor purple tang tried to drive them away by attacking, but each time it hit one manini it just moved a bit and the purple tang found another manini in its path, depleting its pantry of precious algae.[20] On a more general level, George Williams calls this the Saint Ignatius strategy after the bishop of Antioch who adhered to the practice of "being companionable so that evil will befall a companion instead of oneself."[21]

Cichlid fishes in Lake Malawi also mob territory holders.[22] And so do cichlids in Lake Tanganyika.[23] The herbivorous *Petrochromis fasciolatus* forms columns, just like the manini, and swarms the territories of the dominant herbivorous cichlids. The most effective school size falls in the range of about 40 to 150 fish. Among the several species, they select especially the territories of *Neolamprologus moorii*. The reason they are so selective is because the territories of *N. moorii* have almost fifteen times as much algae as that found in other territories.

Even with our limited knowledge, we see again that cichlids are out of the ordinary among freshwater fishes. Several species have evolved distinct feeding territories paralleling those recorded among coral-reef fishes. But some cichlids are not territorial at all. Certain reef-dwelling cichlids in Lake Malawi and in Lake Tanganyika have even given up breeding territories, which means they do not fight over space.

To return to our starting theme, aggression, as manifested in territoriality, arises only when the benefits exceed the costs. For reasons we do not yet understand, some species have found ways to reproduce without establishing and defending territories. The underlying cause is most likely differences in feeding behavior. Algal feeding favors territoriality, and other modes of feeding, here conveniently lumped as carnivory, tend to promote a more mobile lifestyle and hence a reduction in territoriality.[24]

THE FIGHT

The way an encounter develops between cichlids is mirrored in clashes among all kinds of animals. The overture between two hostile animals is remarkably general and applies to cichlids as well. When actual fighting emerges, however, its form tends to diverge across species depending on the weapons with which they are endowed. Here is the way a fight between two cichlids progresses.

Reading the Mind of a Fish

When two cichlids approach one another to contest control over some resource, and they are equally matched in all respects, each is stimulated simultaneously to attack and to flee from the other. The external signs suggest they experience a conflict between aggression and fear.[25] As they start their approach, they seem highly aggressive: The mouth is clenched shut, and the gill covers, or opercles, are flared out from the head. The median fins are spread, but the spines of the dorsal fin may be laid back. The colors are intensified and are rich in contrast.

As they come closer together, signs of fear become steadily more detectable. They slow their head-on approach. Once in close proximity, each turns off, presenting its side to the opponent so the two fish become aligned, side by side, sometimes head next to head and sometimes head to tail. Now all the fins are fully spread. Predictably each fish opens its mouth wide. Concurrently, it passes deep sideways undulations down its body while braking with its pectoral fins to resist the forward force; this behavior is called tail beating, and it may be done in a forceful snap-whip fashion. The result is an imposing display and one that washes the opponent with a hefty flow of water that may even be perceived as sound.[26]

All of this behavior indicates conflict. Each fish has positioned itself so that it is prepared either to flee or to attack. Fleeing, swimming away, would be done by means of powerful body undulations in which the tail swings from side to side. Thus tail beating reflects the urge to escape from the dangerous opponent. At the same time, the fish extends its pectoral fins in a breaking action that counters the undulations, reflecting the motivation to stay put and attack the opponent. During lateral display, the body is commonly bent into an S-shape. The head turned to the opponent indicates a greater likelihood of biting; the head away, a higher probability of giving up. The open mouth suggests preparation to bite, but initially the opponents do not actually bite, probably because of the danger of being bitten back.

The close coordination between the contestants, faintly resembling a dance, suggested to Peter Hurd the possibility that the contestants may actually be, in some unconventional way, cooperating. Two dwarf cichlids in a fight presumably would benefit from positioning themselves so as best to assess the prowess of the opponent.[27] Thus they would be supplying and obtaining valuable information about their opponent while minimizing the

FIGURE 5.1 Two males of the South American cichlid *Aequidens pulcher* lock jaws as the last step in an intense fight.

possibility of incurring costly injury. The same argument applies to further coordination between the opponents, if the fight escalates.

Escalation

After the initial salvo of displays and tail beats, one or both contestants usually decides retreat is the better part of valor. But when each is in the position where the perceived benefits outweigh the risk, or when escape is out of the question (as in an aquarium), the clash may hang at this point briefly. If neither yields, the encounter then escalates to the next stage, frontal confrontation.

Now the fish maneuver to face off. An odd, almost gentlemanly behavior happens at that time. If the two combatants are, say, head to tail, they then turn toward each other to face off. As they do so, they may touch the side of the opponent with their mouth, but, and this is the amazing part, they do not bite their opponent's exposed flank. On the rare occasion when this happens, the bitten cichlid seems startled, as though shocked by the violation of protocol. The bitten fish quickly retaliates by trying to bite its adversary. More often, however, the two fish coordinate their movements with one another to assume a head-on position. That facing orientation is a dangerous escalation and indicates they are preparing to inflict damage.

As the two face off, they jockey to bite the opponent on the mouth. Mouths are opened to varying degrees, widely in some species whose mouth cavity may be adorned with conspicuous color (which in the case of the firemouth, *'Cichlasoma' meeki*, gives rise to its name).

The battle is truly joined when the antagonists finally lock jaws, each holding the other by its mouth. Having coupled, each combatant alternately pushes or pulls the other. At some point, one adversary vigorously jerks its head to the side, ripping the lips of the other. Then they may either resume jaw locking, with its mouth wrestling, or progress to carouselling.

Carouselling emerges as each fish lunges at the side or tail of the other. If they separate a bit, the lunging becomes an accelerated ramming action, the goal of which appears to be to rip the side or face of the opponent with the open jaws. Often they wind up side by side and head to tail, each trying to grasp the fins of the other. That produces the carousel effect as they rapidly circle one another, frantically biting at their opponent. Around and around they go, oblivious to all about them. Now one latches onto any fin of the other and often hangs on. The two fish may lie almost still for moments, each munching on some fin of the other and shredding its membrane as it pulls away with a sideward jerk of the head.

Carouselling and mouth locking alternate. Occasionally elements appear from earlier stages of the fight, such as tail beating. During the advanced stage, individual differences in fighting style are likely to appear. One of our Midas cichlids regularly swam up and then dived down onto the head of the other, biting as it made contact.[28] We dubbed him the "bomber."

Up to this point neither fish shows signs of giving up, or of winning, each maintaining the pretext of being willing to fight to the death. But then the eyes of one of them seem to search the aquarium, seeking a way out. When the other attacks, it nonetheless carries on. Sooner or later, however, that fish suddenly gives up.

The loser backs off with folded fins. Its colors fade, and the markings become soft edged. It moves up and away from the bottom, which is the prize area. If this were happening in nature, the loser would rapidly swim away.

I have misgivings about staging fights between these powerful fish. I am not consoled by having seen serious fights between Midas cichlids in Nicaraguan lakes. The fish that fought in the lab were immediately separated at the end of a fight and put into a recovery aquarium with bactericidal medication. By the next day both contestants were bruised but otherwise recovered, ready to defend their breeding territory and seek a mate. Within days their injuries had healed.

WHAT IS YOUR INTENT?

Nowhere is the difference in thinking about group selection and individual selection more apparent than in the history of aggression.[29] Just a few decades ago, the reason contesting animals display rather than fight was said to be that fighting risks injury, even death. True enough, and that is still accepted. The difficulty arose when answering the question, why should they evade serious harm? The answer would seem so obvious that some biologists wondered why it would even be asked: Injurious combat would be bad for the species. The benefit of not fighting therefore accrues to the group. Unfortunately, no one has been able yet to explain how that could evolve.

When the question was first posed, many ethologists believed that if you analyzed a prolonged combat between two animals you should be able to detect early on, perhaps right at the beginning, which individual was going to win. Fights are so complex and rich in data, however, that few had the patience to gather and analyze the mountains of data that resulted–but Michael Simpson did.[30] The unanticipated outcome of his study on the Siamese fighting fish was that neither the investigator nor, apparently, the fish were able to predict who would be the eventual victor.

A little later, game theorists such as John Maynard Smith provided a better framework.[31] Ethologists had used terms like "bluffing" in reference to hostile encounters, but they had not gone far enough. The argument from game theory extended our thinking by focusing on the interests of the individuals, not those of the species. When two animals fight over some limiting resource they are motivated only by their own selfish Darwinian fitness. In the game, neither contestant should reveal its true state, or intent. Each aspires to "cheat" by "deceiving" the other into believing it is superior. It does this by emitting signals indicating size or prowess beyond reality.

Amoz Zahavi argued, however, that natural selection should favor signals that cannot be cheated.[32] Uncheatable signals carry a cost, a handicap. An inferior contestant would find the costs too great to bear and thus would fail if it tried to signal something that it is not. The huge tail of a peacock, for instance, is a severe handicap and can be borne only by a high-quality male.

This argument was originally meant to apply to female choice of superior males, but it holds just as well for displays exchanged by rival males. In any event, genes for signals that are effective in bluff should spread through the population with the upper limit set by the costs incurred by the exaggerated trait. In the case of cichlid fishes, the upper limit of, say, large fins should be determined by the increasing hydrodynamic drag they create, which is a severe cost.

Theory therefore holds that to prevail, neither competitor should show its hand. Each tries to impress the other with not only its prowess but its persistence in defending the resource. If one contestant reveals that it intends to withdraw, then further fighting would have no point. So, the name of the game is to outbluff your opponent by carrying on this "war of attrition."[33]

MODELING THROUGH

Building on the game-theory model, and based on fights of a South American dwarf cichlid, *Nannacara anomala,* Magnus Enquist and Olof Leimar produced an improved perspective on hostile encounters.[34] They called their model "sequential assessment."

The fictional contestants in the model are assumed to be sufficiently well matched that each is uncertain about its chance of winning. Each probes the other to assess its prowess relative to itself. Hence, when the encounter begins, the contenders try to maximize the information communicated while keeping costs to a minimum, for instance, through displays and tail beating.

Enquist and Leimar made several predictions from their sequential-assessment model. Some of these were insightful, but some were post hoc. For instance, fights proceed in stages, advancing from one stage to the next; students of cichlid behavior knew that already. The insightful part was explaining why. At each stage, the contestants repeat a certain kind of behavior. Each time they do so, less information is communicated, as though the other contestant is saying, "Yes, of course. You've already said that." What each is trying to impress on its opponent, the model asserts, is, "I am bigger and stronger than you, so kindly yield."

If neither yields, they must proceed to the next, more dangerous, and hence costly, stage where new information is obtained about the prowess of the opponent. The more closely the two fish are matched (or "symmetrical"), the more likely they are to keep escalating through successive stages until outright damaging fighting finally erupts.

In nature one seldom sees a knock-down-drag-out fight that progresses to actual injury, though they have been reported,[35] and I have observed a few while diving in Nicaraguan lakes. The reason the staged fights become damaging is obvious. In an aquarium, the fish that senses it is losing has nowhere to flee. Seeing no path for escape, it continues to fight as the only means of protecting itself because trying to flee in an aquarium makes the loser more vulnerable to further attack. Failure to recognize the subtle clues that indicate the fish has lost explains why the final stage of a fight in captivity is so variable and its duration so unpredictable.[36]

Another variable influencing contests is history. When two fish have fought in the past, they may remember how they fared. In matched encounters in captivity, the previous loser is more inclined to give up.[37] Game theorists find more interesting, however, a more universal reason why animals rarely get into damaging encounters: Such encounters are seldom symmetrical. And the sources of asymmetry can be neatly categorized.

MISMATCHED CONTESTANTS

A step forward in theory was the realization that asymmetry can be tied to ownership. The phenomenon was labeled briefly the bourgeois effect, but it is most generally known as prior residency. Its existence was first demonstrated in a platyfish, a relative of the common guppy, and well before the theory had been elaborated.[38]

The resident individual has a substantial advantage over the intruder.[39] The difference is mental and is enhanced by a number of variables such as the quality of the home, the presence of a mate or eggs, the physical properties of the owner, and so on. Theorists refer to the totality of these effects as resource holding potential.

To overcome prior residency, the intruder must be substantially larger than the resident, another fundamental asymmetry, though how much larger varies with species. Studies on a variety of fishes suggest they can perceive a difference in size between themselves and an opponent of about 10 percent, perhaps less.

Body size is one of the most, perhaps the single most, important aspects of asymmetry.[40] Scientists have tested the effect of body size on the outcome of actual fights by fooling the fish into thinking each is the legitimate resident. With the help of William Rogers and Natasha Fraley, we did this with the Midas cichlid.[41] We tried to match the fish by weight, but the match was never perfect. We then put each fish into one-half of an aquarium, separating them by an opaque barrier. The barrier was removed after varying periods of time.

When the barrier was left in place overnight, each fish reacted as though it were the resident. The encounters quickly escalated into serious fights, and the fights were often long and drawn out. The loser was almost always the smaller contestant. A size difference as small as 2 percent by weight usually predicted the outcome. The fish behaved as though they could not assess such a small difference through displays but instead had to resolve the contest through overt fighting.

This experiment also revealed what a difference the prior-residency effect makes. When the barrier was removed in less than two hours, the fish behaved differently, as though they had not yet fully developed a sense of ownership. They seemed uncertain and took longer to reach the escalating phase of the fight. And the fights were much briefer than when the fish were 24-hour residents. Size then played only a minor role.

When prior residency was eliminated as a factor, the most enlightening result was that the fish that bit first almost always won. In contrast, when they had 24 hours to become the resident, the question of who was first to bite played no role. Further, the fish known from previous studies to show the highest readiness to fight its mirror image, thus the inherently more aggressive fish, was predictably the first one to bite after two hours of separation. Among fish that had been separated 24 hours, readiness was of no consequence.

This study told us some things of fundamental importance about the nature of aggression in cichlids, and it may generalize to other kinds of animals. Readiness to attack is highly variable across individuals. That suggests the trade-offs between costs and benefits resulting from quickness to fight are themselves variable. Second, the ability to fight varies only slightly among the individuals. That indicates strong selection for prowess, once committed to a fight.

ARE MALES THE TOUGH GUYS?

In characterizing a cichlid fight, I have taken some license. The model for the fight is that of monogamous substrate-brooding cichlids that are relatively monomorphic. It is a reasonable starting point because such cichlids are probably nearest to the starting condition in the evolution of this family, and they are belligerent cichlids that show the full gamut of aggression.

Monogamous cichlids often face fierce competition for breeding sites.[42] A male Midas cichlid in a Nicaraguan lake cannot alone secure a territory. It requires the combative abilities of a pair to do so. After they have acquired a territory, they remain under siege from other pairs who seek breeding territories through the entire reproductive cycle of five to six weeks. Thus in biparental species, selection favors not only highly aggressive males but also combative females who can withstand the onslaught until the end of the breeding cycle. Consequently, natural selection has produced females who are as aggressively competent as are the males. Some of our findings indicate nonbreeding female Midas cichlids are the equal of a male of the same size.[43] And our recent research reveals that the female African cichlid *Julidochromis marlieri* dominates her male mate even when he is slightly larger.

Differences between the sexes are largely a matter of motivation, as in their readiness to attack, and they have not been studied much in that respect. Jenny Holder, Richard Francis, and I explored this issue by analyzing the way male and female Midas cichlids attack their mirror images.[44] This study was unusual in that so many individual fish were used, 130 females and 132 males, and individuals were tested a number of times.

Males proved more ready to assault their mirror image than females. Also, aggressiveness varied from one male to the next, and appreciably more than between females. Males also varied in relation to their reproductive state. Those closest to spawning, as judged by their genital papillae, were the most aggressive. (The genital papilla is a small flap of tissue right behind the vent of the fish.) Despite the fluctuation in aggressiveness, individual males were more consistent through time than were females. Females were maximally aggressive just before spawning.

Recent research on males of a monogamous African cichlid, *Tilapia zilli,* have shown that the probability of winning a fight also depends on the reproductive state of the male, as estimated from the size of its testes.[45] To understand the aggressiveness in this and possibly other cichlids, a window in the fish to allow viewing the testes would be convenient. In a way, the male's jaw provides such a window, albeit a smudged one: The larger the male's testes the larger the mouth, though the correlation is not strong.[46]

In the highly dimorphic mouth-brooding cichlids, we see a striking difference between the sexes in aggressiveness. Females typically do not hold reproductive territories. The general situation is that of a lekking species. The female approaches the displaying males only to spawn. Then she departs. Males fight over territories. Females do not.

The haremic dwarf cichlids of South America, the *Apistogramma,* differ from both of the preceding patterns. Recall that the males are much the larger sex and are highly ornamented. Their aggression appears to be exceptionally ritualized, consisting of elaborate movements of fins and exaggerated postures, though they do engage in overt combat; however, the extent to which they do in relation to degree of dimorphism needs studying.

The female apistos are differently but brilliantly colored and exceedingly aggressive. They also have a complex system of visual displays and engage in overt combat with other females, contesting for spawning sites and protecting their fry.

We have seen how fighting among cichlids varies in relation to need. The lekking females just pick up sperm from males and then brood their eggs by themselves. They have no need to fight. Monogamous and haremic substrate brooders, in contrast, have evolved to be highly aggressive in order to reproduce.

The major asymmetries influencing which fish is likely to win an encounter are now modestly well established. Residents usually prevail over intruders. When residence is uncertain, the winner is likely to be the larger, more aggressive, and more motivated fish. Further, individuals vary greatly in their willingness to engage in combat. None of this will seem surprising to most of you, but these findings take the conclusions out of the realm of speculation. Without this basis, further consideration of the role of aggression in the life of cichlids might be contentious.

But how and what do they actually communicate when assessing one another? How do they attempt to persuade the opponent to yield? Part of the answer was apparent earlier in this chapter in the description of displays seen during a typical fight. Cichlid fishes have been important players in the study of communication during aggressive interactions. My objective now is to bring you further into the world of fish talk.

≈ *six* ≈

CICHLID SPEAK

Those of us who study fish behavior forget how attuned we become to the actions of these creatures. For years I had a 200-gallon (about 750 liters) aquarium in my office across from my desk. During conversations with visitors, we often sat facing the cichlids, and we were drawn to their behavior. What impressed me was how little the untrained visitors saw. I might say, "Look at that!" and the visitor would typically respond, "Look at what?" The interacting fish were speaking worlds to me, but to the guest they were merely swimming near one another.

Most of us are practiced readers of human nonverbal behavior. We tune in to tone of voice, body posture, subtle movements of hands and feet, skin color, and most importantly, to our agile faces with their expressive mouths and eyes. But consider a cichlid fish, or any fish for that matter. It cannot smile or smirk, or stick out its tongue. Nor can it wink, squint, or raise eyebrows. A fish has no limbs to gesture with. Its body is stiff compared with ours, so it can neither curl up into a ball nor crouch. A cichlid does communicate visually, however, but in ways that are mostly, but not entirely, different from the way we do. It takes practice to learn how to tap the lines of communication between fish.

In a hostile encounter, each fish tries to frighten the other. They do this in a number of ways, two of which I have already mentioned. First, they make themselves look larger to intimidate the rival. Second, they try to impress the opponent with their strength. And third, they demonstrate their growing intention to bite.

A fish cannot smile, but it can do some remarkable things with its face. The mouth can be opened widely, showing teeth and sometimes color inside. It can clamp its jaws together such that its opercles extend out from the sides of the head, increasing the apparent size of the face; many people call this flaring. A cichlid can also lower its throat by means of the long (branchiostegal) rays within; that changes the appearance of its head, sometimes making it seem enormously enlarged. The eyes can be turned ahead a bit in an effective stare when one cichlid faces another, which shows the opponent that it is being measured for a bite.

The body is stiff, but it is not a steel rod. A fish can flex its body laterally into an S-shaped curve, the sigmoid, or into a simple C-shape, to alter its guise. The sigmoid posture coils the fish to dart forward, like a sprinter in the starting blocks. The C-shape reveals the intention to turn toward the opponent. In the lateral display, the fish may deliver mechanical stimulation from currents generated by tail beats. These buffet the opponent with turbulence, sometimes even pushing it away. In this way, the actor shows its strength.

Meanwhile, the median fins are busy. Spreading them increases apparent size. But cichlids do more than just raise and lower the entire fin. The median fins can be partially raised and lowered, and in different sections. Just how the fins are spread or folded says a lot about how the various parts of the display came into being.

The first part of the dorsal and anal fins, usually about one-quarter to one-half the length of the fin base, is supported by sturdy, sharp spines, whereas the rear part of each of these fins is braced by soft, jointed rays. When a fish swims in a cruising mode, the spiny parts of the median fins are folded away. The soft portions are loosely open. Propulsion derives from a lateral wave passing down the body from the head to the tail end, increasing in amplitude as the wave travels rearward. The function of the median fins is to prevent the fish from rolling while undulating. Thus as the fish cruises along, the rearward, soft portions of the fins act like the stabilizing feathers at the end of an arrow. When starting from a sigmoid posture, the spiny portion of the fins is also typically folded.

When turning, or stopping, however, the spiny part of the fins come into play, expanding quickly into keels to brace the fin against the destabilizing turbulence introduced by these maneuvers. Thus the intention of the displaying fish to start, turn, or stop is revealed by the lowering or raising of the spiny median fins. The pelvic fins of cichlids, which have one major spine, are also raised when stopping or turning, acting like anchor points. One can read a great deal about what the fish will do next by closely observing the positions of the fins.

We can put some of these elements together. When a cichlid presents its side and performs tail beating, it is basically swimming in place with exaggerated wide undulations. To stay in place, it has to brake with its paired pectoral and pelvic fins. That creates unsteadiness in the fish's balance, so it raises its median fins.

When fish face off, they often assume the sigmoid position, revealing the intention to shoot forward, smashing into the opponent. Then the spiny portions of the median fins are usually folded. The fish seems to be imagining that it has already launched the forward lunge. The intent of the fish is also conveyed by changes in the face. For instance, flaring the opercles is caused by the fish biting down hard, as though the opponent were already in its grasp.

COLORFUL CONVERSATION

A fish can also enhance or soften its displays. The most apparent way of doing this lies in the dynamics of coloration. Fishes that associate closely with the bottom, and particularly with a broken up environment such as a coral or rocky reef, often have exquisite control over their color patterns and can quickly change them. One reason for doing this is to make themselves difficult to see, either as the hunter or the hunted, through camouflage. In the simplest case, they become pale over a sandy bottom and nearly black when next to a dark lava reef. Among cichlids, fading can take up to half a minute and darkening even more time.[1]

Controlling Coloration

Although some details are still lacking, we have a fairly good idea of how cichlids change color. The skin has pigment cells called chromatophores. The most common and active ones are filled with dark melanin and are called, not surprisingly, melanophores; their actions are the ones best understood. The striking hues, on the other hand, are produced by separate chromatophores that most often contain yellow to orange or red carotenoid pigments, much like those found in carrots. The color emitted by such pigments can be radically altered by attaching them to various proteins.[2] Still other chromatophores are filled with guanine crystals, and these produce iridescence, or structural colors, especially blues and greens.

The ability to change the color pattern derives from the behavior of the tiny chromatophores in their collective action. Under a microscope, the multibranched chromatophore looks like a web spun by an inebriated spider. The pigments can be withdrawn from the branches and concentrated into a dot at the center of each cell. When that happens, that color almost disappears from the fish. At other times, the pigment flows out into the reticulated web. The intensity of the color depends on the degree of expansion. Fully expanded, the mesh of a multitude of chromatophores overlaps, saturating the skin with that color at that spot.

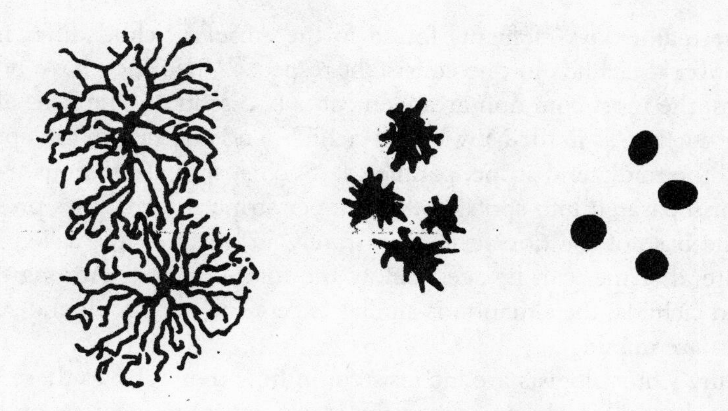

FIGURE 6.1 Schematic representation of melanophores lying in the skin of a fish. The left cell is fully expanded, the right fully contracted, while the middle one is partially expanded.

The chromatophore cells lie in up to four or five outer layers in the skin.[3] Melanophores are found in the first three outer layers, and consequently, when they expand they can mask the pigments contained in deeper-lying chromatophores. Typically, they expand and contract faster than the more inward cells. Speedy change is accomplished by direct neural stimulation.

Moving down into the skin we find two more layers of chromatophores. These are filled with the colorful pigments. Their expansion and contraction is usually slower than that of the melanophores. They are more influenced by hormones, as when the fish start to reproduce.

The chromatophores are organized into elements that are basically linear and give coherence to the color pattern. They remind me of the numbers seen in a digital display on a wrist watch or calculator; those are based on columns and rows of single light cells. Squarish numerals or letters can be produced with one small grid. Organized units of chromatophores differ from this example in that the intensity also varies, as in an analog system. But let us stay with the digital model for a bit.

The Color Code

Analysis of the development and organization of melanophores into units in the African mouth brooder *Oreochromis mossambicus* revealed that this species has available to it no more than fourteen elements, such as bars, stripes, and various spots and facial markings.[4] Other related cichlids, even those with remarkably different color patterns, use much the same fundamental units.[5]

The major dark elements found in the African cichlid differ from the New World cichlids in one consistent respect. When they have horizontal stripes, the most common arrangement is two stripes. One runs along the midbody line, as in the New World cichlids, and another on the upper side, called the midlateral stripe, parallels it. Such stripes are variously continuous or separated into spots. A third upper stripe is sometimes present, lining the base of the dorsal fin. And rarely, a fourth stripe, called here the sublateral stripe, can be seen below the midbody line. Among the New World cichlids, the situation is similar except the midlateral and sublateral stripes are missing.

Many ichthyologists are inconsistent in how they refer to these bold linear markings. Simply put, when the markings are vertical, as on the well-known convict cichlid, they should be called bars; this includes bars on the head and face that recur in different combinations on so many species of cichlids. When the markings are horizontal, they should be called stripes.[6] The bars atop the front end of the head cause some ambiguity because as

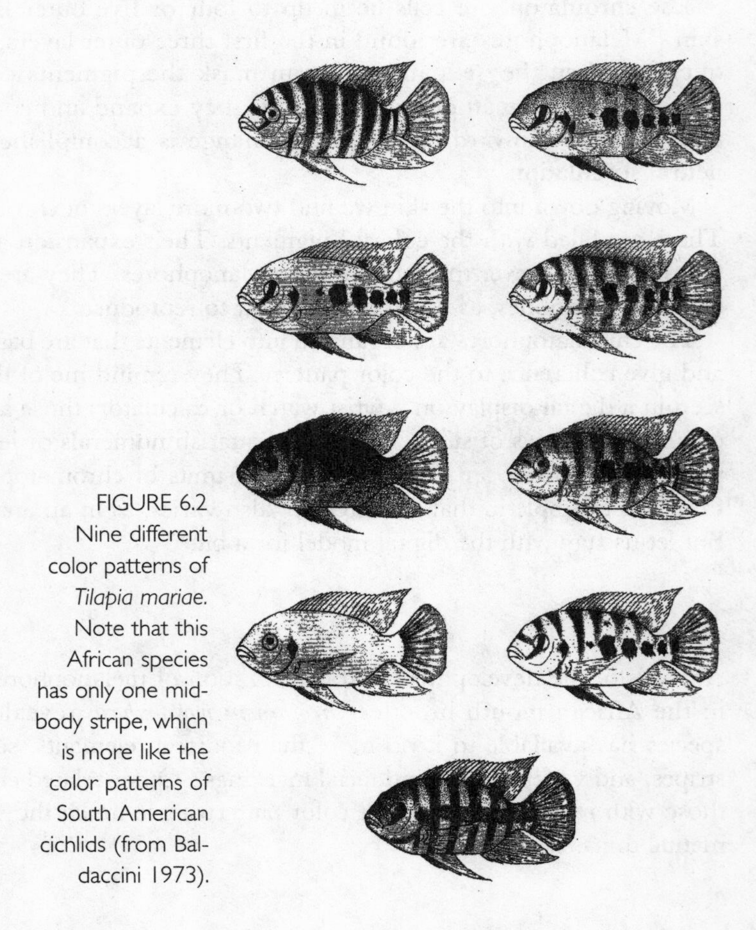

FIGURE 6.2 Nine different color patterns of *Tilapia mariae*. Note that this African species has only one mid-body stripe, which is more like the color patterns of South American cichlids (from Baldaccini 1973).

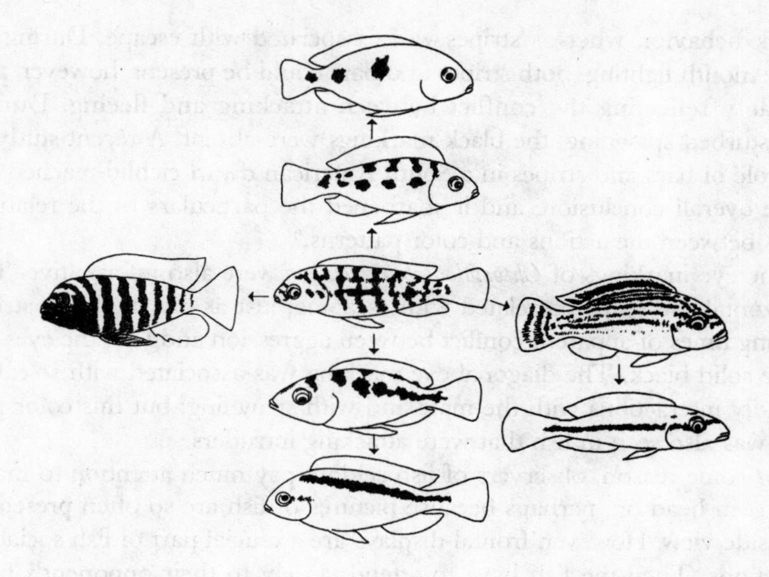

FIGURE 6.3 This scheme shows the connection between the basal condition of bars and multiple stripes in an African cichlid, to derived conditions in other African species. The progression is, to the left, fully barred; to the right, from multiple stripes to a single midbody stripe; upward, to just a few body spots; and downward, to a gradual shift to an oblique stripe by using elements of different bars.

they follow the downward turn of the head, they become oblique; nonetheless, they should be referred to as bars.

The behavior of the Midas cichlid in Nicaragua illustrates the adaptiveness of these markings because this fish alters its markings rapidly as it changes its behavior. When feeding up in the water, it turns on a single body stripe, but if it moves closer to the bottom into aggregations, then the stripe breaks up into a few prominent spots, one at midbody and the other at the base of the tail. Finally, if the fish settles close to the bottom its color pattern becomes one of multiple dark bars. During breeding, the bars intensify, making the fish conspicuous.[7]

Color, Fish to Fish

Getting back on the theme of how cichlids talk with color, Gerard Baerends and his colleagues assessed the relationships between the various elements of the color patterns and the behavior of the fish to reveal their role in communication.[8] Their approach was to quantify which color patterns occurred with which behavioral events. The pattern of bars, perhaps better put, the vertical rows of patches, in *Chromidotilapia guentheri* correlated with attack behavior, whereas stripes were associated with escape. During intense mouth fighting, both stripes and bars could be present, however, presumably reflecting the conflict between attacking and fleeing. During undisturbed spawning, the black markings were absent. A recent study of the role of bars and stripes in a South American dwarf cichlid reached the same overall conclusion, and it sharpened the particulars of the relationships between the actions and color patterns.[9]

The eye markings of *Chromidotilapia guentheri* were also informative. The horizontal eye stripe correlated with escaping, just as did the body stripe. During times of apparent conflict between aggression and fear, the eyes became solid black. The diagonal eye marking was associated with so-called friendly interactions with the mate and with spawning; but this color pattern was also seen in fish that were attacking intruders.

For some reason, observers of fish seldom pay much attention to markings seen head on, perhaps because pictures of fish are so often presented in a side view. However, frontal displays are a crucial part of fish social interactions. Then the fish have to attend closely to their opponent's face, seen head-on. Unfortunately, the analysis of the use of the color patterns on the front end of *Chromidotilapia guentheri* was unsuccessful. I noticed such markings intensifying during hostile encounters among other species. Systematic observations, however, are wanting.

The general pattern among cichlids, at least for the river dwellers, is for the color pattern of a dominant or territory-holding fish to become more intense. The coloration of the subordinate, in contrast, becomes muted, often paler, and renders the fish cryptic. Among many species, dominance produces more pronounced barring, but this is no rule. Virtually the opposite is seen in the African cichlid *Tilapia mariae:* The fish that is territorial becomes spotted, the subordinate barred.[10]

In the Central American cichlid *Neetroplus nematopus,* a dominant fish becomes black with white eyes and one white bar in the middle of its body; the "neutral" color pattern is the reverse–a gray fish with dark eyes and a single black bar. But in another New World cichlid called the Oscar, *Astronotus ocellatus,* the loser of a face-off becomes nearly black with some mottling.[11]

Recent studies in my lab of an African cichlid, *Julidochromis marlieri,* are consistent with the generality that lake-dwelling African cichlids change colors ever so little in the service of social interactions. The black-white contrast in the markings of the winner of a fight, or the dominant fish in an aquarium, do sharpen, and the pelvic fins darken in the winner of a fight and blanch in the loser, but this is a subtle difference to the observer.

Markings on the side of the face also enter in, and some of these are probably meant to be seen head-on. In the African mouth brooder *Astatotilapia burtoni,* a prominent black bar below the eye characterizes territorial males. The presence of that bar, which is switched off in nonterritorial males, is an especially effective marking in communicating ownership of a breeding territory.[12]

In the Oscar, the base of the tail carries a conspicuous eye-like marking, a black disk with an orange ring, and the tail is vibrated during lateral display when the median fins are spread. The orange-ringed disk attracts the biting activity of an opponent, but the significance of this behavior is not understood.[13]

Eye-like spots are called ocelli (*ocellus* = small eye). An ocellus is a disk marking that is dark in the center with a pale band around it, much like the iris of the eye encircled by its white. The common assumption is that such ocelli mimic the eyes, especially when they occur as pairs. Ocelli pop up in all kinds of animals, including cichlids. Among the tilapias, juveniles characteristically have a single ocellus on the trailing edge of the dorsal fin; it probably serves as a signal helping to keep the school together.

In those species that can extend the opercles especially far out from the gills they cover, their tips often bear a large ocellus on each side. This behavior is so striking that it inspired some informative experiments.[14] The scientists gently scraped off the skin containing the ocelli in an experimental group of male firemouths. In a control group, they scrubbed off a similar area of skin next to each ocellus, but left the ocellus itself unaltered.

When one male had ocelli and the other lacked them, the fight escalated faster than usual with much more biting. But about one-third of the way through the fight, the rate of biting returned to a normal level. To force the issue, the experimenters put one each of the two types of male firemouths into an aquarium with a single flower pot, which is a desirable territorial refuge and spawning place. The objective was to force the males to compete for a resource that was important to them.

The male with intact ocelli almost always won the flower pot, even if it was the slightly smaller of the two fish. Moreover, when a solitary male firemouth saw its image in a mirror, those males without ocelli threatened from closer up than did those that had them. These results are consistent with the hypothesis that ocelli function as eye spots, stimulating fear and thereby inhibit attacking in the opponent.

More recently, Matthew Evans and Ken Norris went after the question of whether the red color on the face and inside the mouth communicates to a firemouth that it is facing a formidable opponent.[15] They took advantage of two means available to them to manipulate the red color. The first is the inability of fishes, and of other vertebrate animals, to synthesize carotenoid pigments. And those are the pigments a firemouth deposits in its skin to produce the red color. Second, when bathed in spectrally pure green light, red cannot be seen.

They staged fights in white light between male firemouths, some of which had been fed a diet rich in carotenoids and the others an identical diet, except it contained no carotenoids. The two diets produced males with bright and dull red throats and mouths. When they staged fights between otherwise similar males of the two groups, the redder males won more of them.

One might object that without carotene in their diets, those males were unhealthy and thus unable to fight well. To overcome this possible objection, the two scientists set up encounters between the two kinds of males, but now under green light. This time neither group of males prevailed. Winners and losers occurred in about equal numbers among the bright and pale red males. In addition, the larger male tended to win, irrespective of color. Obviously, withholding carotene did not produce weaker males.

In a yet earlier series of papers on the Midas cichlid, *'Cichlasoma' citrinellum,* I demonstrated a parallel effect of orange color, which I call gold. In nature, just the throat, belly, and eyes of this species are usually, and to varying degrees, yellow through red. Otherwise, a normal fish is typically gray-green with black markings. A totally gold fish, then, would be remarkable.

In the lakes of Nicaragua, about 8 percent of the adults are gold colored. They are that way because they lack dark pigment in their skin; they lose the pigment during development (they are not albinos because the eyes retain their melanin). When the dark pigment disappears, that reveals the underlying pigment.[16] In most instances that pigment consists of carotene and allied molecules, giving the fish its gold coloration.[17] A few individuals lose even the carotenoid-bearing cells, and they become white. Fish that lack dark pigment also lack the ability to change color patterns.

When gold morphs of the Midas cichlid were matched against normal ones of the same size, an incredible advantage emerged.[18] The golds always won. By varying the sizes of the contestants, we discovered that being gold conferred the equivalent of a weight advantage of 17 percent. Recall that a difference in weight of only 2 percent usually determines the outcome of a fight between two Midas cichlids of the same color, whether normal or gold. When groups of gold and normal fish competed for food in a large aquarium, golds had priority of access, which they asserted when they were hungry.[19]

Why do golds dominate? Is it the absence of the species-typical markings, such as pronounced bars? If so, then gold morphs should have no advantage over white ones. But they do. White morphs suffered the same 17 percent disadvantage as normal morphs in encounters with gold ones.[20] Lack of markings is not the answer.

Perhaps the gold color is correlated with a genetic tendency to be more aggressive? Again, wrong. When gold morphs were shown dummy Midas cichlids, they attacked, or not, at the same rate as normal morphs.[21]

The likely answer is the following. In the normal fish, the gold markings serve as a threat display. So, when a fish becomes entirely gold it becomes a supernormal threat display.

WHAT DID THAT FISH SAY?

Mention of dummy experiments brings up a recurring paradox about the role of color patterns in interactions among cichlids. And it highlights the whole issue of communication and how we analyze it. If a given stimulus, here some evolved signal, produces a predictable change, then we can assume communication has occurred. However, it is another thing to say what was communicated. That is where the study of communication becomes more demanding.

You just read convincing evidence in the case of two contesting firemouth cichlids: The red color and the opercular eye spots intimidate the perceiver, apparently by inhibiting attack through fear. That seems to account, too, for the intimidating effect of the gold Midas cichlid. The essential point is that not only was a change produced, but it was in the predicted direction: Aggression was reduced in the perceiver.

Yet in nearly all the many experiments performed on cichlids, and on other kinds of animals, when presented a series of dummies, the one that bears the color pattern associated with dominance evokes the most attacking, not the least.[22] That is a striking and consistent difference between responses to a live opponent and to a passive dummy. Why should such a bipolar response exist, and what does it tell us?

The answer lies in the nature of the experiment, as well as an assumption about the function of the color patterns. As I have presented them here, color patterns found on dominant cichlids evoke fear in the beholder. Actually, they communicate something at a higher level. They signify the most important and potent rival, the one who most needs attending to. Hence, if the fish is a territory holder, it should attack most vigorously the intruder that bears the markings that bespeak dominance because that is the most dangerous rival. Conversely, an intruder into a territory should flee especially from the fish who has the same dominance markings that signify ownership.

Fish in experiments that are shown dummies have prior residence over the dummy, and the dummy does not threaten back. So the resident fish attack most vigorously the dummy that bears the color pattern of the main rival. In the Midas cichlid, that meant the passive gold dummy was attacked the most.

Paul Siri and I tested this hypothesis by arranging to have dummies respond to the subject.[23] When the Midas cichlid attacked at the dummy behind a glass wall, the dummy wheeled and charged back. When the subject withdrew, so did the dummy. The dummy mimicked the behavior of the subject. To the subject, that meant attacking the reactive dummy might carry costs. And when the color pattern of the dummy indicated the possibility of dominance, as in the case of the gold dummy, the subject ought to be cautious.

That proved the case: The reactive gold dummy received the fewest attacks and was avoided the most. However, Midas cichlids are smart. They quickly learned the reactive dummy was all show and no substance. Within a few presentations the reactive gold dummy was approached and attacked the most, as before.

I have been focusing on color, but a cichlid fish has still other communicatory tools at its disposal. One is the pattern of movement and posture. I have already mentioned the actions that are so recurrent across different species, such as approach, lateral and frontal display, and tail beating. But many more specialized movements are found in certain species. Some cichlids almost stand on their tails, either facing toward or away from another cichlid. In another situation, a cichlid poses almost vertically, head down; in many species this is followed by demonstrative digging in the sand or mud bottom.

In normal digging, which appears to be communicative as well as serving to provide a nest pit, the fish tips down, bites into the bottom, then returns to normal and swims a variable distance and spews out the substrate. Sometimes it spits the sand out at a neighbor, "Here's mud in your eye." The functional significance is obvious.

SOUND CONVERSATIONS

Cichlids are no match for birds, but they can vocalize in a soft and crude fashion. You won't be able to hear them do this. A few scientists have listened to them by putting a hydrophone into the aquarium to amplify their sounds, and so far, sixteen known cases of acoustic communication have been reported for cichlids.[24] That is a relatively small number, considering the size of the family.

The sounds heard have been mostly low-frequency grunts, and these are emitted characteristically during hostile displays and sometimes during courtship. During aggressive exchanges, the rainbow cichlid, *Herotilapia multispinosa,* of Central America, for instance, produces what have been described as volleys and thumps[25] or purrs.[26]

The tilapia *Oreochromis mossambicus* has a relatively rich repertoire of sounds, produced by grinding its pharyngeal teeth. Adult males vocalize when they are territorial, and when attracting females. Their fry start doing this when they are only three weeks old.[27] Groups of feeding tilapia also make sounds.

Only one study, done in my laboratory, has attempted to determine the effect of acoustic signals by one cichlid on another cichlid hearing them.[28] Two males of a Central American cichlid, *'Cichlasoma' centrarchus,* were placed in adjacent aquaria where they could display at one another. The air space between the aquaria prevented sounds made by males from reaching one another. Using an underwater speaker, various noises, including sounds recorded from one of their species during hostile behavior, were broadcast to one or the other male. Hearing the sounds made by an aggressive male inhibited the aggressive display of the male hearing them. Apparently the vocalizations act much as do visual displays to intimidate a rival.

Sometimes you can tell when a cichlid is emitting sounds, even if you cannot hear them, by observing trembling of the body and median fins. Exceedingly tame Midas cichlids display and bite at us through the glass walls of their aquaria. Such individuals often tremble, as though tensing their muscles. I suspect they emit sound through that action. Some species, such as *Aequidens latifrons, Geophagus brasiliensis*, and *Julidochromis marlieri,* probably vocalize, too, when they jerk the head to one side in a hostile encounter.

THE LAST WORD

A cichlid fish has lots to say to another adult, and it expresses it mainly through aggressive actions. An observer soon learns to recognize the signals and the context in which they occur and can thereby decipher the language of the fish.

When one compares mating in the substrate and mouth brooders, a fundamental difference is quickly seen. Aggression between male and female substrate brooders is a major issue, and one that has led to elaborate displays by both sexes during courtship. Among polygynous mouth brooders, in stark contrast, males pretty much restrict aggression to other males; females communicate submission when approaching the male to spawn. Consequently, how mates are selected, and reproduction achieved, differs profoundly between the two types of cichlids.

≈ *seven* ≈

BEAUTY IS ONLY FIN DEEP

As a young man I pondered what made some women seem more beautiful to me than others. That is not a trivial question when considered in an evolutionary context. Equally mysterious to me, most men agreed with my judgments, at least at the supreme level of gorgeous movie stars. But why are such women universally regarded as more attractive than others? I gather women generally, but certainly not always, agree among themselves which men are the most attractive.

At that time in my life I speculated that culture must play a central role. The importance of experience, of learning, is indisputable, but it is only part of the equation. Recent research suggests our perception of beauty is also partially intrinsic, transcending racial and ethnic differences.[1] Some core persists despite cultural influences.

This view has been harbored by many for centuries but without any way of evaluating the belief. Scientists are now testing that presumption, and some of their studies burst into the public's awareness with a *Newsweek* cover story in June of 1996. The article started with an account of what females of some animal species find attractive about males: Female penguins prefer the fattest males, jungle fowl (progenitors of barnyard chickens) choose the fanciest roosters with colorful feathers and combs, and female scorpion flies favor males whose wings are most symmetrical. That *Newsweek* should use these examples as a springboard into a discussion of beauty in humans is testimony to the impact of studies of mate choice among animals on attitudes about human behavior.

Psychologists have also been making progress toward understanding beauty. Judith Langlois and Lori Roggman, to cite one example, gave meaning to the expression, "I'm just looking for an average mate." Using computer graphics, they merged faces of people taken from photographs in a college yearbook. The more faces they blended, up to thirty-two, the more attractive the resulting faces were judged to be. Which visages they used made no difference. Any set of photographs produced the same outcome: The most intermediate face was considered the most attractive.[2] As a sobering note, few people have an average face, so average here does not mean the most frequently seen face.

Of course, the most average face is also the most symmetrical one. Some argue that may be an important marker for quality of mate,[3] but others contend the evidence for this is lacking.[4] Current research suggests attractiveness in humans is a multifaceted phenomenon, which I am sure does not surprise you. Ratio of waist to hip, overall height, proportions of the face,[5] voice, and subconscious detection of odor[6] may also be important. The approach used in these studies was anticipated by recent research on mate choice in animals.

How humans choose mates is inherently fascinating to most of us. You might well ask, however, why we should care how an animal, here a cichlid fish, chooses its mate. The immediate answer has two forms. For one, understanding parallel behavior in animals provides us with a model for thinking about our behavior, often in fresh ways. For another, how animals behave, like it or not, is constantly brought up in serious discussions of the evolution of our behavior, the "biological roots" issue. Therefore, having the right information is important because often either the facts get distorted or only the examples that support a particular point of view are put forth.

SEXUAL SELECTION

The role of attractiveness as an agent of evolution started with an 1871 book by Charles Darwin, *The Descent of Man and Selection in Relation to Sex*. There Darwin proposed that natural selection can be viewed more broadly than the conventional considerations of obtaining food, avoiding being eaten, and producing young. He argued that any trait that provides an advantage in mating would be favored. He called this sexual selection.[7]

Sexual selection has much in common with traditional natural selection, so much so that many evolutionary biologists argue that it is only a subset of natural selection.[8] In conventional natural selection, individuals compete for food and refuges, for instance, both among themselves and among other species. In sexual selection, no other species are involved in the competition. Males compete for access to females' eggs. To the male that achieves high status in that competition comes high Darwinian fitness.[9]

While recognizing that sexual selection is a subset of natural selection, I treat mating among cichlids in the framework of sexual selection. I do this because that aspect of theory has assumed such a prominent place in a provocative hypothesis for rapid speciation among cichlids.

Naturally selected traits always have their costs and benefits. What has drawn so much attention to the concept of sexual selection is that in the more spectacular examples, the trait must confer substantial benefits because the sexually selected trait incurs such profound costs. Put another way, the bearer of the trait is handicapped by the very trait females prefer.[10] The huge tail of a male peacock, for instance, hinders flight from a predator or to a place to feed.

The risk of predation incurred by a sexually selected trait was elegantly demonstrated in the calling behavior of the Tungara frog. Males gather in a lek, and each produces a noisy call of chucks and whines to lure in females who are ready to mate. The females need the chuck sound to locate the male, but a bat also uses the chuck call to zero in on the male, who becomes his dinner. Thus sexual and natural selection work in opposite directions.[11]

Any advantage achieved through competition between males, such as fighting, is called *intra*sexual selection. Contrasted with that, female predilection for certain males is termed *inter*sexual selection. When people refer simply to sexual selection they usually, but not always, mean female preference for some feature or features of the male.

The obvious outcomes are all around us.[12] When males fight for access to females, selection favors males who are vibrant and powerful, typically resulting in males that are bigger and more formidable than females. Enormous bull elephant seals are an extreme example among mammals. Larger size in males, however, can be driven by more than fighting–females have often been shown to prefer mating with larger males. Thus female choice can contribute to differences in size between males and females.

The more typical situation is that the female favors ornamentation in males. The result is conspicuous male traits. The most famous product of intersexual selection is the incredible tail of the peacock. Among cichlids, certain haremic and lekking males are renowned for their spectacular coloration and relatively large size.

Returning to Darwin, his writings about sexual selection were not fully accepted at first. Some influential evolutionary biologists argued that sexual selection was basically the same as natural selection. That position has merit and is still held by some serious students of evolutionary theory. What rekindled biologists' interest in sexual selection, however, was the republication in the early 1950s of a book by Ronald Fisher, the noted theorist. He presented a plausible model for how intersexual selection works, one called runaway selection.[13]

Fisher's model proposes the joint evolution of female choice and the exaggeration of the critical trait in the male. If females prefer a certain trait, then males possessing that trait will obviously be chosen by the females. Their male offspring will also possess the trait and will thus be preferred. That will lead to a positive feedback system that simultaneously exaggerates the trait and increases the preference. Ultimately, the costs of the trait to male survival cancel the benefits from attracting a mate, limiting the further elaboration of the trait.

Fisher's model is probably overly simple, and competing models have been presented.[14] Its most radical tenet is that the trait preferred by the female has nothing to do with the genetic quality of the male other than genes for the conspicuous trait.

The good-genes model is a robust competitor. It posits that when the female selects a male with the most exaggerated trait, she is picking the one with the best genetic material for her offspring.[15] As Amotz Zahavi put it, the male that can carry the greatest handicap, and still survive to compete with other males and reproduce, must have the best genes; this is the so-called handicap principle.[16]

A similar hypothesis is that if the female chooses the male most attractive to females in general, then her sons will also be more appealing to females. The result is that her genes will be more widely spread through her handsome sons. This is the "sexy sons" hypothesis, an idea that has been argued pro and con by the theorists.[17]

Sexual selection has also been explained at another level of analysis, that of the sensory mechanisms. Put simply, females are more likely to notice and attend to males whose ornaments best "exploit" her sensory receptors. Assume, for example, that the eyes of a female are most strongly tuned to a certain range of reddish hues. That visual tuning could be an adaptation for efficient detection of prey in the illumination in which her eyes evolved. A male that evolves red ornamentation could then be favored by the female. This is called sensory bias[18] in the female or sensory exploitation[19] by the male.

A major consequence of renewed interest in sexual selection is emphasis on the act of female choice among the males available to her. Until recently, many biologists doubted whether females discriminate among individual males as long as they are of the same species. Now the study of mate choice by females, and even by males, is a vigorous area of research.[20]

In the following, I take up mating in only the polygynous, mostly dimorphic species of cichlids because intersexual selection is most obvious among them; allure reigns supreme, and speed is the key word. For the typically monomorphic monogamous cichlids, initial attraction is also important; the key to their successful pairing, however, as we will see in the next chapter, lies in negotiating what amounts to a peace treaty so the pair can stay together during the long period of parental care.

MATING AS A QUICK PICK

When mating, any animal has to make sure the potential mate satisfies three criteria: It must be the right species, the right sex, and especially for females, the highest quality individual the seeker can get. As humans, we seldom think about the first two prerequisites. After all, who would confuse a chimpanzee with a human? But an animal like a cichlid fish, inhabiting a lake with a multitude of other kinds of cichlids, is regularly faced with the task of discriminating among them. In reality, usually only a few of the many species living together closely resemble one another, reducing the problem substantially.

For the sexually dimorphic cichlids, recognizing males and females ought to be relatively easy because they are so obviously different. This needs some qualification, however. Many African lake species nest on open sand or mud bottom where they are exposed to predation. Because of that, they are more likely to be protectively colored. Such species tend to converge on the same best coloration.[21] Being vulnerable, the differences between the sexes are also less pronounced than in reef-nesting species, though when nesting, the details of the male's color pattern can be subtly colorful and beautiful. The reef fishes can afford to be more colorful because they can hide among the rocks.[22]

In general, the polygynous mouth-brooding species mate quickly, and the females appear to identify their prospective mates before they come close to them. In more exposed soft-bottomed areas, males center their breeding territory on a nest they dig, and it can take various forms. He excavates the nest slowly and laboriously, one mouthful at a time. The distance swum before spitting out varies among species and is one way different shapes of nests are produced. In other species, the male stations himself over a favored rock or near a cave among rocks; even then, the male commonly does some digging, cleaning house. In fact, some males dig in the sand but then deposit the material on a prominent rock, making the nest more conspicuous. The consequence of the variation in digging in many instances is the creation of distinctive nests typical of their species.[23]

Once his nest is built, the male stands ready for females. He inspects cichlids that approach his nest, assessing whether they are of his species, an intruding rival male, or a female ready to spawn. A rival male immediately elicits aggression. But just as quickly, the resident male inhibits aggression toward females and instead reacts with courtship. I stress this happens quickly, as though recognizing species and sex is easily accomplished even at some distance.[24]

Such a mating system would seem to suggest only a weak capacity to recognize individuals. On the other hand, natural selection should favor the ability of neighboring territorial males to recognize one another so they don't have to repeat the initial aggressive interactions, the dear-enemy phenomenon. Females should discriminate among individuals in the sense of choosing the most attractive males.

Generally, the burden of picking the right species for spawning falls not to the male but to the female. That follows because the cost to her of making a mistake is higher: She lays an energetically expensive clutch of eggs and then invests in a long period of care. The cost to the male of a mistake, in comparison, is small. Consequently, where closely related polygynous cichlids occur together, males of related species are commonly distinctively different in appearance, facilitating species recognition, whereas females of different species are drab in the same way and thus are hard to tell apart.

We need to pause a moment to consider the significance of how males of related species living together set themselves apart. When they have been preserved in formaldehyde, which destroys the distinctive hues on their bodies and leaves just the black markings, they are almost indistinguishable. When alive, however, they have brilliant colors. Their color patterns are like flags of nations and mark the species of the males. In this they differ sharply from those species that form pair bonds, whether in monogamy or in harem dwellers. There the differences among related species are more often expressed in the dark stripes and bars, particularly on the head.[25]

Because the males of polygynous species are regularly differently colored but the females not, it follows that the male should not be as good as the female at identifying the species of the opposite sex. That is a widely held belief for many different kinds of animals. Various field observations and one experimental study suggest, nonetheless, that males can tell the difference between males of their own and other species,[26] though males may not be as skilled at that as their conspecific females, and they often behave as though the difference is unimportant.

When the female approaches, the male characteristically swims toward her and displays. Cichlids vary in precisely how they do this, but given the diversity of species, the ceremony is remarkably similar. The display usually consists of spreading the dorsal and anal fins to varying degrees, then swimming back to the nest, an act called "leading" the female. The swimming action during leading may be exaggerated, with wide body undulations of the male, as though swaggering.

And species differ some in the way the male moves about over the nest.[27] Once at the nest, a male *Oreochromis koromo* faces the female, quivers, and does "nose wagging," which is just tail beating but emphasizes the front end of the fish. In another tilapia, *O. variabilis,* the behavior is similar, but the males leave out nose wagging.[28] Among different male haplochromines that lek in Lake Malawi, some circle over their nests, others swim in a figure eight and yet others move in an S-shaped pattern.[29]

A receptive female follows and enters the nest. There she does boringly little in most species, beyond watching the male, but in some species the female may nudge the vent region of the male.[30] In response, he may wash water at her by beating his tail while holding position by braking with his paired pectoral fins. The male commonly rapidly shakes his head a small amount from side to side in a movement that is little more than a tremble and is called quivering. By now, the two fish surely have figured out their species and sex and soon commence spawning.

The foregoing mating behavior is an idealized representation of how polygynous mouth-brooding cichlids come together to spawn. Things get more complicated when we consider some of the substrate-brooding cichlids that are highly dimorphic and polygynous. Take the haremic dwarf cichlids of the genus *Apistogramma,* the apistos. The sexes are extremely dimorphic and differently colored, as befits a polygynous species. And males and females have different behavior and roles in courtship.

Unlike the mouth brooders, the female apistos are not passive. On the contrary, they have an impressive array of displays they present to territorial intruders and to the male with whom they will spawn. Probably the chief reason for the elaborate displays of females is that they have adjacent territories and compete with one another for the nesting sites. In parallel fashion, the males compete with other males. Each male has a territory that embraces those of several females. He visits the clustered females at intervals to see which is ready to spawn and thus in need of his sperm. Consequently, the female needs a means of communicating to the male her reproductive status. Sorry to say, we know too little about the mating behavior of these fascinating tiny cichlids.[31]

By describing the mating behavior of polygynous cichlids in a general way, I have prepared you for the basic question we started with, what do such cichlids find attractive in one another? The first step toward answering this is to explore how they proceed to mate, to garner insight into how they make the necessary decisions.

ARE YOU MY KIND OF FISH?

Clearly, cichlids are able to recognize the species to which they belong. Many of the species aggregate or school only with their own species. In

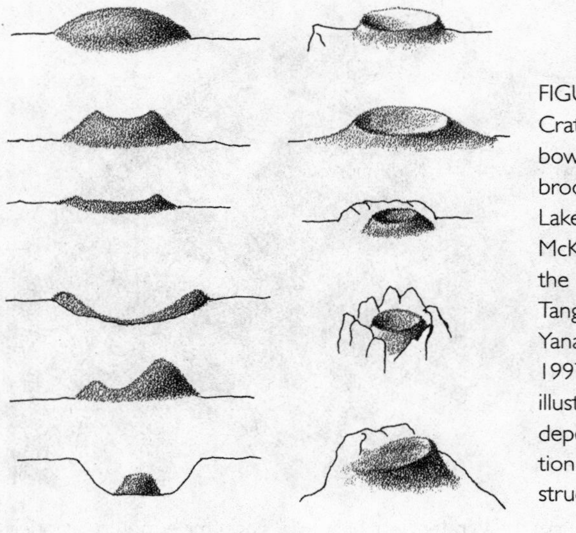

FIGURE 7.1
Crater nests, or bowers, of mouth-brooding cichlids in Lake Malawi (after McKaye 1991) on the left, and in Lake Tanganyika (after Yanagisawa et al. 1997) on the right, illustrating the independent evolution of such structures.

aquaria, the individuals sort themselves by species even when the observer has difficulty distinguishing them.

Just how the species recognize their own kind comes up often in the literature on cichlids because it is pivotal to how they radiate into multiple species. Most commonly, biologists hypothesize that color patterns are the key to species recognition in cichlids.[32] Chemical and acoustic communication get scant attention in this regard. And lest we forget, part of species recognition may lie in such ecological factors as where the nest is placed by the male.

The Nest as a Species-Specific Bower

Indeed, nests may offer females important clues about which males belong to their species. For example, some male tilapias nest in swamps where dark debris covers the bottom. The males clear the debris to expose the white sand underneath, and those nests stand out like bright white discs on the dark bottom.[33] Later, in comparing the behavior of various tilapias, came the realization that the nest differed between species in distinctive features; that led to the suggestion that the structural differences may be used by females to recognize their own species.[34]

Based on underwater observations in Lake Malawi, Ken McKaye extended that insightful conclusion. He argued that the various unique sandcastle nests enable the females to identify the species of the builder; he calls the nests bowers because of the rough parallel to the display arenas built by

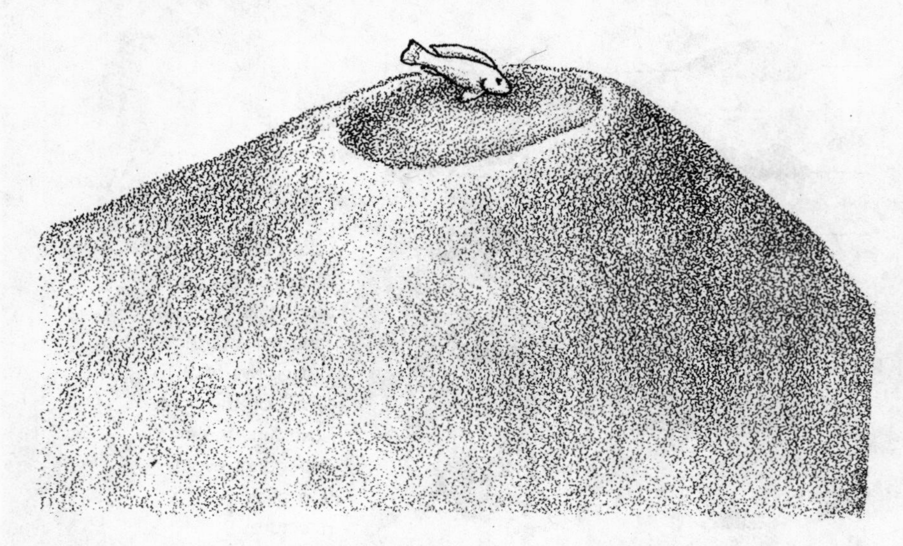

FIGURE 7.2 The immense size of a crater nest, built by a single male mouth-brooder in Lake Malawi (after Stauffer and Kellogg 1996).

bower birds. Some of the nests are huge, compared with the size of the male, and are virtual sand castles.[35] They must be built at an enormous cost in energy.

The same habit of building distinctive nests is seen in cichlids of Lake Tanganyika. There, tiny males of *Cyathopharynx furcifer* build remarkably convergent bowers that resemble volcanoes and are about two feet (77 cm) in diameter at the base and about one-half foot (13 cm) high. Some of the males failed to construct castles, and females spawned only with the males that did. Because females did not choose among the males according to the size of the sand castle, Kenji Karino suggested the "bowers" served only for species recognition.[36] The close parallel with the situation in Lake Malawi is another example of how cichlids independently evolve similar adaptations.

Some bower-building cichlids in Lake Tanganyika have added a twist. The elongate male of an undescribed cichlid, *Enatiopus* sp. "Kilesa," scoops out a shallow nest. He spits out the sand in patches, forming up to twenty or more small sand turrets around the periphery of the nest, and these distinguish his species.[37]

Asking the Relevant Question

Given the significance of the question of how species recognize others of their own kind, surprisingly little experimental work has been done on the question. And I must register one caution here. The few existing experiments on species recognition are of limited use because they fail to ask the question in the context of the potential that the species would meet in nature. The investigators had other reasons for doing their experiments, so their results, albeit of some interest, are not relevant here.[38]

A more germane study is the early report by Sebastian Holzberg.[39] He became intrigued by one of the most common cichlids in Lake Malawi, Africa. The cichlid, then called *Pseudotropheus zebra,* is a popular aquarium fish because it is so colorful and easy to care for. Individual fish come in four different color patterns, termed morphs.

When Holzberg undertook his studies, these color morphs were all thought to be forms of one unusually polychromatic species. Holzberg suspected, however, that they were actually two species. Field and laboratory observations confirmed his hunch. Males of one type feed slightly differently than do males of the other type, and they also have larger territories. The two types of males also do not bother to defend their territories against one another but instead reserve their attacks for males of the same morph.

The clincher was that the two kinds of males court different colors of females. Holzberg's research on species recognition thus established that the name *P. zebra* embraces two different species, now placed in a new genus, *Metriaclima.* One retained the name *zebra,* and the other, which had the informal name "zebra cobalt," became *M. callainos.*[40]

Holzberg's research, however, did not determine *how* the two species recognize their own kind, or the opposite sex, but merely that they do. Holzberg suggested, reasonably enough, that the discrimination is based on coloration, and that hypothesis has been tested recently on some cichlids from Lake Victoria.

These mouth-brooding cichlids, of the haplochromine tribe, present special problems because they have been proposed as examples of how species might arise as a result of female choice of differently colored male morphs of their own species. In Lake Victoria, the males of two species in the *Haplochromis nyererei* complex differ in color, one being predominantly blue and the other chiefly red. In an aquarium experiment, females readily chose the male of their species.

However, females then had to choose between the males when they were bathed in monochromatic light that made the males appear the same color. The females could not discriminate, demonstrating that they recognized their species solely on the basis of male coloration.[41] They did, nonetheless, respond to the males with mating behavior. That proved they could still distinguish the sex of the males in the absence of their brilliant colors.

Choosy Males

Recall that I told you the females are the choosers among the polygynous species. They have the responsibility of picking the right species. But if you were really alert, you noticed that Holzberg discovered that the males of the two species of *Metriaclima* discriminated among the females. Earlier observers also noticed in Lake Malawi that males seemed to swim out to greet only females of their own species.[42]

Now an experimental study has shown that males of three species of *Metriaclima* that live together, *zebra,* "zebra gold," and *callainos,* can distinguish between their females, when the task is not difficult. The females of *M. callainos* differed in color from those of the other two species, and their males courted them. But in the other two species the most common color form of their females looked alike, and the males could not easily distinguish between them.[43]

Natural selection should favor males recognizing females of their own species. Competition for female eggs is intense. The male that sallies forth to meet an approaching female probably has a better chance of enticing her to mate with him than the male that hesitates.

Natural selection should therefore favor those males within a given species who are best able to identify a female of its own species. But pity the male when other selective forces conspire to make females of different species look alike. For instance, exposed to the same kind of predators, females converge on a common cryptic pattern. That is the typical state among polygynous cichlids. Then the burden of species recognition falls mainly on the females, unless their prospective mates have some other way of distinguishing among females of different species, a much neglected question.

So far so good. We have at least some feeling for how species are discerned in the polygynous dimorphic species. But how do the fish know which sex is which?

THE RIGHT SEX?

In dimorphic species, the pronounced differences between the male and the female should suffice for recognition of sex. I have already mentioned that many researchers suspect that dimorphic females recognize both species and sex by shape and coloration. Sex recognition is apparently helped by olfaction as well, as was shown in two experiments, both of which involved lekking mouth brooders.

On the Odor Trail

Males of the African cichlid *Astatotilapia burtoni* (formerly put in the genus *Haplochromis*) were exposed to the odor of females ready to spawn (gravid) and those not (nongravid), as well as to water containing the odor of males, or to water with no odor at all.[44] Surprisingly, at least to me, the males did not respond clearly when first exposed to the water of gravid females. But over a period of a few days they became more obviously ready to mate, as shown by their activities and change in coloration. Thus the odor of females seemed not to play a direct role in recognition but did prime the males for reproduction.

This species is abundant in small pools along the shore of Lake Tanganyika. Conceivably, chemicals exuding from gravid females could be sufficiently concentrated that they would prime males. That could well stimulate the males to commence digging nests. However, that is different from sex recognition per se.

A similar experiment was done on a species of tilapia.[45] Males of *Oreochromis niloticus* responded to their mirror image by displaying at it when they received the odor of a gravid female of their species, less when the odor was from a nongravid female, and not at all to odorless water. Thus chemical cues enable one species of tilapia to recognize the presence of a female ready to spawn.

Color as a Primer

The color pattern of male *Haplochromis elegans* appears to prime their females.[46] As in many other species of haplochromine cichlids, the male has bright spots on his anal fin that resemble the eggs of the species. In this experiment, females were kept with males whose egg spots were unaltered and with males whose egg spots had been removed. The females with "eggless" males produced clutches of eggs at only about half the rate as did females with intact males. Given a choice, the females also preferred unaltered males. Whether that indicates better recognition of sex or of a male of superior quality, however, remains a question.

THE BEST MATE?

How a cichlid recognizes the sex as opposed to the quality of the potential mate is often difficult to separate because the same cues may be involved. Thus in the example just presented, eggs spots on the anal fin of male *H. elegans,* females might recognize males based on the presence of those spots. However, they might also discriminate among males on the size, number, and brightness of the color of the spots.

Another experiment on the egg spots was done underwater at Thumbi Island in Lake Malawi, but using the species *Pseudotropheus aurora*.[47] The males differed in number of egg spots, and one male each was placed in a cage with six females. The results were clear: More egg dummies had a priming effect and resulted in more spawnings by the females.

Quality Males

Information about female choice for higher-quality males is spotty. Among lekking cichlids in Lake Malawi, individual females regularly spawn with several males to fertilize one clutch of eggs.[48] Males in the more central part of the lek, however, appear to fertilize more eggs than those on the periphery.[49] The differential in reproductive success suggests females prefer the central males, and the central males are most likely those of the highest quality.

In addition, males of *Cyrtocara conophorus* that made the largest nest, the so-called sand castles or bowers, were visited the most by females and presumably fertilized the most eggs.[50] In another bower-building and lekking cichlid in Lake Malawi, a species of *Copadichromis,* the males that built the biggest castles were the males with the fewest parasites in their livers; they also fertilized more eggs than did the more parasite-ridden males.[51]

The nest-building behavior of *Cyathopharynx furcifer* has been studied more closely in Lake Tanganyika than in the observations in Lake Malawi, and a revealing finding has emerged. McKaye's thinking was influenced by the finding that males build their own bowers. However, Kenji Karino discovered that among *Cyathopharynx furcifer* in Lake Tanganyika only some of the males build their own bowers. Other males in the lek hold territories without bowers, and when a bower is vacated they take it over.

But why should a male vacate its bower when possessing one assures reproductive success? The reason lies in energetics. First, it takes a great deal of work to build a bower. Second, once occupied, the bower-owning male engages in much more aggressive and spawning behavior than do the other males. When analyzed, bower-holding males had considerably lower fat reserves than did the other males. In short, they run out of gas and have to take time off to restore their energetic reserves; they may also be more vulnerable to predators when over the nest.[52]

One researcher, Mindy Nelson, set about to test the hypothesis that females of a tilapia, *Oreochromis mossambicus,* fancy males who make the biggest nests.[53] If the female tilapia did, she would be indirectly assessing the quality of the male. That would be a nice example of intersexual selection. Nelson realized that life is seldom so simple, however. Suppose larger males make larger nests, and females were simply selecting the larger male. Moreover, females could also be choosing among the males based on the way they interact with one another. So long as the difference is based on female choice, however, all these possibilities would be intersexual selection.

The experiments she did are a pleasing example of how to tease out such differences. Unfortunately, this is not the place to report them in detail. She played off male size against nest size and availability by creating artificial nests. She also analyzed female choice in relation to how the males interacted with one another. She even got gravid females, in the absence of males, to choose the nest in which to spawn. The bottom line is that Nelson nicely demonstrated that females discriminate among individual interacting males of their own species based on the size of the male. Further, with all else equal, females opt for the larger nest when given a choice.

Lopsided Mates

Before closing here, I return to one aspect of mate choice I touched on in opening this chapter. It is the highly controversial issue of the role of asymmetry.[54] The argument is that the individual with superior development, hence the best genetic "machinery," has the smallest differences between left and right. It is called fluctuating asymmetry because the variation varies among individuals. If offered a choice of mates, select the more symmetrical one. Most of this research has been done on animals other than fishes, including humans.[55] The early studies on guppies and a related fish have produced contradictory results,[56,57] so any evidence from cichlids is worth noting.

One Tanganyikan cichlid, *Cyathopharynx furcifer*, has been cited as an example of choice based on fluctuating asymmetry. Its pelvic fins are so long they reach past the rear end of the anal fin, and they are held below the fish in plain view where the female can assess any differences between right and left. Further, the tip of each fin bears a bright yellow egg spot. Females approach most frequently those males who hold castles and who perform the "shaking" display most actively. The observed females, however, did not necessarily spawn with the most vigorously courting males. Apparently, the displaying males simply attracted the female to the nest. When at the nest, she circled with the male, "checking him out." The males with the least difference in length of the two pelvic fins received the most eggs.[58]

A SOUND CHOICE

Phillip Lobel has discovered in Lake Malawi that the nesting males of two lekking species of cichlids, *Copadichromis conophorus* and *Tramitichromis intermedius,* make distinct sounds in response to the approach of females. In an inventive departure from recording sounds in the laboratory, he positioned a hydrophone at the nest of a male in the lake and connected it to a video camcorder a few meters away. The coordinated visual and sound information enabled him to analyze and interpret the meaning of the calls the males produced.[59]

Those calls were emitted simultaneously with the visual displays of fluttering the fins and quivering. The calls consisted of short pulses of low-frequency sound. Through careful analysis Lobel has determined that the duration of each pulse and the number of pulses per call differ slightly but statistically significantly between the two species. The entire call, of which several may be given, lasts only about 0.2 seconds, so one wonders how much information about species differences is contained in such fine features as the individual pulses. He suggested cautiously that those differences might be used in species recognition.

However, calling could serve one or more functions. Because mating is done quickly, it could couple with the visual displays to stimulate the female to spawn quicker. Calling could also assist in recognition of sex, though I suspect sex, as well as species, is determined by the female before she gets close enough to elicit displaying by the male. Rather, the entire display, visual and acoustic, is probably important in mate assessment. Consonant with other kinds of animals, the ability of the male to produce lots of calls during a female's approach could be a valuable means for the female to assess the quality of the male.

PUTTING IT TOGETHER

So, where does that leave us with regard to the polygynous species, and especially the mouth-brooding ones? For these cichlids, initial attraction says it all. Sexual selection is well demonstrated by them. They have evolved a system of mating such that the male and female have no enduring relationship. The males typically differ in appearance from the males of those closely related species with which they might be confused. When the female is ready to spawn, she proceeds to the displaying male of her species and of her choice, enters into brief courtship with him, spawns, and departs. That's it.

The male and female have no reason to be aggressive toward one another. They form no bond. They have no personal, lasting relationship. The situation among the monogamous cichlids is different. Although mate assessment also involves traits such as size and "good looks," the personal relationship is central to forming a pair. And that makes their mating remarkably different from the polygynous species of cichlids discussed in this chapter.

≈ *eight* ≈

MATING GETS PERSONAL

Monogamous, substrate-brooding cichlids are the archetypical cichlids and present the starting condition for the evolution of cichlid reproductive behavior. The most primitive cichlids are found in the Indian subcontinent and in Madagascar, and all of them are substrate brooders. They also rule the rivers of the New World tropics where 70 percent of the known genera of substrate brooders are found.[1] Some substrate brooders, however, are prominent in the rivers of Africa, and the tribe Lamprologinae dominates the rocky shores of Lake Tanganyika.

Any of these cichlids that are monogamous and seeking a suitable mate share a problem with other pair-bonding species, for instance, ours. Reflect on the human condition for a moment to see what I mean. Perhaps you are seeking the ideal person with whom to raise a family, or anticipate that some day you will. The following message is embedded in advertisements in the lonely hearts section of the newspaper, though it is seldom spelled out so graphically:[2] "Single, attractive woman looking for a tall trustworthy man with a secure income. Must be good with children and willing to make a commitment." That covers the main points, and it starts with a self-advertisement of attractiveness.

Perhaps love at first sight does result in a harmonious, enduring union. More often, however, you probably were intensely attracted to a person of the opposite sex and then, after a few dates, lost interest. Worse, you were enamored of someone, but that person rejected you. My life's experiences taught me that most individuals go through such experiences a number of times. Obviously, just being good-looking is not enough. For whatever reason, that person was not right for you. Most of us can guess why, and it usually depends on the interplay of personalities. The prospective mate was physically appealing, but you did not "click" with the personality under the veneer. Monogamous cichlids are like that, though I am not going to claim they have likes and dislikes, or that they have personalities (though they probably do.[3]) An often repeated observation that has been little studied is that when a person puts a male and female cichlid together, more often than not they are not compatible.[4]

When hobbyists or biologists want to breed a particular kind of monogamous cichlid, characteristically they select a male and female that seem ready to spawn. Then they simply put them together in a suitable aquarium. The male predictably starts to court the female, and the female responds positively. However, about half of the pairs proceed to get into trouble.[5] The male starts attacking the female. If she is not rescued, the male may seriously injure or even kill her. Yet both are ready to mate and spawn. What is the matter?

This phenomenon is seen merely as a bother by many researchers, yet it is key to understanding mating in monogamous cichlids. By the end of this chapter, you should have an understanding of what is probably going on. We'll start by recapping how pairing in monogamous cichlids differs so sharply from the polygynous mouth brooders who lie at the other extreme. Note that a female cichlid could have placed the same advertisement in the lonely hearts section, for it says she is available, seeks a quality male, and wants one who will commit to raising the brood. In what follows, we can assume the female is available. However, the key themes transcend those that humans face. Just as in the polygynous mouth brooders, species recognition, sex recognition, and mate assessment must take place before the issue of compatibility enters play. Note, too, that these issues apply to males as well as females in monogamous species. After all, men place personal ads, too.

MATING WHEN MALES AND FEMALES LOOK ALIKE

The male and female monogamous cichlid often look so much alike that at first the fish themselves may not be able to recognize which sex is which. As one example, male and female Oscars *(Astronotus ocellatus)* are notoriously difficult to differentiate. The sexes in some species differ slightly, but seldom conspicuously. Characteristically, however, the male is larger than the female in freely formed pairs.

The growth of fishes has misleadingly been called indeterminate. By that it is meant that an individual fish continues to grow throughout its life, which is only partly true. Fishes do differ from birds and mammals in that they commonly continue to grow after they have become reproductively mature. After maturity is reached, growth is ever slower until the rate becomes almost static. A fish, nonetheless, might increase its weight severalfold after maturing, before reaching its ultimate weight. That is a remarkable difference when compared with mammals or birds.

The consequence of this form of growth for cichlids is a substantial overlap in size between the sexes. If you measure all the males and females of the Midas cichlid in a lake, you find the males are larger on average than the females, and the largest male is larger than the largest female, yet females regularly mate with males larger than themselves.[6]

Why does a pair regularly consist of a large male and a smaller female,[7] whatever their absolute sizes? Can we produce pairs in captivity in which the female is larger than the male? Only a few attempts have been made to produce such pairs.[8] I will suggest some explanations for this relationship when I take up the effect of size and aggression on pair compatibility later in this chapter.

Given that the sexes look alike and that size does not necessarily indicate gender, the initial distinction between intruder and potential mate cannot be based on a difference in size. Until recently, the consensus was that pair formation in monogamous cichlids occurs when a male establishes a territory and a female approaches the male, as in the mouth brooders. If that were so, identification of gender would be easy for females: Just look for a member of your species holding a territory. Males, however, would still have a problem because an approaching fish might be another male set on evicting him from his nest site.

The belief that males independently establish territories originated from watching cichlids in unnaturally small aquaria. Early in my career I became suspicious of this scenario while intensively observing one monogamous species of cichlid, the orange chromide *(Etroplus maculatus),* from India and Sri Lanka. I monitored a group of juveniles that were just becoming reproductively mature in a large aquarium.

At maturity, small differences between the sexes start to appear. The male develops a relatively inconspicuous touch of dark red on the rear margin of his eye, as though someone had touched him with a pointed lipstick. For her part, the female acquires two pale, slightly blue markings on her tail fin, one at the upper margin and one at the lower; here, too, the difference is not obvious.

That community of orange chromides regularly foraged in loose groups. I noticed that when a male and female were side by side feeding, the male sometimes began rapidly trembling his head from side to side, a behavior called quivering. Occasionally the female quivered back. In many but not all instances, the two fish paired within a few days and together set up a breeding territory. Now we know that the general pattern among monogamous cichlids is that the pair forms first, and then together they establish a territory.[9]

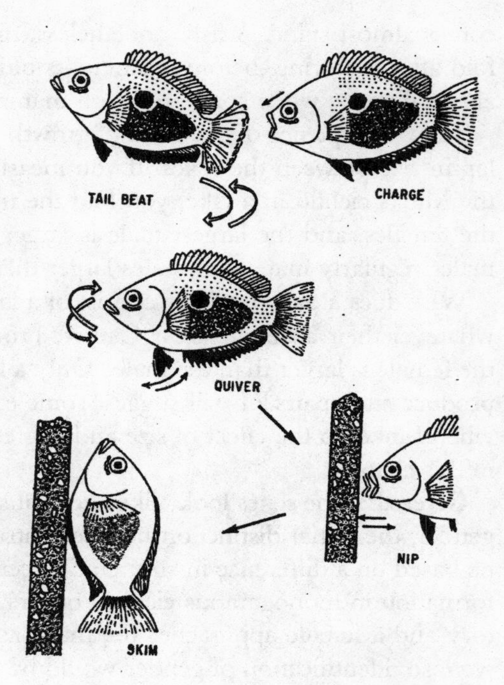

FIGURE 8.1 Courtship sequence in a pair of orange chromides *(Etroplus maculatus).* From the top, the male charges at the female, who turns away and delivers tail beats into the male's face. One down, the female then shakes her head rapidly from side to side (quivering) while flickering the pelvic fins in and out. She then proceeds to nip demonstratively at the spawning site, and finally she places her vent against the spawning site, simulating spawning (skimming). The sequence is the same when the female charges at the male, and he answers with tail beat, quiver, nip, and skim.

After pairing, when the female orange chromide approached the male, he went into frenzied display. This usually started with vigorous tail beats aimed at the head of the female. Then he might suddenly scoop up gravel in his mouth, carry it off, and spit it out (that is how the nest pit gets dug). Or he might quiver rapidly and then begin biting vigorously at the selected spawning substrate, often the vertical wall of a rock or of a terra-cotta flowerpot; that action, called nipping, cleaned the surface to receive the eggs. When the burst of displaying was particularly intense, it usually ended in skimming; in that act, the fish pressed its vent against the wall of the nest and slowly glided along, and that is symbolic spawning. Then it was the female's turn. She performed the same displays as did the male. Thus not only do the male and female orange chromide look nearly alike, they express the same behavior patterns in their mating behavior. This arrangement is representative of other monogamous cichlids.

The lack of qualitatively dissimilar male and female displays is another arresting separation in behavior between the dimorphic and monomorphic cichlids. Among the haremic substrate brooders, and especially among polygynous mouth brooders, males and females behave differently. The females of polygynous mouth brooders direct little distinctive behavior at the male during spawning but show the core behavior patterns when aggressive.

Many variations on the basic theme have evolved, and these are most evident among the substrate brooders. In courtship, the pair-bonding *Juli-*

dochromis marlieri rolls on its long axis, folds all of its fins, and flutters just the rear lobe of its dorsal fin. In lateral display, one fish or the other sometimes extends the elongate pelvic fin nearest the other, as though pointing a finger. They also do a rapid, darting swimming action, accompanied by quivering, when excitedly courting.

In the Midas cichlid *('Cichlasoma' citrinellum),* the behavior is simpler than in the orange chromide, but they engage one another during pair formation in a way I have never seen in another cichlid. In a large pool, but not in an aquarium, the overture to pairing is initiated by the female. She swims directly at the male, head-on, in a game of "chicken." Her opercles are spread as she rushes toward him, and he answers by facing her, opercles spread. At the very last moment, the female avoids collision and slides along the side of the male as she passes, an action termed "slipping." The Midas cichlid is also singular in that it does not perform ritualized skimming.

I am taken by the consistent nature of the core courtship displays seen in so many species of monomorphic cichlids. If you can recognize flaring of the opercles, tail beating, quivering, nipping, digging, and skimming, you can tune in to the cardinal components of courtship.

THE RIGHT SPECIES

All cichlids experience the familiar problems of deciphering species, sex, and quality of potential mates. Among monogamous, monomorphic cichlids, compared with the mouth brooders, we find even less experimental evidence for how the fish recognize their own species. A comparative study, however, suggests that the pattern of bars and stripes on the face is important for distinguishing species; the bars and stripes on the body seem more influenced by ecology, that is, whether the fish lives out in the open or in a structured environment such as a rocky reef.[10]

This makes sense in a mating system in which the male and female are so close to one another that they literally get in one another's face, and take some time doing so. Consequently, the potential mates have ample opportunity to study the other's facial markings. I hasten to add here that by no means are elements of bars and stripes found on the faces of all pair-bonding cichlids. Many cichlids, such as the African jewel fish, *Hemichromis guttatus,* and the South American uaru, *Uaru amphiacanthoides*, lack such facial markings.

One way to investigate the question of how cichlids recognize their own species is to present them with dummy fish whose color patterns depart to varying degrees from the typical pattern. Paul Siri and I were interested in how individual Midas cichlids respond to dummies that lack all species-typical markings and are completely gold in color.[11] That question has salience for the Midas cichlid because it is polychromatic: A small number of fish in nature are totally gold in color. Thus they lack the normal dark markings on a gray background. So the question naturally arose whether such oddly colored individuals can mate with the normally colored ones and whether they would prefer to pair with a fish that shares their color. After all, they might regard such differently colored fish as different species.

The experiment we did with dummies indicated that gold color alters the aggressiveness of the test fish. That might suggest interference with species recognition to some researchers. But in live gold-colored fish, the lack of color does not interfere with species recognition.

In most studies, the more a dummy resembles the species of interest, the more it is attacked or courted, depending on the situation.[12] Those results are useful for informing us about the role in communication of color markings. But to nail down how fishes discriminate between species one has to offer them a choice. The choice should be between two fish, one its own species and the other a close relative with which it might hybridize in a natural situation.

Is One's Species Learned?

If a fish can recognize its own kind, how does it do that? Could the fish learn who it is as a result of experiences with its own kind early in life? If so, then if raised by a different species, or with offspring of a different species, it should come to treat that species as its own. To test this, cichlids were fostered by radically different species, or raised with the alien species' offspring, starting at what amounts to infancy.

As adults, the experimental cichlids in social groups responded differently toward the foster species than did nonfostered young.[13] However, the change in behavior was slight and not interpretable; no interbreeding resulted. Thus the studies said little about species recognition but nonetheless demonstrated some learning had taken place.

In another experiment to explore how cichlids learn how to tell one species from another, we started with wrigglers of the Midas cichlid and fostered them to parents of their own species, but in all color combinations of gold and normal.[14] When they became reproductively mature, we tested their preference of mate by color. Earlier attempts to analyze mate choice had suggested the behavioral responses of the treatment fish could influence choice of the subject.[15] To make sure the subject was choosing solely on the basis of appearance, the treatment fish were presented behind a one-way mirror, preventing them from responding to the subject. The selectivity of the females was not crisp and could be revealed only through statistical analysis.

The issue here was whether the fish showed sexual imprinting. This is the same question asked previously, but now with more careful experiments: Do cichlids have the potential to learn which species they are? We were asking this at a more subtle level: Would they prefer to mate with a fish that shared one or both parents' color?

The females in this situation responded to males of both color morphs. However, the salient point that emerged was that female Midas cichlids are biased toward normal-colored males no matter how they are raised. The normal color pattern is the primitive one. The experiment had revealed an inherent tendency to recognize the basic color pattern irrespective of early experience.

The finding bears on the question of how new species of cichlids originate. In trying to understand how so many different species of cichlids could arise in each of the African Great Lakes, some have proposed that young cichlids of polychromatic species imprint on the differently colored parents. When they become adults, they choose to mate with just that color morph, and, voilà, a new species is started.[16]

Cryptic Species

Jeff Baylis, Jennifer Holder, Ken McKaye, and I have studied a cluster of monogamous cichlids from Nicaragua, an undertaking that included the Midas cichlid. These closely related species are difficult to tell apart and often coexist in a concentrated area. They are so similar in shape and coloration, in fact, that they were long thought to be the same species. In reality, they are sibling species.

From field studies and from pilot experiments in the laboratory, we proved we were dealing with at least three different species. One, the red-devil cichlid *('Cichlasoma' labiatum),* is found only in the Great Lakes of Nicaragua. Another, the Midas cichlid *('C.' citrinellum),* is found in the Great Lakes but also in the adjoining rivers and small lakes, though each lake may contain a genetically different population of them, perhaps even different species. The third and clearly new species, found only in the crater lake Apoyo, I named the arrow cichlid because of its more elongate shape; its scientific name, *'C.' zaliosum,* means wave-borne, in recognition of its habit of swimming in the open lake near the surface. These three species mate only with their own kind in nature, and detailed morphological analysis revealed subtle, consistent morphological differences between the species.[17]

When two of these species inhabit the same lake they must be able to distinguish one another. They do this in the absence of overt differences in shape and coloration. How do they do that? We do not know the final answer, but we have made progress.

In 1976, Jeff Baylis undertook a painstakingly detailed quantitative study of the mating behavior of the three species in our laboratory.[18] Not only are the three species nearly indistinguishable visually, but so are the sexes and their courtship displays. They differ only in the circumstance in which they express the elements of their courtship behavior. Yet, when with members of their own and the other species, they mate only with their own kind.

Using mathematical models from information theory, Baylis analyzed the effect of a behavioral act by one mate on the likelihood that its mate would respond with any given courtship display. Behavioral differences were real, but not obvious to the casual observer. Baylis decided that both species and sex recognition were accomplished through a complex web of communication.[19] He would be the first to admit that although this conclusion is the most plausible explanation, and one I agree with, it still is not proof.

Jennifer Holder followed up on those studies by testing the ability of two species to differentiate their own species.[20] One species was the familiar Midas cichlid, and the other was the red devil, *'C.' labiatum*. The two species occur together in the Great Lakes, but the Midas cichlid occurs additionally in other lakes. When two species are found together, the risk of hybridization is real, and so evolution ought to sharpen the behavioral mechanisms that prevent interbreeding and the dilution thereby of their gene pool, one that has been evolved to enhance their adaptations. But when the species occur in separate lakes, they lack the continuous need of behavioral mechanisms that keep them distinct.

In the test, the subject was given a choice of two chambers that it could enter and view a treatment fish through a one-way mirror. In addition, the odor of some fish could be introduced through a tube in front of the treatment fish that was shown behind the mirror. Thus, a subject could view a fish either of the same or different species, to which the species odor was either matched or mismatched.

For those Midas cichlids living in contact in the same lake, said to be sympatric, both males and females responded strongest to their own species, although only when the visual and odor cues were matched. When mismatched, they gave no sign of discriminating. And they would not respond to the odor if no visual cue was provided. When the fish from one lake were tested against cues from those living in a different lake, said to be allopatric, only the female subjects discriminated.

This experiment made the following points: In these monogamous cichlids, both visual and chemical cues are needed to recognize one's own species in the absence of interaction. Taken together, the experiments indicate that species recognition is based on a multiplicity of cues. That may not strike you as astonishing, but many behavioral studies conclude that just a single cue is used by other animals to recognize one's own species.

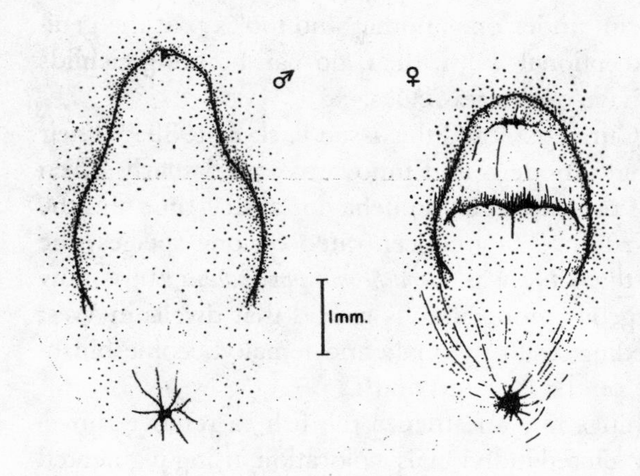

FIGURE 8.2 The genital papillae on the bellies of two Midas cichlids, heads toward the bottom of the page. The more conical papilla of the male is to the left, the more disc-shaped papilla of the female to the right. The vents, from which excretory products are released, are just below (in front of) the papillae, and the genital openings are at the tips of the papillae.

For some animals, such as insects, that can be the case, but monogamous cichlids are more sophisticated than that.

THE RIGHT SEX?

Once a monogamous cichlid has determined that he or she is about to mate with another cichlid of the same species, the task then becomes one of ensuring that the potential partner is of the opposite sex (in reality, these assessments are probably carried out simultaneously). In the previous chapter I complained about the lack of information on sex recognition in the polygynous mouth brooders. The situation is even worse among the monogamous substrate brooders. Unfortunately, scientists like easy problems because the answers are clean and readily publishable. That might explain why the few papers on sex recognition have been mostly on the highly dimorphic mouth brooders, where the answer is known before the test is done.

In the monogamous cichlids, the sexes commonly look much alike, so much so that at times they are almost impossible for us to separate. In my lab, we hold the Midas cichlid upside down and inspect the tiny genital papilla at the vent of the fish. The differences are a reliable means of distinguishing the sexes.[21] In mature fish, the papilla is more elongate and pointed in males but more disk-shaped in females.

I doubt that the fish check one another's genital papilla at the time of pair formation (it swells just prior to spawning).[22] However, a cichlid in Lake Tanganyika, *Julidochromis marlieri,* might provide an exception. The papilla of the male is pigmented and unusually long, so long, in fact, that it can be seen from the side as the male swims by. In at least one other fish, a distantly related labrid found in the Galapagos Islands, the sexes look alike, and during mating they swim under one another and look up at the genital papilla.[23] But this is exceptional. How, then, do pair-bonding cichlids distinguish the sexes? We have only a few clues.

G. K. Noble and Brian Curtis explored this issue back in 1939.[24] Their classic paper was gratifyingly modern and innovative, particularly given the almost amateur level of most research on behavior at that time and the lack of statistical sophistication. They concentrated on one species, the jewel fish, whose name at that time was *Hemichromis bimaculatus* but is now *H. guttatus*.[25] This is a gorgeous monogamous cichlid that dwells in West African rivers. When breeding, both the male and female become blush-like red with blue spangles on the face and body.

The investigators did things like anesthetize the fish to remove movement as a clue to sex. They altered individuals' coloration, using pigmented grease. They also administered hormones that produced deeper-red males, injected saline solution to make females appear full of eggs, and placed tiny blindfolds over the fishes' eyes.

Experienced females unfailingly distinguished the sex of a jewel fish in an adjacent aquarium, but a naive female often made mistakes. When they blindfolded fish, neither males nor females could make out the gender of another fish. But when the blindfolds were either transparent or had a small hole in the center, permitting the fish to see, then the subjects detected which sex was which. From this, Noble and Curtis concluded males and females must move differently. Because sex was not discerned when the test fish was blindfolded, the investigators might also have concluded that odor plays no role.

Mature males of many species of cichlids, both polygynous and monogamous, develop a swollen forehead, called the nuchal hump, or just the hump. The throat also expands, but less noticeably. In the Midas cichlid, the swelling normally occurs during the time of pair formation, then subsides around the time of spawning. The cause of the distension is local edema: Fluid floods the tissues within a day or two.[26] I mention this because others have jumped to the conclusion that the hump is filled with fat;[27] actually, the fat content is diluted by the edema.

Using dummy Midas cichlids, Paul Siri and I established that females respond best to a dummy having a hump that agrees with the normal size, which is relatively small compared with the huge humps some overfed males in captivity develop.[28] Contrary to predictions made by the sexual-selection hypothesis, the largest humps drew the least response from females. Thus the nuchal hump is one way the male reveals his gender to the female, and at the time it is needed most, right at pair formation. For his part, the male Midas cichlid responded most to the dummy that lacked a nuchal hump. Thus the male might be using lack of hump as one clue to femaleness.

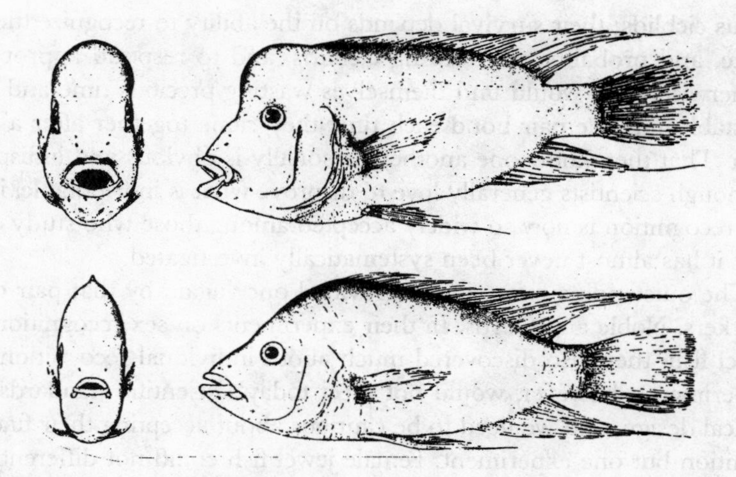

FIGURE 8.3 The upper Midas cichlid is a male ready to mate. He has an expanded nuchal hump and throat. The frontal view reveals how the swollen head and throat extend out to the side as well, although the opercles in the drawing are broader than normal, an accident of preservation. The lower figure is that of a nonbreeding fish.

So far I have concentrated on visual cues in sex recognition. Other senses, such as olfaction, might also be important, but they have been remarkably little explored. One study in our lab on the role of odor led to ambiguous results.[29] The evidence that sound may be involved is a bit more substantial but still scant.

In my laboratory, we examined the question of whether sound could be used to enable sex recognition, employing the Central American cichlid *'Cichlasoma' centrarchus*. They make low-pitched grunting sounds that are audible to us only when amplified. Males produce more sound when they see a female but reduce the grunts if the female shows courtship behavior.[30] That lends credibility to the hypothesis that sex recognition might be helped by sound signals, adding another dimension to the abundant cues. Yet, the finding remains only a tantalizing possibility. The Midas cichlid produces similar sounds in a social setting, but their function has not been revealed.

GETTING PERSONAL

Hobbyists often insist that a particular pet cichlid distinguishes between various humans; that is, they recognize people as individuals. If they can, that wouldn't surprise me, particularly if the pet were one of the monogamous cichlids; their survival depends on the ability to recognize their own mate, and probably neighboring cichlids, and to respond appropriately. Otherwise, they would find themselves wasting precious time and energy reestablishing the pair bond each time they came together after a separation. That they know one another personally is obvious and indisputable. Although scientists generally love to disprove what is indisputable, individual recognition is now so widely accepted among those who study cichlids that it has almost never been systematically investigated.

The outstanding exception is provided once again by that pair of early workers, Noble and Curtis. In their experiments on sex recognition in the jewel fish, they also discovered much about individual recognition. Their experiments, however, would not meet today's scientific standards of statistical design, and we need to be cautious about accepting their findings. I mention but one experiment: Female jewel fish could not differentiate between their mate and another male when both males were anesthetized and moved artificially, but they could tell them apart if the males had on blinders and moved slowly. Thus for mate recognition, they concluded, appropriate responses from the male are not necessary, but life-like movement is.

A recent careful study of the Lake Tanganyika cichlid *Neolamprologus pulcher* confirmed that the female recognizes the video image of her mate.[31] The Midas cichlids that I studied for so long are able to recognize their mates and tank mates with no trouble. This applies even to the all-gold morphs that lack any distinctive marking. Taken together, these kinds of observations reinforce the idea that cichlids discriminate among individuals based on many, and sometimes subtle, attributes.

Anecdotal observations on different cichlids over the years, however, makes me suspect that both individual and species recognition in pair-bonding cichlids is facilitated primarily by the face of the other fish, seen from the side. I once observed a breeding pair of green chromides, *Etroplus suratensis,* who were on the alert for predators on their fry and were totally at peace with one another. The aquarium was partially divided by two tiles, stood on end to create a low wall. In the middle of the wall was a small vertical opening, not large enough for the parents to pass through, but large enough to see another fish on the other side.

Whenever the parents were on the opposite side of the wall and happened to pass the slit at the same time, they spotted one another. Each immediately wheeled and went to the window to investigate. Face on, they tried to attack one another through the slit. They became excited, and one or the other soon darted up into the water, over the wall, to the other side. There it found its mate, equally excited. They greeted and acted as though they were saying to one another, "Did you see that intruder?" Clearly, they could not recognize one another from just the front view of the face. Re-

Paratilapia polleni, from Madagascar, tending her eggs (courtesy of Paul Loiselle).

A pair of *Etroplus suratensis* from Sri Lanka. The female, in the flower pot, quivers her head in courtship while looking at her mate.

A male jewel fish, *Hemichromis guttatus,* from Africa (courtesy of Paul Loiselle).

A male tilapia of the species *Oreochromis macrochir*, in breeding coloration (courtesy of Paul Loiselle).

Two male firemouth cichlids, *'Cichlasoma' meeki,* in frontal display with their branchiostegal membranes lowered and opercles, with false eyes, widely spread (courtesy of Ad Konings).

Female convict cichlid, *'Cichlasoma' nigrofasciatum,* examining her clutch of eggs.

A large specimen of the piscivorous peacock cichlid, *Cichla ocellaris* (possibly *monoculus*) (courtesy of Ad Konings).

A pair of South American dwarf cichlids, *Apistogramma nijsseni,* nicely demonstrating the differences between the sexes in size and coloration; the female is the smaller, yellow fish (courtesy of Ad Konings).

A male of the Central American cichlid, *'Cichlasoma' managuense* (courtesy of Paul Loiselle).

A female of the Costa Rican cichlid, *Herotilapia multispinosus,* with some of her young (courtesy of Jeffrey Baylis).

An unusual cichlid, *'Cichlasoma' minckleyi,* from northern Mexico in the desert springs called Cuatro Cienegas. In this haremic species, the male becomes black and the female white. Here is a pair at the mouth of the cave they have excavated for spawning (courtesy of Juan Miguel Artigas).

A pair of Mexican cichlids, *'Cichlasoma' labridens,* with their offspring. The female presses her belly against the soft bottom and passes rapid, shallow undulations down her body to stir up the detritus, from which her fry feed (courtesy of Juan Miguel Artigas).

A pair of Midas cichlids, *'Cichlasoma' citrinellum,* from Nicaragua. This is the first day of swimming for their school of fry. The female is colored gold while the male bears the normal coloration. The polychromatism is found in both sexes.

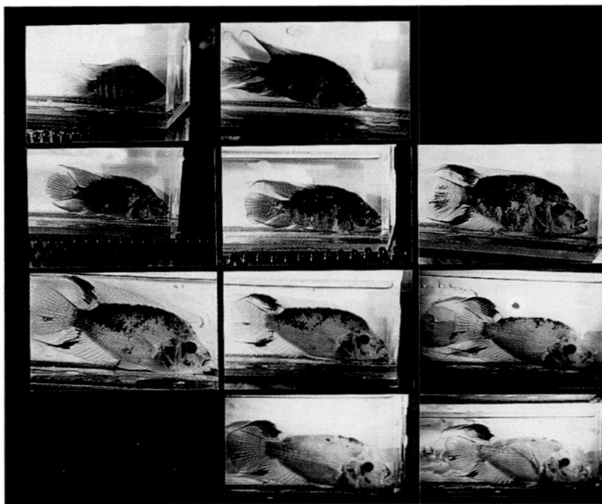

A series of photos over time of a Midas cichlid, showing how it gradually changes color from normal to gold to white as it loses chromatophores (courtesy of Michael Dickman.)

A pair of South American earth eaters, *Satanoperca leucosticta,* in the act of spawning. The female, in front, is laying eggs (courtesy of Wayne Leibel).

One of the goby cichlids, *Eretmodus cyanostictus,* from Lake Tanganyika. They live in the shallow surge zone (courtesy of Ad Konings).

A male of a widely distributed mbuna in Lake Malawi, *Metriaclima barlowi* (courtesy of Ad Konings).

The most well known species of mbuna from Lake Malawi, *Metriaclima zebra,* here represented by one of its more colorful manifestations, a male OB morph. Note the egg spots in the anal fin (courtesy of Ad Konings).

Another Aufwuchs-scraping cichlids from Lake Malawi, a male of *Labeotropheus fuelleborni.* Note the egg spots in the anal fin. This species is widely distributed in the shallows (courtesy of Ad Konings).

Ophthalmotilapia ventralis is an open-water cichlid that comes to the reefs in Lake Tanganyika to spawn. Here, the female responds to the egg spots at the tips of the male's anal fin (courtesy of Ad Konings).

The piscivorous *Cyphotilapia frontosa* of Lake Tanganyika. This haremic species inhabits relatively deep water where it forms harems. The female in this photo has a mouthful of eggs (courtesy of Ad Konings).

A pair of *Julidochromis marlieri*, with the female in front. This particular color pattern is found at Rutanga in Lake Tanganyika (courtesy of Ad Konings).

In this remarkable photo, a family of *Neolamprologus pulcher* (formerly *brichardi*), from Lake Tanganyika defends against the predaceous *Lamprolongus elongatus* (courtesy of Michael Taborsky).

A tiny pair of *Neolamprologus brevis* have stationed themselves before their 'home,' an empty snail shell on the sand bottom of Lake Tanganyika (photo courtesy of Ad Konings).

The most extreme example of sex dimorphism among vertebrate animals is shown in this photo from Lake Tanganyika. The huge male of *Neolamprologus callipterus* approaches one of his tiny females (almost under his head), who occupies one of the many empty snail shells (courtesy of Ad Konings).

Another snail-shell dweller from Lake Tanganyika. This group of *Neolamprologus similis* lives as an extended family, like N. multifasciatus described in the text (courtesy of Ad Konings).

An inhabitant of Lake Bermin, *Tilapia bemini* (courtesy of Uli Schliewen).

The tiny crate lake, Bermin (courtesy of Uli Schliewen).

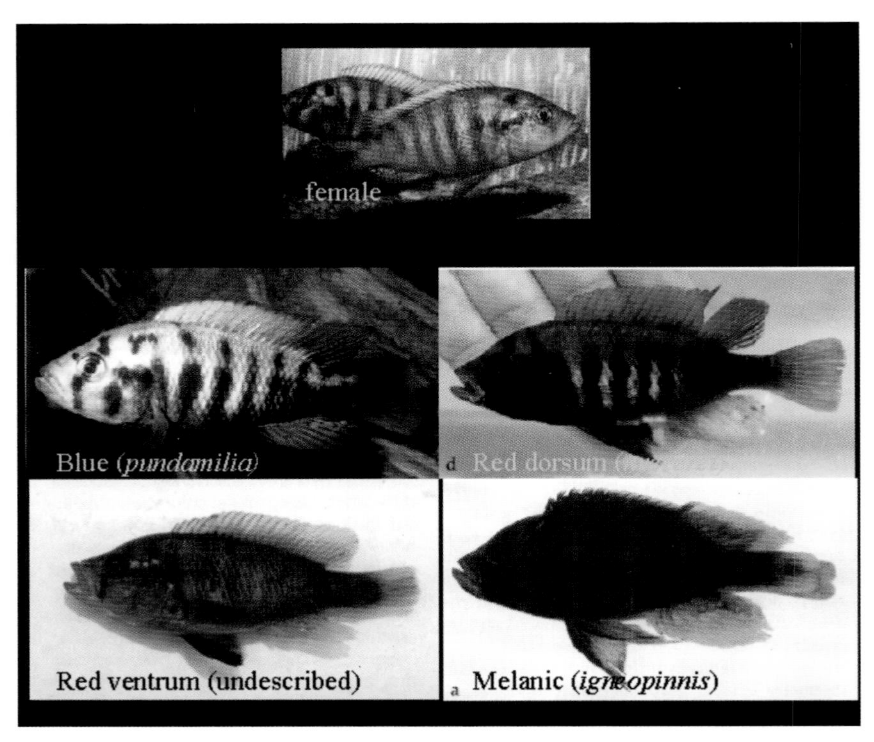

A composite set of photos showing the differences in coloration of a group of closely related *Haplochromis*-type cichlids from Lake Victoria. Note particularly the drab females at the top, and the blue, red top, and red bottom males (courtesy of Ole Seehausen).

cently, I saw the same thing happening within a pair of Lake Tanganyika cichlids, *Julidochromis marlieri.*

That the side of the face is so important is revealed by a test conducted by Noble and Curtis of a small monogamous cichlid from Central America, a species they called *Cichlasoma cutteri.* Think of their experiment this way: We can recognize a friend from a considerable distance by his gait and general appearance. However, we sometimes make mistakes. Imagine approaching such a friend from behind and greeting him. But when the person turns, he has a huge handle-bar mustache, which your friend lacks. Whoops! Mistake. Not really. The friend has fooled you by putting on a false mustache.

In essence, Noble and Curtis performed the mustache experiment by putting a black marking under the eye of the male mate, a marking he normally lacks. That caused his mate to attack him.[32] Thus, cichlids recognize individuals visually through many aspects of their appearance, but facial marking are especially salient. Other kinds of sensory information are also involved.

Stephan Reebs has shown that female convict cichlids use smell to recognize their individual mate.[33] He tested several females who were guarding their eggs at night by confining them in the dark to small plastic boxes with a male of the same species, either their mate or a stranger. The females attacked at their own mates a little, which they commonly do when their mates approach the clutch. But the strange males were attacked much more. From that and other aspects of his findings, Reebs concluded females recognize their mates individually on the basis of close-range chemical cues.

HOMOSEXUALITY, AGGRESSION, AND SEX RECOGNITION

The orange chromide stands out for the ease with which females form what some would call homosexual pairs;[34] I prefer to refer to them as isosexual. When animals have access to members of the opposite sex, homosexuality is virtually unknown in nature, with some rare exceptions among primates. Humans are a well-known exception, and some of the great apes engage in bisexual behavior.[35]

At one time, same-sex pairs of gulls on one of the Channel Islands off Southern California attracted attention in the national media. However, the female pairs laid fertilized eggs, so they had copulated with males.[36] A later study revealed that females outnumbered males in that population, leaving some females without mates. Heterosexual females merely joined forces to cooperatively defend a nest site and raise their chicks.

Female pairs of orange chromides regularly occur when they are kept in all-female groups. They even spawn together, each laying eggs. Sometimes threesomes form, but one female usually gets excluded at the moment of spawning. The same thing had earlier been seen among jewel fish when kept in all-female groups.[37] Males of Tilapia mariae have been induced to pair, but only after each had already spawned and was guarding a school of its young.[38] Male orange chromides, in contrast to their females, are unable to form isosexual pairs; I observed occasional attempts in all-male groups, but the tryst always succumbed to fighting.

The key seems to lie in how orange chromides assess one another, and this is relevant to how cichlids recognize and interact with the opposite sex. In a test situation, visiting six other fish of the same or opposite sex, females were stimulated to court more toward other females than toward males, even though they prefer males. Males courted more to females, too.[39] Thus females are more attractive to both sexes.

Both females and males, however, also attacked more at members of their own sex. Males are too aggressive ever to come to terms with another male. But if two females can resolve their aggressiveness, as when with another female of the same size, they can mate. This is consistent with the notion that an important element of recognizing maleness by females is aggressiveness, as Holder's experiment with the Midas cichlids suggests.[40] Years earlier, Bernard Greenberg suggested the same with regard to isosexual pairs of jewel fish.[41]

Our knowledge of how monogamous cichlids distinguish species and sex remains thus a largely unplowed field, even among those species that show reversed sexual dichromatism. Perhaps because sexual selection is in fashion, a bit more is known about the next level, that of mate assessment.

WHICH IS THE BEST MATE?

Sexual selection most clearly comes into play through initial attraction. Monogamous cichlids present a special challenge here because both the male and female should be choosy. The reason is that each makes roughly the same investment in raising the offspring.[42] Put another way, the male and female share the same fitness as measured by the number of young they raise.

Their roles, however, differ depending on the species of monogamous cichlid, and I'll treat that in more detail in a subsequent chapter. A characteristic pattern among New World cichlids is that males expend more energy in courtship and defending the territory, and females exert more effort in laying eggs and caring for the eggs and fry, especially in the early phase of their development.[43] The orange chromide, on the other hand, is unusual because males and females share caretaking about equally.

In either case, monogamous female cichlids should prefer the largest, most aggressive males, especially in species that contest for breeding territories,[44] as they generally do. The largest and most aggressive male should have the best chance of securing a territory.

For his part, the male should also select the largest, most aggressive mate available, and for the same reasons. Playing "chicken" may serve this decision. The male has an additional incentive, however, for getting a large mate: The larger the female, the more eggs she can produce.[45] That is a consequence of constant egg size. A larger female can pack in more eggs.

Experiments to test this set of hypotheses have agreed with the premise, for the most part, most notably for females. Male choice has either seldom been tested, or investigators have regarded failure to show a choice as failure of their experimental design and hence not reported the research.

In at least one experiment, male cichlids exercised choice of female. Male subjects of the orange chromide, *Etroplus maculatus,* visited two treatment females who were confined behind a transparent screen.[46] One female was about 5 percent longer than the male, and the other female was 5 percent shorter, which is not a large difference for that species. The males clearly courted more to the larger females, as theory predicted. When females were tested, the courtship they expressed did not differ much between the larger and smaller treatment male. However, the females attacked at the smaller males much more, which is a backward indication of preference for the larger male. A later and more elaborate experiment produced similar results.[47]

Female convict cichlids select the larger of two males when given a choice.[48] We did a similar but more elaborate experiment on the Midas cichlid.[49] For females, the outcome was as predicted: Female Midas cichlids responded most to large males, and in a separate test, most to the more aggressive males. As shown in a subsequent experiment, large male size assured the female of a territory, and aggressiveness predicted the willingness of the father to defend his young.[50]

We also tested mate choice by males. To our consternation, the males chose at random and seemed reluctant to court in the artificial setting. Later experiments again found the male reluctant to perform in a situation where the female could not see him and where she was thus unable to respond to his behavior.[51]

One difference between the sexes, common to cichlids, might be clouding the issue. Males can spawn promptly when presented a female ready to lay eggs; males have sperm at the ready. Females, in contrast, need time to manufacture the large yolk-rich eggs. In nature, males may be interested only in those females that are close to spawning. That would reduce the time and energy spent in courtship.

But how would a male know that? For one, the genital papilla of the female swells as she approaches spawning, and then the papilla is visible from the side.[52] The amount of male courtship has been shown to correlate with the size of the female's papilla in the African cichlid *Tilapia mariae*[53] and in the Midas cichlid,[54] but correlations are not proof.

Conceivably, males can detect some other change in females associated with approaching spawning. She could be releasing chemical signals, pheromones, but that has never been investigated. The girth of the female does increase noticeably as she fills up with eggs, and males should be able to notice that.

Daniel Nuttall and Miles Keenleyside decided to explore the role female girth plays in mate choice, using the convict cichlid. The results were illuminating. Choosing through glass, the males spent the most time with the female with the greatest girth, that is, the most gravid. If the females were equally gravid, the males stayed with the larger female. If the smaller female was gravid but the larger one not, then the males chose the smaller female. Thus males discriminate among potential mates both on size and proximity to spawning.[55]

MATING AS A PEACE TREATY

The findings on size and girth of female made me even more baffled by the failure of male Midas cichlids in our experiments to discriminate among females, especially when the females differed in size. I thought aggressiveness might be the key.

The role of aggression in pair formation tells us something important about how a pair bond is formed. Konrad Lorenz, in his book *On Aggression,* defined pair bonding as the moment when the male and female turn their aggression away from one another and instead cooperate in directing it toward other individuals.[56]

The Midas cichlid is a particularly aggressive species, probably because they nest colonially and so have to fight a great deal to obtain and keep a territory. And the pair must cooperate to do this. The males fight other males, and the females fight other females, though that is not an absolute rule. Thus one of the main purposes of pair formation and early courtship should be testing to determine whether the potential mate is sufficiently aggressiveness. If they are to prevail in holding a territory, each member of the pair needs a big aggressive mate.

The color polymorphism (polychromatism) of the Midas cichlid provides a natural experiment for exploring the role of aggression in mating. As a quick review, the color of the infrequently occurring all-gold individuals acts like a supernormal threat display, though the gold and normal morphs are equally aggressive.[57] Being gold confers an advantage in a hostile encounter equivalent to being about 15 percent heavier than the normal opponent,[58] which is a huge advantage.

In the aggressive Midas cichlid, as in so many monogamous cichlids, pair formation looks much like a fight. The male threatens the female, and she responds with threat display toward him. This is all part of testing the opposite sex and just the kind of interaction that is excluded in aquarium experiments. There, mate choice is examined by offering treatment fish behind some barrier that prevents overt fighting.

When members of a potential pair are not separated by some screen, they test one another directly and often lapse into a real fight instead of pairing. To prevent that, we placed a screen barrier between pairs, using all four combinations of sex and color; the males weighed about 15 percent more than the females.[59] So arranged, all pairs spawned, on average, in two weeks.

In the next experiment, we removed the barrier and monitored their free behavior. In all groups but one, about half the pairs failed to form, and the female had to be rescued. The exception was among pairs with a gold male and a normal-colored female, in which the failure rate was much higher: 84 percent of the pairs were incompatible. That was a stunning difference.

I hypothesized that to the normal-colored female, the combination of a large and gold-colored male was too intimidating. Conversely, the smaller normal-colored female was not much of a threat to the gold male. If the female were larger, increasing her perceived potential for aggression while diminishing that of the male, then the pairs should be successful.

So we repeated the experiment on the critical color combination, gold male and normal-colored female, but now the proffered normal-colored female was the same weight as the gold male. Mating success climbed to an unusually high level of 60 percent. That outcome agreed well with the argument that the female cannot mate if the male is too intimidating or, seen the other way round, if the female is not imposing enough. In effect, making the normal-colored female the same size as the gold male was equivalent to a normal-colored pair with the male 15 percent heavier than the female.

An unrelated experiment adds further support. Baylis forced two sibling species to hybridize. Males of the aggressive species *'Cichlasoma' zaliosum* and females of the less aggressive *'C.' citrinellum* could form pairs only if the female was almost as large as the male.[60]

The findings were starting to converge. We knew females preferred large, aggressive males, and we suspected the same applied to male choice of females. But the males refused to cooperate unless they could interact with the females. In particular, the males seemed to require a certain aggressive potential in the female, expressed either through size or aggressiveness per se.

The last experiment marked a change in direction. The goal was to have interaction a part of the test situation while retaining some control over it. The analysis focused on the difference in aggressiveness between the male and the female. That appears to be the key to detecting what is going on.

The smaller normal-colored female Midas cichlid could visit two larger normal-colored males of equal size. Each male was at first confined to the rear half of his chamber by a large-mesh screen. That allowed the male and visiting female to do everything through the screen except get past it. They could even bite through it, if the fish on the other side cooperated. Shortly before starting to observe the behavioral interactions, we counted how often each fish attacked its mirror image and used that as a measure of their aggressiveness.

In the previous studies I described, when a female was separated from two or more males by a screen she regularly ranged back and forth between them, seldom making an all-or-nothing choice. In the experiment at hand, in arresting contrast, most females expressed a strong preference for the more aggressive male of the two, ignoring the other one. The next day the screen barrier was removed. Then things went wild.

Without the barrier, the male and female interacted differently and at a high rate. With that male, it was make or break for the female. Despite the clear preferences shown by the females, 43 percent of them either could not or would not pair with the liberated chosen male and had to flee his chamber through a slit too narrow for him to follow. Of those who fled, 29 percent were able to mate with the other, less aggressive male. Being unable to pair with the first male, therefore, did not mean the female was not ready to mate. She was simply more compatible with the second-choice male than with the first.

The important message here is that female choice did not necessarily determine with whom she mated. The preferred males rejected the females about half of the time. This is a clear instance of conflict between the sexes. Male choice and female choice did not coincide, and the males prevailed.

I was nonplused by the data from the first day of the experiment, the one with the screen in place. No relationship whatsoever emerged when I analyzed the scores for male-female differences in aggression. The male and female in each compartment obviously and vigorously responded to one another. But they were not paying attention to what the other was doing. Each did its own thing. They were not interacting in the real sense of the word. They were "talking past one another."

This says worlds about the body of experiments on mate choice that have precluded direct interaction. It does not invalidate them. It says merely that different things are learned depending on whether interaction is allowed or prevented.

I analyzed the results for the successful pairs on the second day, when the screen had been removed. The picture was clear: The more the aggression score of the female exceeded the male's, the more the pair courted and the less aggression they expressed. Size difference was not a planned variable, but enough variation in relative size existed for an analysis. The outcome was essentially the same: The larger the female relative to the male, the more they courted.

For the females, initial attraction and ultimate pairing did not necessarily agree. Only half the time when a female chose a male based on appearance could they mate. And then mating was achieved when the female had a higher aggressive score than the male. In addition, the male was favorably disposed to a large female. Size and aggressiveness substitute for one another.

That females choose large aggressive males was already known. But males prefer the same traits in a female, at least in a relative sense. This results in a conflict of interest between the male and the female, and the male wins. The males get their way because they pair with large, aggressive females, but females may be rejected by the males they choose if the females are not aggressive enough. Thus mate choice here is indirect.

UNANSWERED QUESTIONS

The last finding raises more questions than it answers, as a good experiment should. It suggests that in nature, aggressive females pair with relatively less aggressive males, which I doubt. Is the most aggressive male in a population unable to breed? The most dominant hen in a flock often does not get fertilized. And why is the male of a pair always larger than the female, at least in nature, even when the overlap in sizes of males and females is sufficient for a smaller male to mate with a larger female? I suspect that larger females are taken by the larger males, and males run larger than females. Also, the role of the male as defender of the territorial boundary probably has been selected for large, tough males.

Can we draw a message for us from all of this? Only in the sense of posing the question. I do not expect the study of cichlids to shed light on spousal abuse in humans, but then it might. In the larger sense, monogamous cichlids seem a good analog to mating among humans. Initial attraction is there, but then the personalities of the male and female must mesh in order to develop and sustain a pair bond.

We might also debate whether monogamous birds or cichlids provide the better model for comparisons with humans. Songbirds have been the popular model, but their position has been questioned because cuckoldry is so common among them.[61] But perhaps they are the better model because cuckoldry is an integral part of human behavior,[62] as any soap opera on television vividly demonstrates. This has led to a lively literature on sperm competition in animals and humans.[63] The possibilities for sperm competition in cichlids are there, but little has been reported. That is one of the issues I take up in the next chapter, where I describe how cichlid eggs get fertilized.

≈ *nine* ≈

HOW GAMETES MEET

An individual is created at conception, but that is not the time life begins. Life is continuous. Each organism alive today has a heritage of uninterrupted life extending back on the order of a billion years or more. For creatures like mammals and fishes, sperm and eggs are but two alternative forms that life takes.

Sperm and eggs, collectively called gametes, have lives of their own. Each is an independent organism, although one with an excruciatingly short life. Sperm and eggs differ from one another in several ways, which I described earlier. The pertinent difference here is that the multiple diminutive sperm swim and the relatively few large eggs do not. From that follows some distinctive cichlid adaptations in the eggs and sperm, and in the way they are brought together.

Both kinds of gametes originated in an aquatic environment, and they remain adapted to a fluid medium. Even in such highly terrestrial organisms as mammals, sperm still swim in a fluid that is mostly water. The male nurtures and delivers his sperm in a specially favorable solution of various salts, hormones, and complex sugars. To receive the sperm, the female mammal creates an aquatic milieu of just the right composition for the sperm to survive and swim to the egg.

Some fishes, for instance, guppies, procreate in a fashion similar to that of mammals and birds. The sperm enter the reproductive tract of the female and combine with her eggs, typically on the inner wall of her hollow ovary. Cichlids, for all their evolutionary success, have never accomplished internal fertilization. Well, at least not in the genital tract of the female.

As with most, but not all, aquatic animals, the eggs and sperm of cichlids are shed into the water around them. The typical life span of fish sperm in water is about 30 seconds, whether in the ocean or in freshwater.[1] Water can be a hostile medium because its chemical composition can vary widely;[2] some waters are acidic, other basic, and the amount of dissolved solids ranges from almost distilled water to nearly brackish. Thus the gametes of each species must be adapted to survive in the water in which the cichlids live. Cichlids are freshwater fishes, but some flourish in marine water and can spawn there;[3] some species even spawn in inhospitable soda springs.[4]

Beyond the ability of the sperm to survive and move on their own, the behavior of the parents plays a critical role in bringing the gametes together. Much of this chapter is about how adult cichlids prepare for spawning and coordinate the meeting of the gametes. I will also tell you about the nature of the eggs themselves, where considerable differences are apparent. I start with the basal monogamous cichlids, say a few words about haremic species, and finish with the spectacular spawning behavior of the highly evolved mouth brooders. For dessert, I will tell you about the issue of sneaker males.

WHERE TO PUT THE EGGS?

The first and crucial step in the enterprise of spawning and nurturing the eggs is choosing where to put them. The spawning site is the focus of the territory the pair establishes. The common feature of the site is a firm, clean substrate onto which the parents can glue their eggs. It should also be a site that the parents can defend against predators intent on devouring their eggs and young. The most common location is a small cave-like hole in a rocky area of a stream or lake. Typically, the fish digs out a cave under a rock or removes the sand and debris from a cave.

Each species chooses a cavity appropriate to its body size. This may seem too self-evident to mention, but it has meaningful consequences when competition for breeding sites emerges in a community of cichlids of different species.[5] Other aspects of the cave may also be important.

In one experiment, pairs of convict cichlids, *'Cichlasoma' nigrofasciatum*, were given a choice between different artificial caves in an aquarium. In the first test, the caves varied in light intensity. In the next test, they were offered "caves" with either one or three entrances. The cichlids chose the dark cave over the bright one, and the one with a single opening over the others. Apparently this behavior guides the parents to a dark cave that is easier to defend because it has only one entrance.[6]

Eggs are usually placed on the ceiling or sides of the cave. That positioning probably reduces the smothering effect of silt falling onto the eggs.[7] However, a few cichlids show little preference and spawn on any surface in their cave, including the floor.[8]

FIGURE 9.1 A pair of the African species *Thysochromis ansorgii* spawning in a crevice. The female is upside down, laying eggs on the ceiling, while the male waits his turn to invert and fertilize the eggs.

Some species burrow deeply into the clay banks of streams and rivers to make their nest. Ron Coleman was in Costa Rica as I prepared this chapter. He wrote, excitedly, about *'Cichlasoma' nicaraguense*. Their tunnels penetrate well into the banks of rivers and provide a refuge. Secure in such deep holes, their eggs don't require adhesive threads to hold them in place; such threads might even be detrimental in a potentially oxygen-deficient cavity.

But how do they excavate such deep tunnels in firm clay? The fish, absorbed in their task, tolerated him watching at close range. Both the male and female participated. The female was out of view in the burrow ahead of the male, working on the end of the burrow, probably a chamber. Coleman was amazed to see the male had mud stuck to his back and dorsal fin. The male repeatedly jammed his body into the initial opening, wriggling around and even rotating like the business end of an auger. Later, Coleman's flashlight revealed the fish had produced a perfectly circular horizontal tunnel into the bank. A cichlid fish has evolved into a drill and bit!

In Lake Tanganyika in East Africa the water is so alkaline and full of dissolved salts that all inanimate material, such as rocks and snail shells, becomes coated with minerals. Sandy bottom soon becomes encrusted, too, giving it a pavement-like surface. One small cichlid, *Neolamprologus tetracanthus,* takes advantage of this to make a secure nest. It bores a deep tunnel into the bottom, spawns there, and then shepherds the fry.[9]

In Panama, I watched a pair of large *'Cichlasoma' maculicauda* tending a nest that resembled a posthole dug into the bottom of Lake Gatun. The shaft was in water only about six inches (15 cm) deep, which was barely enough water to cover the large parents. The depression itself was also around six inches in diameter and roughly one foot (30 cm) in depth. When I put my hand into the shaft I was surprised to discover that the walls of the shaft were covered by tiny hair roots from the trees along the shore. The female had stuck her eggs onto the hair roots.

While scuba diving in the lake, I saw another pair that had made a similar nest, but at a depth of fifteen feet (4.5 m) in a lush bed of the alga *Charra* that carpeted the bottom in a layer about one foot thick. Remarkably, the shaft in the *Charra* was about the same shape and size as the one in the mud bottom, but the parents had carved it out of the plant material.

Since then, I have found other hole-nesting cichlids that attach their eggs to hair roots. In the interior of Belize, in White Water Lagoon, the distinctive piscivore *Petenia splendida* spawns under water lilies. They dig a small pit in the sandy bottom, exposing the hair roots of the lilies, on which they spawn. Such behavior is probably common among cichlids that spawn in rivers where only soft bottom is available.

The orange and green chromides (*Etroplus maculatus* and *E. suratensis,* respectively) live in brackish coastal lagoons in Sri Lanka. The bottom is soft and sandy, and hard surfaces for attaching eggs are scarce. The green chromide rips sea grass out of the sand bottom in shallow water and further digs away the sand. The result is a shallow pit about four feet (1.2 m) in diameter. The eggs are laid on the exposed roots along the perimeter of the pit. The truly remarkable discovery Jack Ward and Rick Wyman made was that the eggs almost perfectly match in size and color the nodules on the roots of the sea grass. The orange chromide makes a much smaller pit, around eighteen inches (around 45 cm) in breadth, and places its eggs on blades of grass, roots, algae, or twigs.[10]

Many monogamous cichlids spawn out in the open. Those species, such as the African jewel fish, *Hemichromis guttatus,* and the South American geophagines, as well as some members of the genus *Aequidens,* commonly select a flat rock as the egg platform. That exposes the clutch to predation but the flowing water may help keep the eggs oxygenated and free of silt. Better sites, however, may not be available to them, and they need to attach their eggs to something to keep track of them.

One of the more remarkable adaptations for a spawning site is one I discovered in Panama. A stream-dwelling species, *Aequidens coeruleopunctata,* lays its eggs on a loose leaf about the size and consistency of a leaf from a rubber tree.[11] Leaves were abundant in the stream, but most went limp when picked up. The few stiff ones I found were already taken by nesting cichlids, so I presume the leaves are a limiting resource.

I never saw the *A. coeruleopunctata* put to the test, but the adaptive significance of the leaf was evident. Well over my head, as I stood on the bank, was flotsam stuck in the branches left there when the river flooded. After a heavy rain, the stream swells and becomes a raging river. That makes life difficult for a mother, the usual caretaker of eggs, if she stays with her eggs in the middle of the stream. Having laid the eggs on a leaf, however, she can tug it into the quiet water at the edge of the stream.

While wading, I approached a female guarding eggs. She grasped the leaf in her mouth and backed into the protection of the undercut bank. When I turned a leaf of another female over, placing the eggs on the underside, she promptly returned and righted it. Other species in this genus, *A. vittatus,*[12] *A. paraguayensis,* and *A. pulcher,* [13] have subsequently been shown also to place their eggs on a moveable leaf.

Females of these species of *Aequidens* use the leaf in courtship in a way so far unparalleled by any other genus of cichlids. When an attractive male approaches the female, she picks up the leaf and swims with it, and she often tugs at the leaf during courtship, demonstrating to the male her readiness to spawn. She provides the nuptial couch. Much less often, the male enters into the act by tugging at the leaf, and he occasionally moves the leaf when it has eggs on it.[14]

Thus, monogamous cichlids have their own version of foreplay, involving the preparation of the place to spawn. You already know the major features of courtship from the previous chapter. As spawning nears, however, courtship changes. The pair proceeds through a quantitative shift in the kind of behavior expressed. The male is at first the more active partner, but gradually the female becomes the most animated as the time to spawn nears.

WHEN AND WHERE

Within the seasonal phenomenon, certain times may be more propitious than others. Spawning in some cichlids tends to be in phase with the moon, perhaps so the fry will disperse during the dark of the moon when they are less vulnerable to predators.[15] A more subtle suggestion is that when the females spawn in phase, they discourage their male mates from deserting because other females would already be engaged in mating.[16]

Better understood is the task faced by the female in synchronizing the release of her eggs in relation to the presence of a mate. She can't just squirt them out. She needs a brief spell to marshal her eggs for spawning, and that may be an important function of courtship. She has to shed the eggs into the lumen of her ovary and prepare them to pass down the oviduct to the exterior. Not to overstate the synchronization component, female cichlids in aquaria are notorious for spawning in the absence of the male.[17]

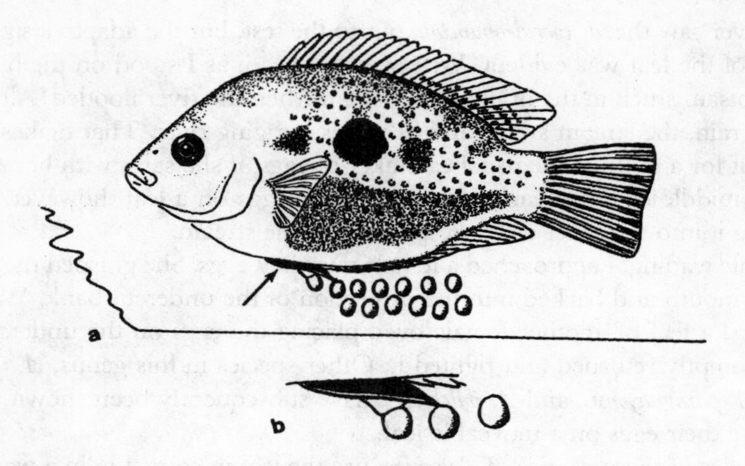

FIGURE 9.2 A female orange chromide, showing how the pelvic fins sense the position of the eggs she has laid (from Ostrander and Ward 1985).

As spawning approaches, the genital papilla of the female swells to several times its normal size,[18] and it often becomes reddened by the now visible blood vessels. The papilla is noticeable from the side and might be evident to the male. The genital papilla of the male also swells, usually presenting a small conical extrusion that typically hooks forward ever so slightly.

The male, sensing that spawning is near, increases the amount of digging around the chosen spawning site. He is excavating a pit next to the place where the eggs will be placed. While doing this, he frequently quivers and nips at the spot chosen for the eggs. The female also quivers, digs, and nips. But as the great moment approaches, quivering fades away. Increasingly, she nips the spot where she will place her eggs and follows that by skimming, the act she uses for laying the eggs. Progressively, this becomes a female show, with the male standing back. He may skim from time to time, though much less than the female.

For their part, males are ready to spawn right from the outset. I have seen a male orange chromide *(Etroplus maculatus)* ignore a female's preparations for spawning, and then immediately fertilize the eggs when they appeared.

Finally, the first eggs are laid. The tiny pearls are only about 1.0 to 1.5 millimeters in diameter, and they are placed in neat rows. The female slowly lays down one or several strings, and then the male briefly replaces her, skimming over the eggs while releasing invisible sperm. She continues spawning. Usually, the affair lasts around an hour. As the female approaches laying her last few eggs, she starts fanning them in brief bouts. The male may do the same. That signals the onset of parental behavior.

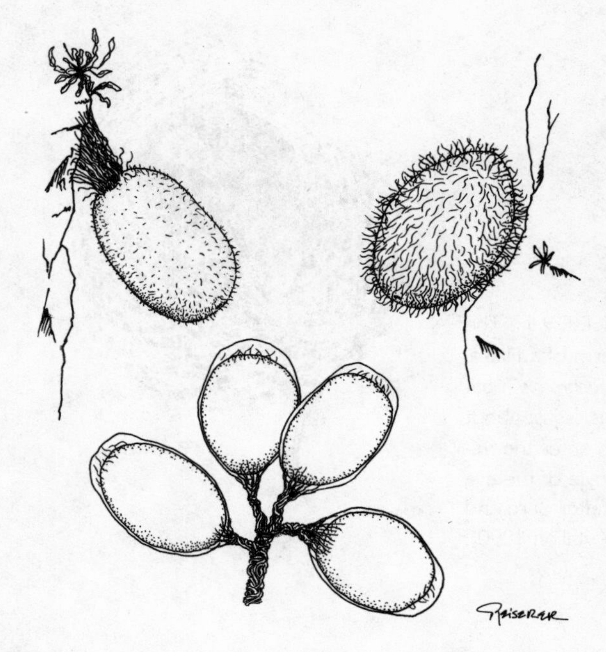

FIGURE 9.3 Three different ways in which the eggs of cichlids are attached to the substrate by adhesive threads. The typical situation for most African and New World species is shown at the upper right, in which small adhesive threads cover the egg. The upper left depicts an egg of the orange chromide; its threads form a stalk. At the bottom is the grape-like arrangement seen in *Paratilapia polleni* from Madagascar (after Stiassny 1993).

In most, but not all, substrate-spawning cichlids, the eggs are laid in a remarkably neat, closely packed patch. Seldom are they laid on top of one another, although some species, such as *Tilapia tholloni,* do that.[19] How the eggs are arranged in such orderly fashion, one immediately next to the other, is a mystery. While depositing the eggs, the female's eyes are in the wrong position to see what she is doing. Therefore, she must deftly feel where the already laid eggs are.

Gary Ostrander reasoned that she senses the eggs through her pelvic fins, which are right next to her genital papilla from which the eggs are issued. He anesthetized some females and surgically removed the pelvic fins. After recovering, those females behaved normally, mated, and spawned. Their eggs were scattered across the nest, demonstrating the role of sensory nerves in the pelvic fins in guiding the precise positioning of the eggs.[20]

MEET THE GAMETES

The eggs of most substrate spawners are slightly elliptical, though some are oblong, like tiny fat sausages. Almost all have adhesive threads that glue them to the substrate, though the details of how this is done differ. The primitive species of the Indian genus *Etroplus* and the Madagascan cichlid *Paratilapia polleni* have a structure for attaching the eggs that diverges from the more ad-

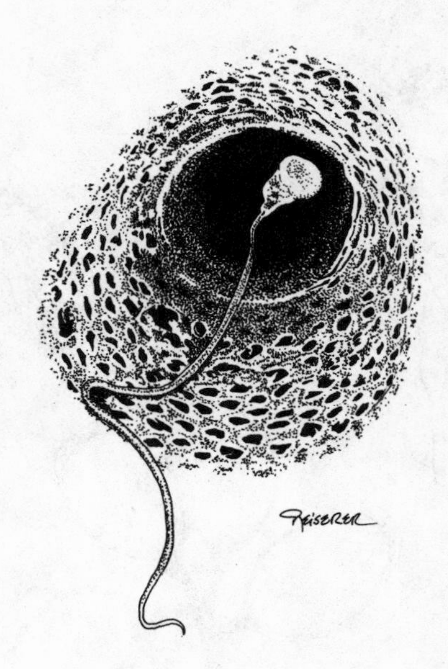

FIGURE 9.4 The sperm of a tilapia, *Oreochromis niloticus,* is just about to enter the micropyle of the egg (after Bern and Avtalion 1990).

vanced substrate-spawning cichlids. In these basal species, a bundle of thread-like filaments extends out from one end of the egg. The micropyle, the pore through which the sperm enters, is situated at the opposite end of the egg.[21]

The eggs of *P. polleni* are relatively spherical. They are apparently unique in that the filaments form an interconnecting network, joining the eggs of the clutch together like a bunch of grapes, three to four eggs deep.[22]

The bulk of the more modern substrate-spawning cichlids, in both the New World tropics and Africa, have distinctly different eggs. They are characteristically only modestly elliptical. They are covered with tacky mucus and sticky, short microscopic filaments that vary in their distribution on the eggs. However, as more species of substrate spawners are examined, more keep popping up that have eggs with filaments attached at one end only.[23]

The sperm of cichlids resemble those of other vertebrate animals. Fish sperm differ in that they have blunt rather than pointed heads.[24] Sperm find the entrance provided for them by the egg, the micropyle, in less than one second.[25] They crowd into the micropyle in a heap. Only one, however, gets in. The others are subsequently evicted as a mass of frustrated dead or dying half beings.

Because sperm perish quickly, the male should do everything possible to release them as close as he can to the virgin eggs. At least among the monogamous species, the male places his vent right on the fresh eggs and releases his sperm. As one consequence of this highly efficient procedure, males of monogamous species have the smallest testes among the different mating systems. Males of polygynous species, who spawn often and more or less continuously, have the largest testes.[26] That would lead one to believe that the few sperm of monogamous males in particular would neither survive long nor travel far. Eggs at some distance would be out of luck. But that is not necessarily true.

I first discovered this when observing my old friend, the orange chromide. I had separated a male and a female by an open-mesh screen. The female spawned on a flower pot about one fish length away from the screen. Despite the separation, the eggs were fertilized. I was not there at the moment of spawning, but the male obviously had to have released sperm that swam to the eggs. They were all fertilized. That made me wonder if the sperm were able to detect eggs and swim toward them, or whether the male simply released so many of them that enough reached the eggs. Later I discovered that at least a few Midas cichlids can do the same, though some males never figure it out.

The males of haremic species are sometimes too large to enter the hollow where the female spawns. When that happens, the males come as close as they can; they release sperm at the entrance to the cavity and leave it up to the sperm to find their way to the eggs. This mode of spawning is common in males of the highly dimorphic dwarf cichlids of the genus *Apistogramma* and has also been reported for the snail-dwelling haremic lamprologines of Lake Tanganyika.[27]

Females in a harem are usually out of phase with one another. The male visits the females periodically to see which is ready to mate. When a female is full of eggs, courtship is relatively brief. After spawning, the male resumes patrolling his harem. Thus in this polygynous mating system, males must be constantly ready to fertilize eggs.

Males must also be constantly ready in the lekking, mouth-brooding species because a given male may receive several females in a day. Females also spread their gametes around. Characteristically, a female wanders through the lek, inspecting the males. She descends to spawn a small portion of her clutch with one of them, then moves on to another. Hence her eggs are fathered by up to four males in this polygynandrous mating system.[28]

DON'T RUSH OFF WITHOUT THE SPERM

I sketch the spawning act of mouth brooders again here because it is useful when following the story. In the prototypal situation, once the female has committed herself to spawning with the male, the pair circles in the nest. Reciprocal snout nudging of one another in the genital region is common. At some point the female drops a few eggs.

The male then passes over the eggs and releases sperm while quivering. As they complete the circle, the female returns to the inseminated eggs lying on the bottom of the nest pit and takes them into her mouth. Then she drops some more eggs, the male fertilizes them, and so on. When she has finished spawning, she departs to gestate the eggs in her mouth away from the spawning area. This basic pattern is seen in the tilapias such as *Oreochromis mossambicus*.[29]

Female *O. mossambicus,* however, sometimes make what might seem to be a mistake in this well-organized ritual. After she releases a few eggs, she may turn in a smaller circle than the male and pick them up *before* he can fertilize them. In many of the African mouth brooders, that is a regular event. But then how do the sperm reach the eggs?

Those "error-prone" female *O. mossambicus* provide a clue. When she picks up the eggs too soon, she places her lips against the genital region of the male for a moment.[30] The white genital papilla of the male stands out against his black belly. Subsequent research has shown that female African mouth-brooding cichlids suck sperm into the mouth to unite with the unfertilized eggs. Moreover, males have evolved various ornaments around their genital region, apparently to direct the females to the source of sperm.

Thus, cichlids have indeed evolved internal fertilization, but it takes place in the female's mouth instead of in her reproductive tract. Many writers refer to this as buccal fertilization; in Latin, *bucca* refers to cheek or cavity. Different cichlids have achieved this end in different ways, though the process has much in common across the species. Let's start with genital tassels.

A pioneer of the biology of tropical fishes, Rosemary Lowe-McConnell made the seminal (no pun intended) observation on the role of tassels in two species of lekking tilapias in swampy but clear water in Africa. She wondered how sperm could get to the eggs and noticed that the females often mouthed the genital tassel of the male. She suggested oral fertilization of the eggs but never pursued the matter.[31]

In most cichlids, the genital papilla is small and inconspicuous. But in the two tilapias that Lowe-McConnell studied, the papillae of the males have evolved into elaborate structures that hang down from the vents of the males. When the fish are not breeding, the tassels regress. Tilapia males that develop genital tassels are united in the subgenus *Nyasalapia*.[32] No other group of cichlids shows this remarkable feature.[33]

Males of *Oreochromis koromo* have spectacular large genital tassels consisting of two branches. The tassels are of the same bright orange color as that on the edges of the median fins of the male. Perhaps more meaningful, the tassels approximate the color of the eggs. The tassels are also nodular, like strung fresh garlic, further enhancing the resemblance to eggs. The male's body is blue-gray with black dots. Against that color, the genital tassel is "very striking."[34] Given the similarity to eggs, the tassel might trick the female into responding to them as though they were her own overlooked eggs.

When the female approaches, the male does something with his tassel that at first seems odd. He drags it across the surface of the nest he has dug.

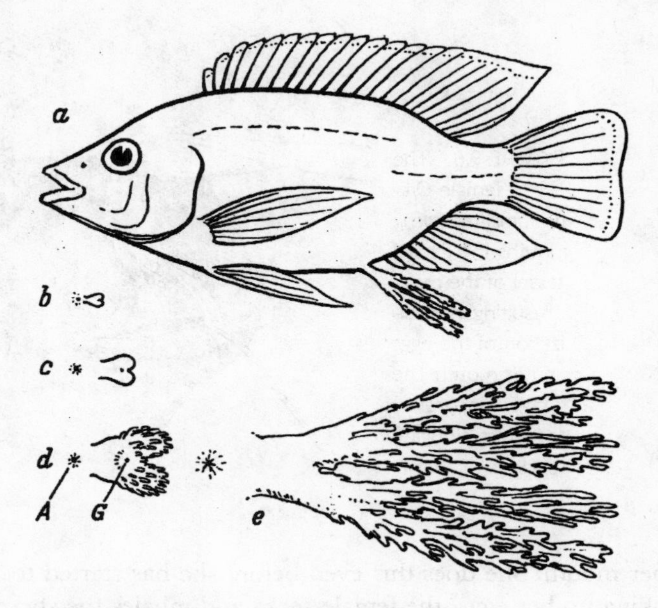

FIGURE 9.5 (a) The male of *Nyasalapia rukwaensis,* with enlarged genital tassel, thought to mimic a cluster of eggs. (b-e) The tassel becomes larger as the fish grows and reaches maturity. (A) The vent. (G) The genital opening (from Wickler 1962b).

He continues doing that after the female has started spawning. Lowe-McConnell hypothesized that the female picks up sperm by mouthing the bottom where the male touched it. She saw no evidence of sperm or packets of them, but microscopic examination later proved that sperm packets had indeed been dropped by the male.[35] Some females entered a nest, went through repeated spawning passes while picking up milt (a term for male sperm), but never laid eggs. Apparently, she briefly stores sperm for later matings.

The case of *Oreochromis macrochir* is worth examining in some detail because most of the speculation here, and below, is about the role of adornments such as tassels in mimicking eggs. In this species, however, the genital papilla seems to mimic not eggs but parcels of sperm.[36] At the start of courtship, the male *O. macrochir* presents his tassel to the female and then drags it over the nest, a shallow pit in the bottom. Right after that, he moves up in the water a bit and proceeds to back up across the nest and over the female. At that moment, he ejects a long, 0.5-mm-thick whitish thread that is sticky at one end. These delicate bombs adhere to the bottom briefly when the male makes contact with it, but within a few seconds they start to drift off.

In Wolfgang Wickler's words, "The female follows the male and grasps the spermatophore-like thread between her lips and inhales it. Very often she not only takes the thread but also the whole genital tassel, or parts of

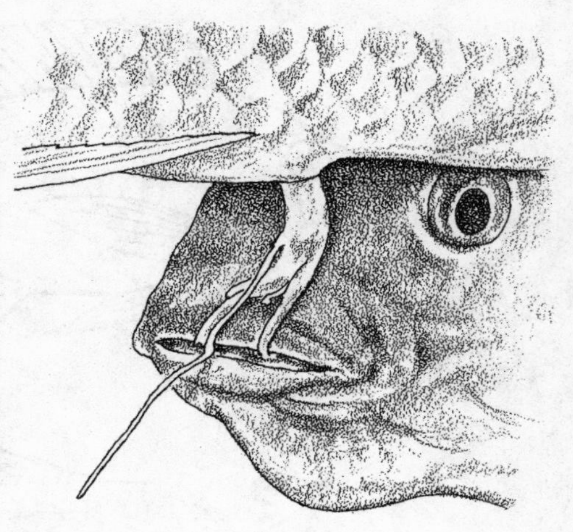

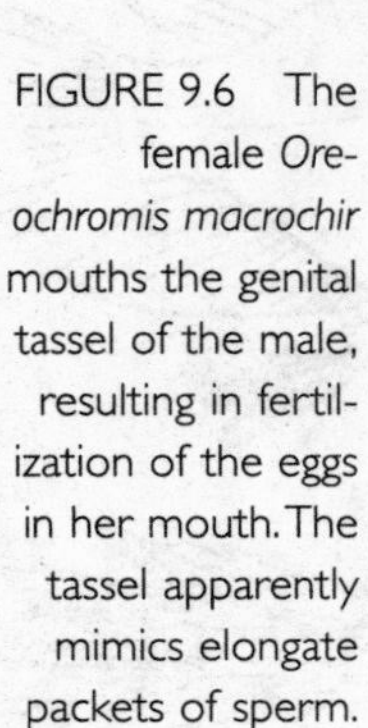

FIGURE 9.6 The female *Oreochromis macrochir* mouths the genital tassel of the male, resulting in fertilization of the eggs in her mouth. The tassel apparently mimics elongate packets of sperm.

it, into her mouth. She does this even before she has started to lay eggs." After picking up her eggs, the female seeks and inhales the sperm threads that are sticking to the bottom or drifting away. Wickler concluded that the function of the male's tassel is to attract females to the source of sperm. The tassel "resembles a large number of whitish sperm-threads . . . and . . . is treated by the female as such."

Subsequent research on other tilapias, *Oreochromis aureus* and *O. hornorum,* revealed that the packaged sperm are in a colloidal suspension of glycoproteins.[37] The males are putting out bundles of inactive sperm that become activated when the colloid dissolves in the mouth of female. Doing so reduces the dilution of sperm in the water that could happen without the packaging. The significance of this adaptation lies in its parallel with internal fertilization in noncichlid fishes where sperm are inserted into the reproductive tract of the female.[38] Those species also concentrate their sperm in a packet that bears the musically polysyllabic name of spermatozeugma.

Those masters of evolution, the cichlids, did not rest on their laurels when they evolved tassels for more effective fertilization of the eggs. Another lineage of African cichlids, the haplochromines, independently evolved another way of enticing the female to take sperm from the male. Those trailblazing investigators of cichlid behavior, Gerard Baerends and Jos Baerends-van Roon, noted in their 1950 monograph that *Pseudocrenilabrus multicolor* engages in oral fertilization, although they lack a genital tassel.[39]

Based on later observations of male *Haplochromis desfontainesii*, Rosa Kirchshofer pointed out that spots on the anal fin seem to mimic eggs.[40] Stimulated by this report, Wolfgang Wickler examined the related haplochromines, *Haplochromis wingatii* and *Astatotilapia* (formerly *Haplochromis*) *burtoni*.[41]

He discovered that the female follows the pattern of wheeling about right after releasing eggs and picking them up in her mouth before the male has had a chance to pass over them. With his head up, the male reclines on one side while spreading his anal fin on the bottom of the nest, all the while quivering. Thus he presents to the female what appear to be eggs on his anal fin. She tries to inhale the errant eggs and sucks up sperm in the process.

Wickler became fascinated with the idea that the anal-fin spots were actually mimicking the female's eggs.[42] To test this proposition, he compared the egg spots on the anal fins of several species of haplochromine cichlids. In those cichlids with colored disks in the anal fin, the coloration was the reverse of that of an ocellus: A thin, dark band at the perimeter of the spot encircles the colorful large central disk bright. That indicates such spots do not mimic eyes. Moreover, the central disk often matches the size and color of the eggs of that species. Because in the cichlids that Wickler studied the spawning female snaps at the spots, the conclusion was obvious. The anal spots are fake eggs. When the female tries to pick them up, she is "deceived" into sucking up the sperm. The inhaled sperm fertilize the eggs already in her mouth.

Wickler noticed that the egg dummies are actually more conspicuous than the real eggs, which are relatively dull and hard to see in the nest.[43] The fake eggs glisten, and their dark ring makes them conspicuous. One might call them supernormal stimuli. But if they were more attractive to the spawning female, then she should pick them up in preference to her own eggs. She never does. Only after the real eggs have been secured does she mouth the egg spots.

ARE THEY REALLY FAUX EGGS?

The story is persuasive. However, aquarists have reported many examples of cichlids, notably haplochromines, in which the female merely nibbles at the unadorned vent of the male after picking up the eggs.[44] Some scientists, too, remained skeptical,[45] and others set about to disprove the two

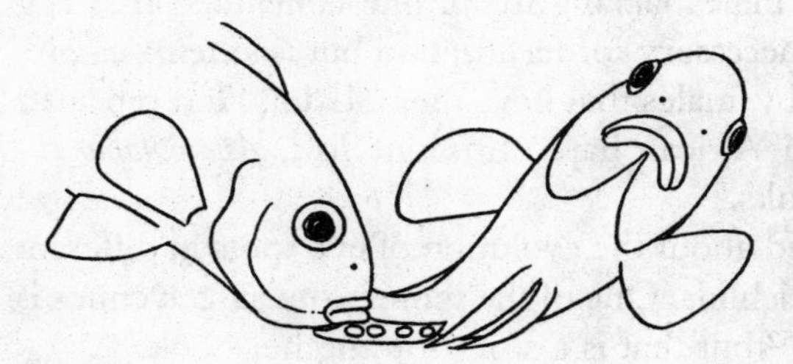

FIGURE 9.7 The male *Astatotilapia burtoni* lies to one side, presenting to the female the egg spots on his anal fin. She mouths the faux eggs and thereby fertilizes the eggs in her mouth (after Wickler 1962b).

major hypotheses here.[46] The first hypothesis is that the eggs are fertilized in the mouth, not outside of it, though Wickler noted that on occasion the male clearly fertilizes some of the eggs before they are engulfed by the female.[47]

Using the African haplochromine *Pseudocrenilabrus multicolor*, Wolfgang Mrowka, sought to determine if the eggs really are fertilized in the female's mouth, outside of it, or both. Through skillful manipulations, he proved that 30 to 50 percent of the eggs are fertilized before the female grabs them. Additional inhaling of sperm by the female resulted in nearly all of the eggs becoming fertile. Then the female was forced to spawn and pick up three batches of her eggs while the male mated behind a glass barrier. Right after that, the female was given access to the male, and 65 percent of the eggs were fertilized. Therefore, in *H. multicolor* part of the clutch is fertilized before being inhaled, but fertilization in the mouth is also essential.[48] This is probably true of other cichlids whose eggs are fertilized in the mouth of the mother.

The more challenging questions, certainly the harder ones to disprove, are, first, whether the anal-fin spots actually mimic eggs and, second, whether their function is to guide the female to the source of the sperm. Often egg spots lie at the far end of the anal fin, well away from the source of sperm, the vent.

DO FAUX EGGS TURN FEMALES ON?

Eva Hert, a colleague of Wickler, asked whether females of *Haplochromis elegans* prefer males with egg spots.[49] She divided a group of males into two, the experimental males from whom she erased the anal-fin spots (by touching the skin briefly with metal that has been dipped in super-cold liquid nitrogen) and the control males whose spots remained intact.[50] Females found males with egg spots more attractive than those lacking them.

In a subsequent experiment, Hert simply placed females in aquaria with males who either lacked or possessed egg spots. She wanted to determine whether the egg spots prime the females to produce more eggs. They did. Females placed with males showing eggs spots produced twice as many eggs as females placed with males lacking them. She concluded that egg dummies are not absolutely necessary for fertilization but that females prefer, and are more stimulated by, males that have them. Later, Hert repeated this experiment with another African haplochromine fish, *Astatotilapia elegans* with much the same results.[51]

Much remains to be learned about the evolution of eye spots in different lineages of mouth-brooding cichlids. One of the more promising avenues is their role in sexual selection,[52] but that is a story for another book.

WHEN FAUX EGGS HAVE BEEN LOST

Even beyond the tilapias, many cichlids have no structure that might be mistaken for egg dummies but nonetheless engage in oral fertilization. Among those in the Great Lakes of Africa, some species probably never had egg spots, and others probably once did but lost them in the course of evolution.

The first example is of two cichlids in Lake Tanganyika.[53] They and other members of their lineage probably never had egg spots, but females nonetheless get their eggs fertilized in their mouths. The small species, *Paracyprichromis brieni*, feeds on plankton in open water, although near reefs and especially those with precipitous faces. Lekking males hold territories over rocks, and the focal point for their spawning is a vertical section of rock; it's not a nest, but for want of a better word I'll call it that. Neither the anal fin nor the pelvic fins of the males have any marking resembling an egg.

When a female approaches, the male quivers and leads her to the nest. There he does a motionless headstand, with his vent close to the rock. The female nuzzles his vent with her mouth. They trade places, the female now head down, and the male nuzzles her. He moves away, perhaps a foot or more. The female drops one egg and quickly backs upward to catch the falling egg in her mouth. She spawns a couple of eggs, and the male returns. The female nuzzles him again, and her mouth can be seen opening and closing, most likely inhaling sperm.

Nearby, another open-water cichlid, *Cyprichromis microlepidotus,* is mating in a similar fashion. However, it does something remarkable. Every now and then the female swims to the male and bumps his vent. Then she returns to feeding on plankton a small distance away. This may go on for as much as eight minutes before she finally lays the first of her small clutch of eggs. And after she has laid all her eggs, she may visit another male and prod him, as though obtaining yet more sperm.[54]

The next example is the Lake Tanganyika species well known to most hobbyists, *Tropheus moorii.* The sexes differ little in color, being dark with a bright girdle of color around the midriff. They often spawn up in the water over a small pinnacle where the female drops one large (about 5 mm) egg at each fertilization. The egg is immediately snapped up by her. She then proceeds to the genital area of the male where, unguided by any egg-like markings, she sucks in the invisible sperm.[55]

In a few species of *Tropheus,* including some populations of *T. brichardi*, the males have poorly developed egg spots at the extreme rear margin of the anal fin. In other populations of this species, the males also have tiny egg-like markings at the tips of the spines in the front half of the anal fin. I suspect the ancestors of *Tropheus* had egg spots that were later lost during the course of evolution. But then I have not even mentioned how the egg spots were thought to have originated.

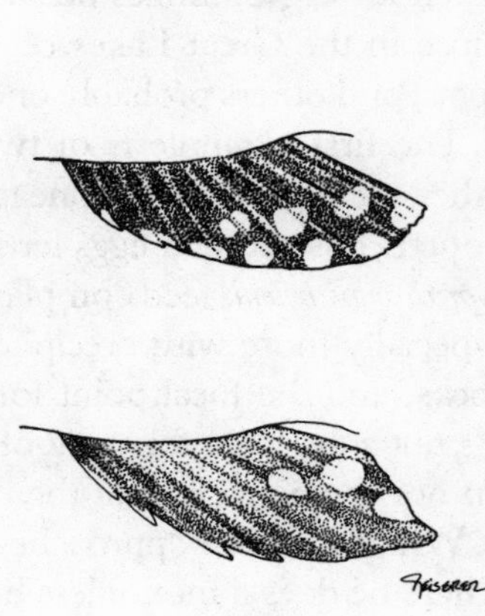

FIGURE 9.8 Anal fins of different species of mouthbrooding cichlids in Lake Malawi. From top to bottom, they depict the progressive change from simple spotting to large spots that resemble eggs.

WHENCE THE FAUX EGGS?

From a comparative study, Wickler developed a plausible progression in the evolution of egg spots.[56] In many haplochromine cichlids, the dorsal and anal fin of both the male and the female have faint reddish to orange lines. In yet other species such lines are broken up and appear as freckles. Other species have yet larger spots in their freckling. In yet others, the spots become bordered with black, producing an egg spot. The spots vary in size and position, depending on the species, but the most highly developed ones closely resemble eggs.

Some of the nonhaplochromine cichlids have what looks like an egg dummy not on the anal fin but at the tips of the elongated pelvic fins. Species in the genus *Ophthalmotilapia*, from Lake Tanganyika, provide several examples. The tip of each pelvic fin is drawn out, almost into a filament. At its extremity is a bright yellow to orange tab that approximates an egg in size and color. As with the anal-fin egg spots, the tip of the pelvic fin is not located right at the vent of the male; instead, it is normally placed toward the rear end of the anal fin. When spawning, however, the male holds the pelvic fins out, away from his body. He drags the tips, with their egg spots, across the floor of the spawning nest, right in front of the female's face.

The male probably ejects sperm in such a way that they follow a course laid out by the filaments of the pelvic fins, and thus into the mouth of the female. But perhaps not. Consider the case of another cichlid in Lake Tanganyika that has similar but not quite such well-developed yellow tips of its elongate pelvic fins. These cichlids are in the genus *Cyathopharynx*. Ad Konings has observed and described their spawning.

The lekking male drags his pelvic fins through the crater-like nest at the top of his sand castle. Konings suggested he may leave sperm there, but the evidence is so far lacking. The female follows, lays some eggs, and then makes room for the male, who fertilizes the eggs in the nest. Only then does the female pick up her eggs.[57] Thus the putative egg spots on the anal fins may play no direct role in fertilization.

In yet another Tanganyikan genus, *Cyprichromis,* all the rays of the pelvic fins are long, but not so long as in *Ophthalmotilapia.* Instead of a tab at the end of the fin as in *Ophthalmotilapia,* the entire tip is yellow. The yellow coloration probably guides the female to the male's vent to fertilize the eggs. This genus of open-water plankton feeders is closely related to *Paracyprichromis brieni,* mentioned above as lacking any signs of dummy eggs.

The original explanation for the evolution of egg dummies seemed simple enough: When the female picks up the eggs in her mouth before they can be fertilized, the egg spots guide the female to the source of sperm. That is probably correct, at least in the species with conspicuous egg spots. But as we have seen, other explanations have been proposed, and for good reason. Part of the problem is that more than one adaptation may apply to a given species. Thus, even when the eggs are fertilized in the female's mouth, some of her eggs are also fertilized while still lying in the nest,[58] probably because excess male sperm are present there.

Likewise, egg spots appear to have additional functions. For one, males with conspicuous egg spots may be more attractive to females; then they serve as a typical sexually selected character. For instance, males of species that live in dim light in Lake Victoria have larger than expected spots, indicating an advertising function.[59]

The egg spots also appear to stimulate the reproductive physiology of the female, though the experiments showing this had the females confined with the males.[60] We need to know whether females in nature, free to approach or avoid males, are actually stimulated by the males to produce more eggs. Do free-ranging females associate more with those males having more spots, and by doing so do they increase the Darwinian fitness of the males? We just do not know.

One critical test of the sperm-guiding hypothesis for egg spots would be to go further than Eva Hert when she removed the egg spots through cold branding.[61] The next step should be to affix faux egg spots to various locations on a male whose spots have been erased. Would the female follow those spots and try to pick them up if they were, say, on the middle of the body of the male? And if so, would fewer of her eggs get fertilized?

Oral insemination of eggs illustrates the evolutionary capacity of cichlids. When confronted with predation on the eggs, of the type described in an earlier chapter, cichlids responded by minimizing the time the eggs spend on the bottom. That meant the female often got the eggs into her mouth before they were fertilized. The solution was straightforward: Draw sperm into the mouth to finish the job. But that has evolved in so many ways.

In some of the *Oreochromis* no special feature has evolved in the males, except to have a brightly colored abdomen. Other *Oreochromis* have a brilliant genital papilla, and yet others evolved tassels, ranging from simple to elaborate. Mostly, the tassels tend to mimic eggs, but in one species they resemble the remarkable sperm packets that species has evolved.

At least one mouth-brooding South American cichlid, *Geophagus steindachneri,* has independently evolved oral fertilization. In that species, the female picks up the eggs quickly. Then the male positions himself at a right angle in front of the female with his genital region directly before her mouth. He quivers, indicating release of invisible sperm, which are then apparently taken up by the female.[62] Thus the capacity to evolve oral fertilization of eggs exists in widely unrelated cichlids that engage in mouth brooding, with or without egg spots.

WHO IS THE FATHER?

One area of research on fertilization has attracted a lot of attention of late. It is the issue of sperm competition and cuckoldry. It has been studied mostly in birds and mammals, including humans, where the possibility exists of more than one male inseminating the female.[63] Female poeciliid fishes, such as guppies, typically mate with several males, and sperm competition is known to happen in those fishes.[64] However, even among those fishes that have external fertilization, more than one male sometimes manages to release sperm near the eggs.[65] Trout and salmon, for instance, are known for what have been called jacks, tiny males that sneak up on a large spawning pair and release sperm.[66] Sneaking fertilizations by the subordinate males is sometimes termed parasitic spawning.

When at least two males release sperm over eggs, the sperm compete with one another to fertilize the eggs. The male having the larger testes is usually, but not always, the larger male. He has a statistical advantage because his sperm outnumber those of the male with the smaller testes. Beyond that, the dominant male, the one that holds the spawning territory, drives the sneaker male away as quickly as he can. Thus the controlling male has more opportunity to wash the eggs with his sperm.

In reviewing sneaking among fishes, Michael Taborsky counted the number of species in each family where parasitic spawning, as he called it, occurs.[67] The highest number of species for a single family has been reported for those marine relatives of cichlids, the wrasses, with twenty-five species known so far. Cichlids, however, were second with sixteen, and

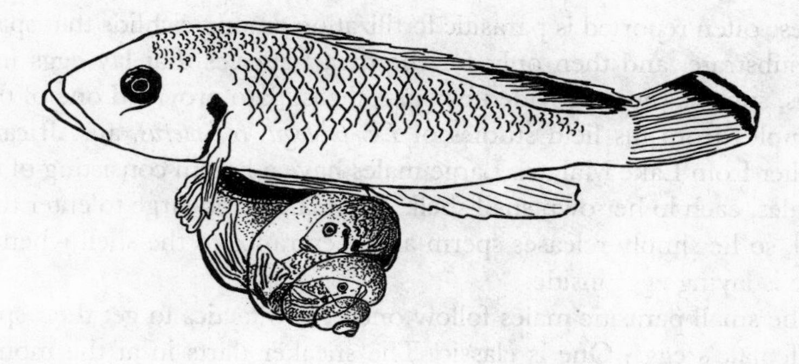

FIGURE 9.9 At spawning, the tiny sneaker male of the Lake Tanganyika cichlid *Lamprologus callipterus* insinuates himself deep inside the snail shell, where he releases sperm as the female lays her eggs. The large dominant male cannot enter the snail shell but releases sperm into the shell from outside (courtesy of Michael Taborsky).

more are bound to emerge. Sneaking was discovered in mouth brooders some time ago,[68] and now many cases have been reported.

A clear example is that of the riverine cichlid *Pseudocrenilabrus philander*, a haplochromine mouth brooder from South Africa.[69] It is representative of an early stage of mouth brooding and probably of sneaking. The female deposits several eggs in the nest and leaves them there a relatively long time, about three minutes, before picking them up. During that time, nonterritorial males may dart into the nest and ejaculate before fleeing. The smaller of the sneaker males resemble females. Sneaking among the mouth brooders is particularly common in captivity, and especially when the fish are crowded together.[70]

All things considered, sneaking has been reported for relatively few species of mouth brooders. It may be absent or exceedingly rare in many other species. For instance, A. Parker and Irv Kornfield observed *Metriaclima zebra* underwater in Lake Malawi for over three hundred hours and never saw parasitic fertilization.[71]

On the other hand, Haruki Ochi recorded sneaking frequently in a lekking mouth brooder in Lake Tanganyika, *Paracyprichromis brieni.* You may recall this lekking species spawns next to steep rocky bottoms. Such places provide hiding places for the sneakers, most often juvenile males, to lurk nearby. Then when a pair spawns, the sneaker darts to the female and releases sperm. Ochi pointed out that parasitic spawning was never observed on the nearby leks of *Cyprichromis microlepidotus,* which spawns in open water. He thought the presence or absence of sneakers could be attributed to the presence or absence of hiding places, respectively.[72]

Less often reported is parasitic fertilization among cichlids that spawn on the substrate, and then only for the dwarf species that lay eggs in small caves, such as those provided by snail shells. Sato provided one of the best examples from his field studies of *Lamprologus callipterus,* an African snail dweller from Lake Malawi. Large males have a harem consisting of several females, each in her own snail shell. The male is too large to enter the snail shell, so he simply releases sperm at the entrance to the shell when the female is laying eggs inside.

The small parasitic males follow one of two tactics to get their sperm to the female's eggs. One is classic. The sneaker darts in at the moment of spawning and ejects sperm at the entrance to the snail shell. The other pattern is more refined. The sneaker male mimics a female and is then tolerated by the territorial male; he is small enough to enter the shell with the female, and does, and there he spawns.[73]

Perhaps you have noticed that one group of cichlids has not been mentioned in this account of parasitic spawning. So far, sneaker males have not been seen in any of the biparental monogamous cichlids, and they have been observed a great deal. The close attendance of the male mate, and having the eggs where the male has direct access to them, is associated with the absence of sneakers.

The reason is because those males invest heavily in raising their offspring. If cuckolded, a male loses some or all of his parental investment, and so the male would do better by deserting and finding a new mate. Thus the female's interests are also served by preventing other males from fertilizing her eggs because she needs the male's help.

In contrast, where parasitic spawnings occur, the males provide no care for their offspring. At the most, they lose some fertilizations to sneakers, though relatively few compared with their own success. To the female, multiple paternity may be a desirable outcome, as suggested by polygynandrous lekking. But in the discussion of parental investment among monogamous cichlids, which we'll turn to next, cuckoldry is not an issue, at least not so far.

≈ ten ≈

FAMILY PLAN

In modern societies, humans commonly plan the size of their families. The fundamental decisions are relatively few: How many children do we want? When in our lives should we have them? How should we space them? The answers depend on several things, though some people give this process little thought: How much is each child going to cost? When will we have enough money to provide them with the advantages we deem important? Should we have fewer children so that we can afford to send them to college?

Animals also confront this issue. Of course, they are not going to send their kids to college, nor are they going to reflect on these decisions. Their decision-making processes are built in, having evolved from ancestors who made the "right" decisions.

For them, it boils down to four questions, the first of which human parents thankfully don't have to consider: How can one avoid being eaten by a predator when reproducing? Second–and this will resonate with human parents–at what age should reproduction be initiated? Third, how many offspring should be produced in any given episode of reproduction? And finally, how can the spacing between reproductive cycles be shortened? I'll not take up the questions in this order, however. The last three questions, taken together, define how many young a potential parent may produce. Or, put another way, how should an animal allocate its energy to maximize its reproduction?[1] We'll turn to this question first.

CICHLIDS AS INVESTORS

The primary task of a young cichlid is to obtain enough food to grow into an adult capable of reproducing, so it spends much of its time eating. The goal is to grow as large as possible, as fast as possible. If it is a female, the larger it becomes the more eggs it can lay. If it is a male, ever larger size brings improved competitive ability and hence the chance of getting a breeding territory and a big fecund female mate to boot.

Achieving ample size is done at a cost and carries some risk. One indisputable cost is time. If the cichlid delays reproducing until it has grown into a strapping adult, it may have missed opportunities to breed. Besides that, our cichlid is constantly hunted by predators. Each day could be the last one of its life; it could die without procreating. Of the up to thousands of baby cichlids in a family, only a few survive to raise their own families.[2] Natural selection should therefore favor an optimum size and age to start reproducing. The right time should balance fecundity of a female, or competitive ability of a male, against future life expectancy.[3]

How Much to Invest?

Once our cichlid starts procreating it has a whole new set of evolutionary decisions ahead of it. These turn mostly on how much to invest in the current offspring. Parental investment is any investment by the parent in an offspring that increases the offspring's chance of surviving to reproduce at the cost of the parent's ability to invest in future offspring.[4]

A particularly unpleasant decision confronts our cichlid when the number of offspring it has falls below some critical level. At that point, the numbers game may favor abandoning, even eating, the progeny and starting over;[5] for one mouth-brooding cichlid, that threshold is crossed when the clutch is reduced by 80 percent.[6] The core idea here is that having a batch of young is not only a matter of the cost of the current young but of how that cost reduces the possible investment in future offspring.

This is comparable to the decision we make when deciding between an expensive private school versus a public one for our children. Sending the first child to a high-priced school might mean reducing the number of children in the future. Enrolling the child in public school with no direct outlay of money, on the other hand, might free up enough resources to have, and to educate, an extra child or two. Eating our children is not an option.

The Common Currency

Investment in youngsters by humans might seem easier to measure than in animals because the common currency is money. How much does it cost to provide medical care, to clothe and feed each child, to provide an education? One could also bring in time spent with the child, or even the energy expended in taking care of the child. Those sacrifices could be trivial or they could be appreciable. Of course that will depend on whether the outlay in time and energy reduces the time and energy that could be applied to earning more money.

Quantifying investment for any animal is difficult. The problem is the common currency. What should it be? Some have suggested energetic expenditure, as measured by metabolic rate. That is a reasonable way to estimate the conversion of energy obtained from food into action or into eggs. Even if that were agreed on, though, metabolism is difficult to calculate when the animal is actively engaged in various complex aspects of parenting, the more so when they are mixed in with other activities such as feeding.

The ideal common currency is the number of offspring that are raised. But even here we encounter problems because the number of progeny may not reflect actual parental investment. The progeny may not reproduce, or they may reproduce less than the fewer offspring left by another pair that invested more in each of their brood. And getting such data in the field is also a formidable task.

Another crucially important cost of parenting is exposure to predation. The term "invest" does not adequately convey the notion of risk (unless you play the stock market). If parenting becomes too risky, not only might future reproduction be eliminated, but the current young might be lost if they lack the protection and nurturing of the parent.

Risk of disease also arises. The parent may grow weak from the expenditure of time and energy needed to feed and protect its young, which may limit its own feeding. Also, if the parent fights to defend the breeding territory, it may be injured and exposed to debilitating infection.

Balancing the Budget

Parental investment, therefore, is often estimated through secondary measures, such as how much time a parent spends doing something important to the well-being of the offspring. This includes investment before the fertilized eggs appear on the scene, such as diverting energy to the production of eggs by the female, or by the male burning up calories as he digs a nest. Another indirect approach is to quantify the loss of body weight during a parental cycle. Thus, appraising the size of the investment is usually done with some index of total expenditure. The choice of how to measure investment is hence to some degree intuitive.

With this synopsis of parental-investment theory behind us, let's start by examining biparental substrate brooders, where the bulk of our knowledge lies. Differences between how caring is done in monogamous and opportunistically polygynous species are sometimes small, so for the most part I will combine them when recounting the more general features of parenting. Haremic species differ in some significant ways, however, and they are given separate treatment where appropriate.

THE WELL-GROUNDED FAMILY

In its structure, a family of substrate-brooding cichlids is comparable with that of humans. A pair cares for the eggs and then the young for an extended period of time. That it lays eggs on the bottom is what we expect for a fish. But outside of birds, prolonged biparental care of mobile young is unusual among animals.

Despite variation in parental care, a common theme runs through it all. The pair forms, establishes a breeding territory, and then cleans the spawning site in preparation for the eggs. A few days later, the female spawns the sticky eggs onto a prepared substrate, such as a rock or submerged log, or hair roots of plants.

Then the pair commence guarding the eggs and ventilating them through a behavior called fanning. During the fanning period, one to several pits may be dug, depending on the species. When the embryos hatch into helpless larvae, called wrigglers, they are placed into a pit. There the parents continue to protect and fan them. About one week after spawning, the baby fish, now called fry, start to swim. Up in the water, they are much more vulnerable to predation than they were as eggs, and the parents defend their offspring even more fiercely than they did before. After a highly variable period of roughly one month, depending on the species and the presence of predators, the fry leave their parents and thus end the parental cycle. Also depending on the species, the parental cycle may be repeated within a few months or done just once in the proper season.

DIVISION OF LABOR

Our society has seen a shift in care of infants during my lifetime from an almost exclusively maternal one to one in which the father shares many of the duties–the nurturing father. Likewise, in some substrate-brooding cichlids the males may be either deserters or nurturing, depending on, say, level of predation on the young. When parents stay together, roles are relatively firm. But just how parents divide up the chores varies across cichlids. It ranges from nearly equal fanning (an action that ventilates the eggs) and caretaking to exclusively female nurturing, with the latter being the more common.

In the typical Central American cichlid, the mother carries out most of the direct care and fanning of the eggs and wrigglers, and is the more active partner in attacking small fish who try to prey on her offspring.[7] The father spends more time further away from the offspring. His role is to intercept other adults of their species who might try to take over their nest site.[8] The same roles recur in some of the African cichlids, such as the jewel fish *(Hemichromis guttatus),*[9] *Tilapia mariae,*[10] and other species.[11]

TIMING THE FAMILY

Orange chromides have two spawning seasons per year, coinciding with the two monsoonal periods.[12] A given individual, however, may spawn only once in the entire year.[13] When they breed, the many pairs tend to do so synchronously.

Pairs of orange chromides that spawned out of phase with the majority were less successful.[13] That could happen for a number of reasons. For one, spawning all at once overwhelms the predators on their fry in the sense that the predator can only eat so many of them; this is sometimes called "swamping" the predator.

The Midas cichlid, *'Cichlasoma' citrinellum,* also has two breeding seasons each year, though a few pairs may be found breeding at any given time. Biennial breeding occurs, too, among the cichlids in Lake Malawi.[15] Numerous species, however, have only a single breeding season during the year. Yet other species may breed sporadically year-round.[16]

Unlike fishes in temperate zones, where temperature and day length vary so greatly with time of year and thus impose seasonal breeding, the timing of reproduction among tropical cichlids appears to be driven by one of two pervasive factors. The first pertains to cichlids that dwell in rivers or streams. There, water velocity plays a profound role because the tiny fry of cichlids are delicate and weak, unable for the most part to resist strong currents. A striking exception is the Central American cichlid *'Cichlasoma' tuba.* It lays relatively huge eggs, 2.4 mm diameter, that produce big fry who are so strong they can hold station in a rapidly flowing river.[17] In general, however, riverine cichlids tend to reproduce during the dry seasons when water flow is mild.

The second factor is access to food, hence energy, for the production of eggs, and it applies more to lake-dwelling species. Availability of food depends on such things as weather, for instance, the pattern of winds that promote rain. Rain increases the movement of water from the land to rivers and on into lakes, bringing with it nutrients or even insects. Wind and rain also promote upwelling in lakes, and that carries to the surface deep water that is rich in nutrients, resulting in algal "blooms."

A few species of fish sift out the plant plankton (phytoplankton), but most do not. Animal plankton (zooplankton) eats the phytoplankton, and the zooplankton is eaten by the cichlids.[18] In addition, the algae attached to material surfaces prosper from the nutrients dissolved in the water, and their growth feeds the herbivorous cichlids. The energy from the food allows the cichlids to invest in the many costly aspects of reproduction.

Competition among species for suitable breeding sites also plays a strong role and, in at least one circumstance, can lead to a form of time-sharing of the available caves for spawning. Dominant species of cichlids take precedence and force subordinate species to reproduce at less favorable times.[19]

CARING FOR THE EGGS

After the eggs have been laid, they are closely attended. In the orange chromide both the male and the female fan the eggs they have laid on a vertical surface. To fan those eggs, a parent positions itself sideways next to the clutch, swinging the near pectoral fin forward, a few beats per second, to propel fresh water over the eggs. With each pulse of this asymmetrical fanning, the generated wave of water ripples the stalked eggs, like a gust of wind stroking a field of wheat, bringing fresh oxygen-bearing water to the developing embryos and washing away their metabolites.

Cichlids that lay their eggs on the bottom, for example, *Aequidens,* fan them symmetrically as they hover over the clutch. The angelfish, *Pterophyllum scalare*, fans both ways, in the style of orange chromides–that is asymmetrically–when next to a leaf with eggs, but symmetrically when it faces them.[20]

Occasionally the parent stops fanning and inspects the eggs. An egg may be gently taken into the mouth, as though testing it. Some eggs die and change color; perhaps they had not been fertilized. Some of those eggs become infected with filaments of fungus. The parents of many species appear to remove and eat such eggs.[21]

The parents in the orange and green chromides take turns fanning the eggs.[22] When the reliever approaches, the relieved fish briefly directs what appears to be a low-level threat display at the reliever, who may answer in kind.[23] If the pair is bonded you might ask, why should the interaction so resemble the start of an aggressive encounter?

In romantic novels and movies, married couples are often portrayed as devoted to one another, and even more so to their children. In real life, this is not always the case. In studying cichlids, Jürg Lamprecht concluded that after the eggs are laid, the male and female cichlid are no longer so devotedly bonded to one another. Instead, each is committed to the offspring and merely tolerates its mate. Thus, he reasoned, when the pair hovers with the school of fry, they avoid facing one another but rather face away from one another, often tail to tail; he mentions six other species that do this.[24] As an alternative or supplemental explanation, facing away could be the best arrangement for the parents to watch for predators on their fry.

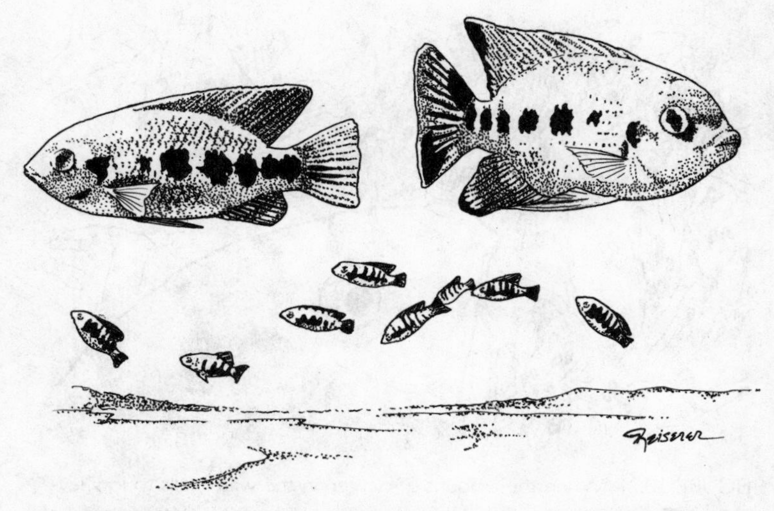

FIGURE 10.1 A monogamous pair of the African cichlid, *Tilapia mariae*, guards a school of relatively large offspring. Each adult, female to the left, faces outward from the center of the school (after Lamprecht 1973).

As we have learned, in cichlids such as *Aequidens* and many of their relatives, the female does most of the direct care of the eggs and larvae. The males are more involved in protecting the breeding territory from potential usurpers. These maternal cichlids continue caring for the eggs at night.[25] Observations through a "snooper scope" revealed how this is done by female convict cichlids, *'Cichlasoma' nigrofasciatum*, and rainbow cichlids, *Herotilapia multispinosa*. The mothers actually fanned the eggs more at night than during the day even when the lights were left on, though they fanned yet more when the lights were off. Thus nocturnal fanning is controlled both by darkness and the mother's biological clock.

The eggs hatch after around two to four days in biparental cichlids, depending on the water temperature and the species. At the time of hatching, the parents become agitated, often inspecting and mouthing the eggs. Apparently the mother and father help hatch the eggs by tugging at them with their mouths. When the parents are not present at this time, the eggs take longer to hatch.

BABY IS AN EMBRYO

Wrigglers are embryos without egg shells. Development of organs such as eyes and ears, for instance, has not been completed. Substrate-brooding cichlids nurture and protect these helpless wrigglers day and night. How they

FIGURE 10.2 When the amount of oxygen in the water falls to low levels, the Central American cichlid *Herotilapia multispinosa* 'pastes' its larvae onto plants near the surface of the water, where higher levels of oxygen exist (after Courtenay and Keenleyside 1983).

do this, however, varies between species. Orange chromides start by putting the wrigglers into a pit they have dug next to where the eggs were laid. There they fan the embryonic wrigglers and gently mouth them. The off-duty parent, especially the male, digs one or more new nests. Each day the wrigglers are moved to a new pit.

Other cichlids, such as those that nest among rocks, commonly keep the wrigglers in one place inside the cave where the eggs were laid. While fanning the eggs, they clean out a depression in a crevice on the floor of the cave. The newly hatched wrigglers may work their way back into small spaces where they are difficult to see and out of reach for most predators. The female guards and fans them, and after about a week they start springing up just hours before commencing to swim.

Yet other cichlids "paste" their wrigglers onto objects rather than deposit them in a pit in the bottom. The rainbow cichlid, for instance, may move its wrigglers from the place they were hatched and spit them out near the surface of the water onto leaves of underwater plants;[26] three related species apparently do the same.[27] The angelfish, *Pterophyllum scalare,* lays its eggs on plant leaves, and later pastes its wrigglers there,[28] as does the discus fish, *Symphysodon discus*.

But why do cichlids stick their wrigglers up onto objects instead of into pits in the bottom? That must be done at some cost, such as exposing the wrigglers to predation. For angelfish the answer is obvious: They lay their eggs on plants[29] in deep water far above the bottom.[30] Putting wrigglers into pits is out of the question.

The rainbow cichlid, however, places its wrigglers either into pits or up onto plants. Pasting the wrigglers near the water surface, it turns out, is a response to low levels of oxygen in the warm swampy water where they reproduce.[31] In an aquarium, when the water was rich in oxygen, the wrigglers went into a pit. When the experimenter drove the oxygen to low levels, the parents stuck the wrigglers onto plants closer to the more oxygenated surface water.[32]

The wriggler is a captivating little creature. It resembles an egg as much as an embryo, with its bulbous yolk sac protruding from its belly. That "belly pack" is the food supply, and it will be absorbed over the next several days as the wriggler's bodily structures develop. At this time, the primordia of its eyes and ears are only crude capsules; the skin's pigment cells are incompletely developed, so the wriggler is nearly transparent; the paired fins are mere stubs; and the dorsal and anal fins are simply folds that lack fin rays and spines.

Each wriggler has three pairs of adhesive glands atop its head. These look like minuscule low pimples.[33] A pore in the tip of each gland secretes a sticky mucus. The mucus glues a wriggler by its head to anything solid, such as the bottom, plants or their roots, or even to one another.

The wrigglers stand on their heads, tails pointed up. Each undulates its body continuously, hence the term wriggler. Because they are in a tight clump, their collective activity creates an upward flow of water.[34] The rise of this water causes water to flow to them from the sides, providing oxygen and washing away waste products such as the carbon dioxide they have produced. The plume of water ascending from the wrigglers must be rich in chemicals excreted by them. Those chemicals can be detected by the parents (and by predators such as catfishes); I'll return to this further on.

Although this is the prevalent pattern, the wrigglers of some species stand out because they are so different that they bring into question the use of the term wriggler. I have recently been observing reproduction in the lamprologine cichlid *Julidochromis marlieri* from Lake Tanganyika. Eggs are placed on the ceiling of a cave, and when they hatch, the larvae just hang straight down, motionless, from the ceiling. There they dangle by their heads like Christmas tree ornaments and are not rearranged by their parents. Rarely, a larva undulates. Parental care is reduced, consisting only of brief inspections in the nest with an occasional perfunctory burst of fanning. But the parents vigorously attack any small potentially predatory fish that approaches. Apparently many of the lamprologine cichlids in Lake Tanganyika express similar minimal care of their brood.[35]

Wrigglers in general take five to seven days to metamorphose into swimming fry. On the last day of wriggler life, some of them begin to swim, taking off from the bottom then dropping back. Those who know cichlids well call such young "hoppers." As more and more of them spring up, they recall images of popcorn popping up from a hot surface. Within hours they form a tight school of fry, hovering in their pit. Then they move up as a group of fry and out into their perilous new world, with their parents in close attendance.

CRADLED IN THE MOUTH

Jurupari lurks near paths twisting through the dark jungles of Amazonia, ready to gobble up travelers. So say the Tupi Indians. They believe Jurupari was born to the first people, though he himself is not really human. Dwelling with his human kin, Jurupari went with a party of village children into the forest. When the children ate forbidden fruit, he exploded in rage and invoked a storm. To escape the lightning, the frightened children took refuge in a cave, but too late they discovered the shelter was Jurupari's open mouth. When he returned to the village, he regurgitated the children before the astonished elders.[36]

Keen observers, the Indians must have noticed parental cichlids in the clear local streams taking their young into their mouths when alarmed and later releasing them. The natives must have assumed the fish were swallowing their own babies and then regurgitating them. Did their belief in the

FIGURE 10.3 An artist's portrayal of the myth of the Jurupari, drawn from an Indian mask.

FIGURE 10.4 The ancient Egyptians revered the mouth-brooding tilapia, bolti, as a symbol of rebirth. Here they have combined the bolti and flowering lotus, another symbol of rebirth (after Dambach and Wallert 1966).

evil Jurupari lead them to name the fish after him, or did their observations of the fish produce the myth? I suspect the latter.

The nineteenth-century ichthyologist Johannes Heckel heard about the evil spirit of the Tupi Indians and incorporated it into the name of a new species of eartheater from that part of the Amazon Basin. The fish is the mouth brooder *Geophagus jurupari* (now named *Satanoperca leucosticta*). The name of another eartheater, *Satanoperca daemon,* is a continuation of that practice; ironically, this species is not a mouth brooder.[37]

The ancient Egyptians were aware of the mouth-brooding behavior of the local tilapia.[38] I was reminded of this when visiting an exhibit of Egyptian art. There I saw a small fish-shaped palette made from stone that had been used for preparing cosmetics around 5,000 years ago. It was unmistakably a tilapia, most likely *Oreochromis niloticus*.

The Egyptians called the fish Bolti. It was a symbol of rebirth, probably because they were touched by the parent spewing forth baby fish in the net when caught from murky water (they might not have seen the fry being taken into the mouth). Its association with beautifying cosmetics is therefore understandable. Tilapia recurs in Egyptian art, frequently with lotus blossoms emerging from its mouth, as in decorative ladles associated with the dead, and is thought to signify resurrection.[39]

Mouth-brooding behavior is a remarkable phenomenon, no matter what the interpretation. Many kinds of fishes other than cichlids are mouth brooders [40], but unlike cichlids, the male in those fishes is the usual and only mouth-brooding parent. Some frogs have similar behavior. One frog even swallows its eggs, which then develop in the gastric juices of its stomach.[41]

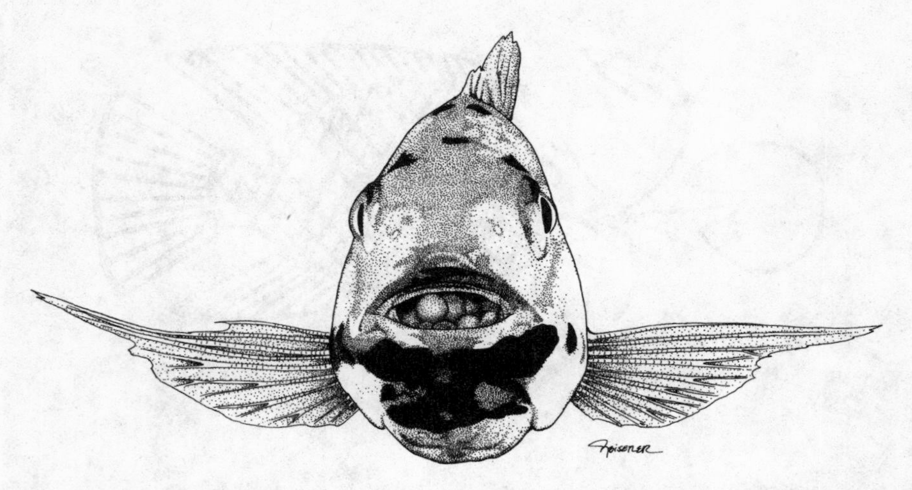

FIGURE 10.5 In this male blackchin mouth brooder, the eggs he is carrying can be seen inside his mouth.

Different lines of cichlids have become mouth brooders, but to do this they all had to overcome the problem of asphyxiating the offspring in their mouth. Developing embryos and larval young quickly use up the oxygen around them. The evolutionary solution upon which the different lineages have converged is twofold.

First, the eggs must not stick together, so natural selection eliminates the adhesive threads, though one may still see vestiges of threads in some species.[42] Adhesive eggs would clump into a ball and the central individuals would rapidly die, fouling the entire brood. The second universal adaptation is the mumbling movements of the parents. They churn the free eggs in a circular motion, toward the rear in the floor of the mouth and then back along the roof of the mouth toward the front, to assure good contact with the oxygenated inflowing water.[43]

How long the eggs/larvae remain in the parent's mouth varies enormously, and that highlights one of the major reasons mouth brooding evolved. In the mouth, the clutch is freer from predation than when exposed as eggs on the substrate or as swimming fry. As in most animals, the younger the animal, the greater the risk of predation. The corollary is that evolution should favor hurrying through the most vulnerable period of an animal's life. The mouth brooders try to get around this by holding their young in their mouths as long as they can to produce large young less vulnerable to predators.[44] The in-mouth time increases in proportion to the size of the egg.

The larger the egg, the larger the hatchling, but also the more time that is needed until the parent can release the fully developed young. Comparing different lines of mouth brooders, a pattern emerges. At one extreme, such as in the tilapiine *Oreochromis,* the female lays some hundreds of eggs, and they are consequently small. The in-mouth time is relatively short, and after that the mother nurtures her fry for a week or two, taking them back into her mouth when danger threatens. At the other extreme is *Cyphotilapia frontosa*. This cichlid lays a few immense eggs (6.7 mm) and keeps the offspring in its mouth for about three months; when released, the large capable young immediately strike out on their own.[45] In general, mouth brooders lay much larger and many fewer eggs than do the substrate brooders, whom we'll return to now.

LIFE OF THE FRY

During their first days, the tiny fry of substrate brooders stay in a tight school near the bottom. But appearances are deceiving. The newly swimming fry are weakly developed when compared with the much larger fry that emerge from the mouths of the typical tilapia type of mouth brooder. This can also be seen in their behavior.[46] They need some object to fix on to maintain a coherent school. When fewer than fifteen such fry are put together without a parent, they do assemble but tend to wander apart. Strong schooling does not develop until several days later.[47]

The parent hovering over the fry provides a focal point, keeping the cluster of fry together. When an individual wanders off, the parent swims to it,

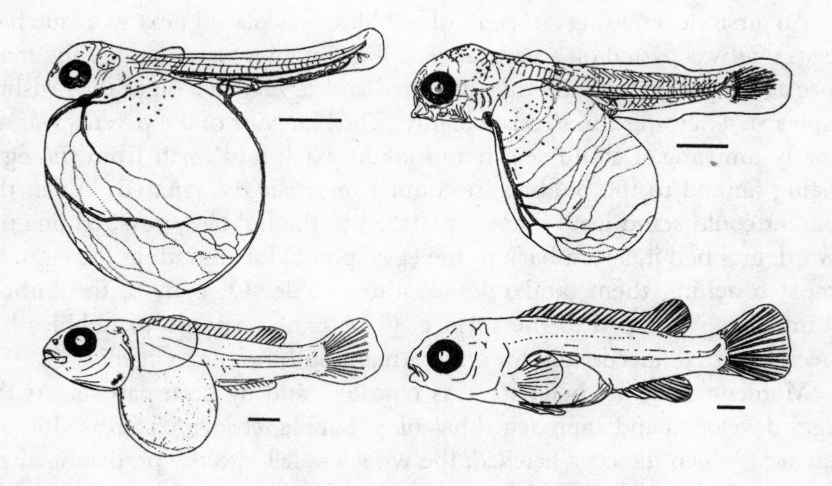

FIGURE 10.6 The eggs of *Labeotropheus trewavasae*, from Lake Tanganyika, hatch into nearly embryonic larvae with huge yolk sacks (upper left). As they grow, the yolk sack shrinks ever more as the organs take shape (upper right, then lower left), until the fully developed young are released. Adhesive head glands are not present. The short black lines indicate the scale of one mm (after Balon 1977).

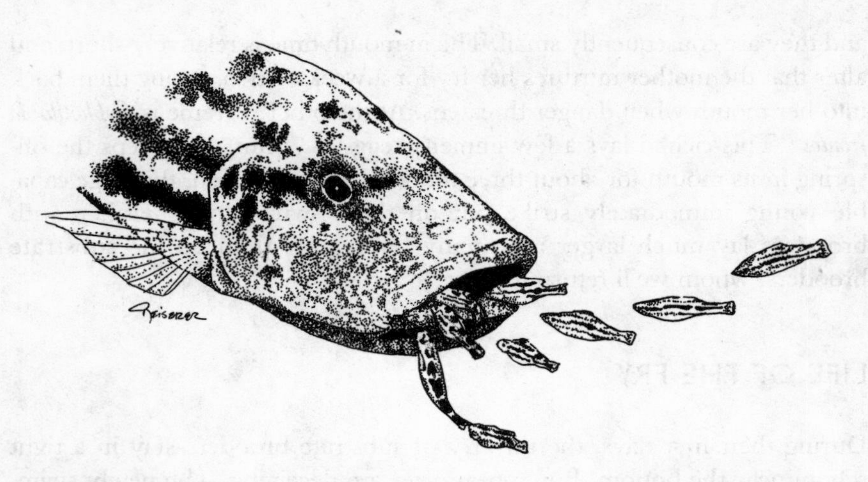

FIGURE 10.7 A female mouth brooder accepts her young back into her mouth.

sucks the vagrant into its mouth, and spits it back into the school of siblings. Retrieving behavior disappears after about the first week as schooling improves.

A pair of orange chromides, *Etroplus maculatus,* revealed how the retrieving response emerges in the parent even before the eggs hatch. The observation was fortuitous, though I saw it repeated a number of times, and it was shown only by pairs breeding for the first time.

An airstone, emitting a torrent of bubbles, was placed next to a clutch of eggs that was pasted on a vertical wall. The bubbles were roughly the same size as the pair's eggs, and they glistened and zigzagged a bit as they rushed upward. They were an irresistible lure. The behavior of the parents was actually amusing. I could see them looking back and forth from the eggs being fanned to the bubbles, becoming increasingly agitated. When the parent could stand it no longer, it rushed to the bubbles, pursued one upward, grabbed it, swam back to the eggs, pointed its mouth at the eggs, almost touching them, and released the bubble. Of course, the bubble immediately escaped to the surface. (The same response to bubbles has been reported anecdotally for a maternal mouth-brooding cichlid.)[48]

Moments later the behavior was repeated, and by both parents. As the eggs developed and approached hatching, bubble retrieving increased in frequency. When the eggs hatched, the wrigglers fell into the pit below where the eggs had been laid. The parents retrieved bubbles even more frequently. But now they tipped down to spit the bubble into the cluster of wrigglers. The behavior persisted until the wrigglers changed into swimming fry. Then the parents retrieved the fry and stopped being drawn to the bubbles.

The parents obviously confused the bubbles with errant young. What this observation beautifully reveals, however, is that retrieving behavior, which is normally seen only after the fry swim, is already present and realizable during the egg stage; it lacks only the proper stimulus. Moreover, the wrigglers of the pair were never moved to a different pit. Ordinarily, parents dig several pits and move the wrigglers to a new one each day. The constant retrieving of bubbles had apparently exhausted the behavior of transferring the wrigglers between pits. These observations provide a method to study quantitatively the dynamics of parental behavior and the underlying physiological changes that prepare the adults for parenting.

Little notice has been given to the fry's response to being retrieved. I have observed this closely only in the orange chromide. Their fry have a highly predictable and stereotyped reaction: When spit out, the fry darts straight down to the bottom. There it lies still for several seconds before becoming active and rejoining its siblings. I presume this is an antipredator adaptation. As the fry grow and become more independent, the parents retrieve them less and less, and the diving response of the fry disappears.

The reaction of the fry of angelfish is also highly specialized. Angelfish in nature live among vegetation in deep water. If a fry were to dive to the bottom when it was retrieved, it would be committing suicide. Instead, the fry spins around at once, dashes to the parent, and bumps into its side where it remains momentarily. The adaptiveness is obvious.

The lamprologine *Julidochromis marlieri* of Lake Tanganyika is again exceptional. When the fry first start to swim they do not school. Rather, they become solitary, keep close contact with the bottom, and retreat into crevices. Almost from the first day, they fight among themselves as though to space out and thus disperse. The parents only rarely retrieve a few fry during the first day. Remarkably, other adult "julies" are tolerant of the fry, who are even permitted to enter their burrows. Parallel behavior has been seen in two distantly related, "almost" lamprologine fish, the river-dwelling *Teleogramma brichardi*[49] and *Leptotilapia tinanti*.[50]

Fry Are Tasty Morsels

Of a thousand or so young that a pair of Midas cichlids brings into the world, only a handful remain by the time they are ready to leave their parents,[51] and this has been demonstrated in other cichlids.[52] Many Midas cichlids lose all their offspring before the end of their parental care. At one extreme, predators have been seen to reduce the number of fry in a family by roughly half every three days.[53] Although the rate of loss may vary appreciably, this situation is typical. Life is harsh.

Those not familiar with fishes might think of predators on the young in the model of the big bad wolf taking the little lamb; the parents are not only powerless but themselves in danger. With cichlids the affair is radically different. The tiny fry, taken one by one, are hardly worth the trouble of a fish predator as large as the parents, and certainly not ones larger than the parents. After all, eagles don't feed on sparrows. The predators, therefore, are smaller than the parents.

Often the predators are the juveniles of the same species, hence cannibals, or those of other cichlids, or are adults of much smaller species of cichlids.[54] In addition, cichlid parents in the New World tropics are plagued by small characins during the day and by catfishes during the night. Similarly, in Sri Lanka, the smaller species of *Etroplus,* the orange chromide, is a major predator on the offspring of the larger species, the green chromide. Cannibalistic orange chromides also prey on the offspring of their fellow orange chromides. In response, the breeding pairs of those species often form colonies where their collective repulsion of the predators is thought to improve the survival of their fry.[55]

Jack Ward and J. I. Samarakoon reported a noteworthy difference between the two species in defense of young. The male and female of a pair of orange chromides, as mentioned, alternate caretaking. The relieved mate not only defends the periphery of the territory, from where it can scurry home if needed, it also has the opportunity to feed. This is a great energetic plus. In contrast, both the mother and father in the green chromide remain with their fry constantly and take no time off to feed.[56]

The reader might well wonder at this point why the fry should remain in a school. Wouldn't that present to the potential predators a sumptuous buffet of baby cichlids? And wouldn't it then be better for each fry to run away from the school and hide? Actually, no. Predators have difficulty catching a fish in a school. Sure, they do catch some of them, but only with difficulty compared with an isolated fish in the open.[57] The same principle applies to a baby cichlid in a school. Think back to the cichlids in a school overwhelming the defense of a dominant cichlid protecting its feeding territory. It is the Saint Ignatius strategy again.[58]

As an aside, I should mention that the bedding down of fry is a charming sight. As night approaches, the family becomes less active. The adults hover just over the bottom, and the fry gather under them in a tight school then gradually descend into a pit where they lie in a dense heap of sleepy fry.

And yes, fish do sleep. I found the easiest way to collect Midas cichlids in Nicaragua was to dive in a lake at night. Using a flashlight and a hand net, I just picked up the sleeping fish. People often ask me how a fish can sleep when it can't close it eyes. I always reply, you can sleep without closing your ears. But back to parenting cichlids.

Family Communication

The pair of biparental cichlids, and the single-mother haremic females, control the behavior of the fry to some degree. One of the more remarkable but subtle aspects of this behavior, however, is *not* communicating with the fry. Gerard Baerends and Josina Baerends-van Roon described this nicely in their memorable monograph of cichlid behavior.[59] When one parent departs from the school of fry, it does not just meander off. If it did that, some of the fry might follow it and get into trouble. Instead, the outbound parent darts off, precluding pursuit by the fry. In that way, the departing parent avoids giving a "follow me" signal to its young.

As the fry grow and become more mobile, the parents guide them across the bottom, often leaving their home territory.[60] To do this, one or both parents swim a few body lengths in the intended direction, stopping near the edge of the school.[61] The fry respond by moving toward the parent, who then swims on a short distance, and so on.

On the other hand, the mass of fry often drift off in one direction or the other on their own, searching for food. Then the fry seem to control the behavior of the parents, who go along with this, unless the fry start to intrude into the dominion of another family. Then one sees striking communicatory behavior because the disturbance accompanying territorial interactions with a neighbor family brings with it the danger of attracting predators. The parents signal their young to gather under them and either follow the parents to safety or hug the bottom. The various species have different ways of performing this alarm signal, as it is sometimes termed.[62]

Among the Central American cichlids, the alarm signal consists of the parent jerking its head slightly and smartly to one side, generating a shallow body undulation that passes almost imperceptibly tailward. At the same time, it snaps the dark pelvic fins downward and the pectoral fins forward. The movement is conspicuous and easily seen by the fry. It may also swirl the water around the fry, stimulating them mechanically. The motion is called jolting and appears to have evolved out of the conflict between a sudden start in the direction of the danger, and a breaking action.[63]

The pelvic fins of the parents are emphatically colored in many species, enhancing the signal. In *'Cichlasoma' alfari,* for instance, the fins are bright yellow. Ron Coleman tells me that when he snorkels in rivers in Costa Rica, the yellow pelvic fins are all that he can see of the parents in turbid water. Likewise, the pelvic fins of the African "krib," *Pelvicachromis pulcher,* are maroon and rich in contrast.

The interpretation of jolting as an alarm signal gains strength from an experimental analysis of its contagious effect on other parents. Pairs of convict cichlids with young viewed a dummy convict cichlid that was equipped with moveable pelvic fins, either in the normal ventral position or on the back.[64] When the parenting pairs saw the dummy flicking its ventral fins, they signaled alarm, too. But when the fins were on the back of the dummy, they reacted only weakly.

In the African jewel fish alarm evokes opening and closing the dorsal fins. This is done rapidly and repeatedly, producing a fluttering action. The South American angelfish also signals its fry by flickering its dorsal and anal fins.[65] Such movements are visually stimulating, but because the fins move in and out, they probably do not function as a mechanical signal since the motion does not produce much turbulence.

The alarm call in the orange and green chromides, *Etroplus maculatus* and *E. suratensis,* respectively, differs from jolting.[66] The parent does not jerk its head. However, as in jolting, the black pelvic fins are snapped down and raised again, but the motion is different. It is a quick flickering performed in several rapid bursts. I often wondered how that could be a signal to the fry because in each species the pelvic fins are black and so is the abdomen. The signal lacks contrast and thus should be difficult to detect, or so I thought.

One day it occurred to me to view the signal from where the fry do, under the parents. When the parents were near the front glass I positioned myself below the level of the bottom of the aquarium and looked up, while someone slightly frightened the parents for me. When the parents flickered their pelvic fins, I then saw the fins do not move just down. They also open out from the center line. And there, between the pelvic fins, is a shiny silver patch. When the pelvic fins are flickered, the effect for the fry is that of a semaphore signal, the silver blinking on and off as the black fins open and close. The motion has been well designed through the process of natural selection: It signals the fry but minimizes detection by predators.

The fry respond to alarm signals from their parents by gathering under them, often sinking to the bottom. In the green chromide, the response can be dramatic. In nature, they go to the bottom and hide among the leaves of the plant *Halophila,* which commonly carpets the bottom in estuaries in Sri Lanka. The leaves are small and mossy green. The fry of the green chromide are similar in color and size to the leaves. When hiding among the leaves, each fry stands, head up next to a leaf and sways back and forth with the leaf as the water moves it. Thus the fry mimic the leaves and are exceedingly hard to detect.[67]

In the Pursuit of Food

The objectives of the fry at this time are simple: Avoid being eaten and eat to grow. On their own, the fry are nearly defenseless against predators, hence the necessity of protective parents. With only a few known exceptions, such as some mouth-brooding cichlids, cichlid fry eat only tiny animals in the beginning, although they may become herbivorous as adults. This is the general situation for most fishes. Herbivory is a specialization that has evolved relatively recently and independently in many kinds of fishes, most of them living in the tropics.[68]

This probably explains why young fish usually start out as carnivores, feeding on microorganisms. At first, they have short intestinal systems, an adaptation to carnivory. As the herbivorous species grow, they gradually include more plant material in their diet; their guts lengthen *pari passu,* a requirement for the slow process of digesting plant material.[69] Cichlids follow this basic pattern.[70]

The school of fry of typical biparental cichlids faces into any current (even lakes have gentle currents). Each fry watches for tiny zooplankters such as copepods and rotifers, which they snap up. (Aquarists hatch the eggs, more properly cysts, of brine shrimp, *Artemia salina* to feed to the fry. The nutrition is provided by the yolk sacs, not by the nauplii themselves.) The school of fry also spread out over the bottom and search the substrate intensely. There they find other invertebrate animals, such as bottom-dwelling copepods.

The fry of some species have developed special feeding behavior. On several occasions I have seen a young orange chromide pick up an object from the bottom and swim straight up with it. Soon it pauses, poised, virtually standing on its tail, appearing to balance the object before letting it drop. If the object is light, it drifts among its siblings, who gather around and may feed on it. If it is heavy, the object drops to the bottom, sometimes drawing other fry to it. This behavior seems to solicit the other members of the school to help pull apart a potential food object that is too large for the offering fry to handle.

Fry of the green chromide, *Etroplus suratensis,* do something even more remarkable. As with the orange chromide, this behavior is also elicited by a food object too large to handle–in this case, intact tubifex, a small worm about the same length as the fry.

A fry picks up a worm in its mouth, but it is too long to swallow. The solution? Swim to its parent nearby. The apparent goal of the fry is to position itself such that the worm trails between itself and the body of the parent. If the worm is on the side of the fry away from the parent, it ducks under, or swims over the parent to get the proper alignment. Also important to the success of this maneuver, the fry approaches the parent from the rear. Recall that cichlids have ctenoid scales; their rear margins present a sharp edge. The fry zooms at the side of the parent and hits it with a glancing blow. The worm is caught between the bodies of the parent and the fry. The action shears off the trailing edge of the worm. The fry swims off, munching down the piece of worm while its siblings dart after the leftover falling piece.

The behavior of the parents has also evolved to help feed their young. Some of the behavior is relatively simple. For instance, while the parent is guarding the fry, it may tilt down, plunge its snout into the bottom, take up a mouthful, and then spit it out among the fry. This is the equivalent of a hen scratching for her chicks to expose food items. In some species, the parent may even swim away, find some food, swim back to the family, and spit the chewed-up food out in front of its offspring.[71]

A parent convict cichlid may also "plow" the bottom for the fry,[72] and the female usually does this more than does the male.[73] The movement, sometimes called fin digging, starts just like regular digging but the parent then drops its belly down onto the bottom. It pushes forward with the tail and backward with the pectoral fins as it slowly moves forward. That action opens and stirs up the bottom and ends with the parent spitting out the stuff in its mouth. The fry rush in to find the exposed delicacies.

The parental green chromide does something similar to plowing. I saw one parent grab with its mouth the edge of a sheet of algae covering the bottom. It pulled the sheet up, letting it fall back among the eager fry. The parent then graciously backed away, making room for the fry to feed from what it had stirred up.

In some species, parental fish flip over objects on the bottom. Orange chromides do this by hooking the lower jaw under the edge of objects like clam shells and then swimming up and forward. Parental convict cichlids, especially the mother, turn over leaves to expose prey for their fry.[74]

The Mouth as a Café

The female mouth brooder provisions each of her offspring with what amounts to a large picnic basket, the yolk in the egg. The yolk must provide all the nutrition the youngster will get until it leaves the mother's mouth. Like her offspring, the female does not eat until the young are liberated. She lays down a store of fat before she spawns and then lives on that until the fry exit her mouth. That arrangement is characteristic of many mouth-brooding cichlids, but now mothers in several species have been found to take in food to varying degrees.[75]

The real trick for the parent is to separate an edible item from the young in the mouth because they are both tasty and commonly about the same size. Mothers in some species have that ability.[76] Most examples, so far known, are from cichlid fishes in Lake Tanganyika. An early stage is demonstrated by the big piscivore *Cyphotilapia frontosa*.[77]

More advanced are maternal *Tropheus moorii*. The mother does not feed while the embryos in her mouth are still in the egg stage. But when they hatch, she begins taking in some food, threads of algae, which shows up in the guts of the larvae in her mouth, the more so as they age.[78] (This is a rare instance of larval herbivory.) Maternal females of the herbivorous *Tropheus duboisi* continue to hold a territory, and they feed at about 80 percent the rate of nonmaternal females. Their stomachs are filled to about the same degree with algal filaments, compared with nonmaternal fish.[79]

The highest level of development is shown by an unnamed species of *Microdontochromis,* which might be a member of genus *Xenotilapia.* These silvery cichlids live in large aggregations over open sand bottom in Lake Tanganyika where they feed on plankton. It is an intriguing species for several reasons. First, after brooding the eggs the female hands over, or rather spits over, a large portion of the larval brood to the male. Second, the size and number of its eggs do not fit the general pattern for cichlids. Not only are the eggs smaller than those of any other mouth brooders, being only 1.9 millimeters in diameter, they are smaller than the eggs of many substrate brooders.

Normally, small eggs mean more eggs, but here the number of eggs is greatly reduced, a third peculiarity. As a consequence, the mouth is not packed with offspring and has lots of spare room, making it easier for parents to ingest food and separate it from the clutch. Both parents eat copepods, small planktonic crustaceans, and at the same rate as nonparental fish. The young stuff themselves on the free lunch flowing in to them; that enables them to grow well beyond that allowed by the paltry amount of yolk in the egg; their increase in weight by the time they leave the mouth is about seventy times the weight of the egg.

For their part, the parents continue to grow, and the female produces more eggs. Thus the parents make up for a small brood by being ready to produce another brood when they finish the one onboard. This is a remarkable evolutionary achievement by a mouth-brooding cichlid fish.[80] Other species in the genus *Xenotilapia* may behave similarly.

Munching on Mom and Dad

People interested in natural history are by now familiar with the plight of the male of some spiders: After copulating, the female has him for lunch. We seldom think of animals like fishes doing that sort of thing, but cichlids do something similar albeit not as dramatic. The more apt comparison is with lactation in mammals.

In many species of biparental substrate brooders, the fry eat from the slime-covered body surface of their parents. Most likely they are getting calories from the mucopolysaccharides, sugary substances that are abundant in that mucus. The behavior has evolved independently a number of times. It starts so simply that one can readily understand how natural selection could improve on its humble beginnings.

The fry of most cichlids do not ordinarily graze on the parent. However, if the fry want for food and the opportunity exists, they may begin to nibble on their parents. I have seen hungry fry of the convict cichlid, for instance, nibbling the mucus coating of their parents. This is called facultative feeding by the fry and is seen only under unnatural circumstances, such as famished fry in an aquarium where the parents cannot avoid them. Even though not natural, such observations reveal how more advanced provisioning by parents might evolve.

In the Midas cichlid, so much studied in our lab, we noticed that fry regularly feed from their parents,[81] a behavior we called contacting. A fry swims right up to a parent, appears to place its mouth on the parent's side, and then swims away (young fry are so tiny that one cannot be sure they bite). Contacting is not an artifact of captivity. We observed this behavior underwater in a lake in Nicaragua. Contacting is, however, to a degree facultative. Well-fed fry contact little, but when they become hungry, they quickly turn to their parents for nutrition.[82]

The amount of contacting also varies with age. During the first week of swimming, the fry seldom contact. But by the second to third week of swimming, they feed vigorously from the parent. By that time, they have developed a preference for the male over the female. Thereafter, contacting falls away again, apparently coinciding with the time in nature when the fry would begin to leave their parents. (This pattern differs from our published observation and rests on subsequent more carefully controlled experiments.)

If left in an aquarium with their parents, however, within another week the incidence of contacting increases rapidly. The fry behave as though they go through a process of turning off contacting, but then, left with an ever-full larder, they seem to learn anew that they can eat their parents. And eat they do. They open wounds in the caretakers' skin. The parents seek to avoid their offspring, which they would do in nature. They do not, however, try to hurt the fry or to defend themselves from their children, except to fold their fins and hide in a corner. After discovering this, we took care to separate older fry from their parents.

In some respects, provisioning the fry by the caretakers is analogous to giving milk in mammals.[83] The mammary gland evolved from sweat glands, so secretion of fluid was already present. Milk is like sloughed-off skin in a liquid medium. Milk production occurs when the mother gives birth, and it is stimulated and sustained by mechanical stimulation from the baby.

When the Midas-cichlid fry start contacting, microscopic examination revealed that goblet-shaped mucus cells on the surface of the parents swell up and start secreting more mucus, just as nursing by baby mammals stimulates the breast to produce milk. The parents' skin becomes thinner.[84] We presume that contacting by the fry stimulated the mucus glands to secrete. The skin of other fishes is known to produce more mucus cells when irritated.[85] Of greater relevance here, the more a given Midas parent was contacted by its young, the more mucus cells it had in its skin.[86] But the analogy to giving milk goes further.

Right after the birth of a mammal, the mother secretes a special milk called colostrum. It is thin and watery compared with normal milk, but it is a rich source of special nutrients that are extremely important to the baby in starting its immune system working. That has a rough parallel in the Midas cichlid.

The mucus the fry eat contains chemicals that are not a source of calories but rather are important as nutrients. These are growth hormone, prolactin, and a form of thyroxine. Unidentified proteins are also present in the mucus. When the fry were raised without parents, their growth was accelerated by adding growth hormone to their food.[87] The study of skin grazing in cichlids is in its infancy, and much remains to be done. Two further cases make that point.

The first one has a special place in my heart. It is, again, the orange chromide, *Etroplus maculatus*. Recall that this is a primitive cichlid, so contacting in it may have evolved early in the history of the Cichlidae. At first, we weren't even sure the fry were feeding from the parent because the contacting changes form as the fry grow. We thought it might simply facilitate schooling around the parents.

To test whether the fry were actually feeding from the parent, Jack Ward and I rolled several parents in carbon dust, which briefly adhered to their mucus coating. After the dusted parents were put back with their young, we waited a while and then removed a few fry. Sure enough, they had particles of carbon in their stomachs.[88]

Even though they independently evolved the trait of feeding their young from their skin, the orange chromide and the Midas cichlid converged on the mechanism to do this. As in the Midas cichlid, the skin of the parental orange chromide develops more mucus cells when its young are contacting it.[89]

The fry of the orange chromide take mucus from their parents mostly during the first five to nine days of their life. This behavior, called micronipping, reaches a peak when the fry are about one week old, and subsequently declines. Then micronipping is gradually replaced by a movement called glancing.[90] In this latter movement the fry swim at the parent and chafe off of them in a glancing motion, but without taking mucus. Oddly, and unlike the Midas cichlid, the rate of micronipping[91] and of glancing[92] is low before feeding and, as the fry grows older, becomes ever more frequent after taking a meal. Glancing carries over into adulthood. Mated pairs often glance off one another.

We found that raising fry in the absence of parents during the first nine days after starting to swim was difficult,[93] which is not the case with most

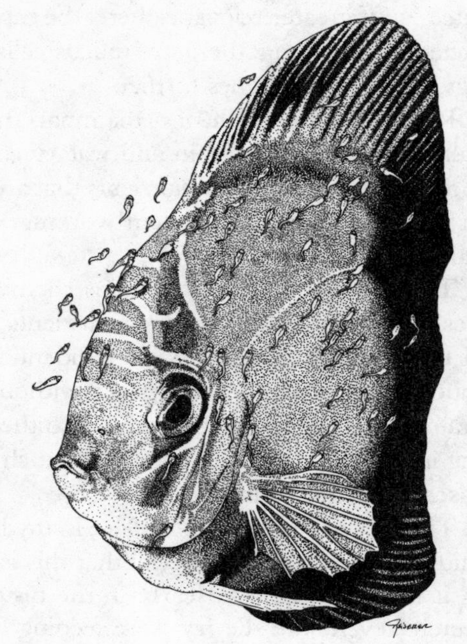

FIGURE 10.8 The discus fish, *Symphysodon discus*, of South American provides a broad surface on which its young cling as they feed from it.

cichlids. Apparently, the fry are getting something beyond calories from the parents' mucus, but that has not been investigated in orange chromides.

The last case is the most spectacular and well known, though much remains to be learned. The beautiful discus fishes, *Symphysodon discus* and *S. aequifasciatus,* from the Amazon Basin have by far the most highly developed provisioning of their offspring. The fry swim to their pancake-shaped parents and cling to them, browsing on their skin. Evidently they require that food to survive and grow, especially during the week or so after commencing swimming,[94] so the behavior is said to be obligatory instead of facultative.

The skin of the parents develops secretory cells that are obvious when examined under the microscope; they are also present in a reduced state in nonbreeding adults.[95] The specialization of provisioning in discus fishes, however, has gone beyond merely secreting mucus. Recent work even suggests that skin mucus is of lesser importance to the fry.

Heinz Bremer and Ulrich Walter suspected the fry were getting more than just mucus. They carefully analyzed the parents' skin and also the contents of the fry's guts.[96] Using electron microscopy to study the skin of *S. aequifasciatus*, they tracked the development of specialized cells called secretocytes. They also demonstrated that the outer layers of the parents' skin becomes much thicker by the addition of more layers of flattened cells.

What Bremer and Walter discovered is amazing. The layer of mucus on parental discus fish is thin. The secretocytes are released intact as free cells onto the surface of the parent where the fry graze on them like tiny hors d'oeuvres. They believe that secretocytes are an entirely new type of cell that discus fish have evolved to feed their fry. Now the skin of other cichlids, whose fry feed from their surface, needs reexamining to see if they have secretocytes, or at least their precursor.

The skin of parental discus fish also has microorganisms in it, such as diatoms and bacteria. Both were found in the guts of the fry. Bremer and Walter suggested that the bacteria eaten by the fry are important in establishing the flora of their guts, much as humans eat yogurt to restore their intestinal flora after taking antibiotics.

The study of what cichlid fry ingest when they graze off their parents' skin is in its infancy (again, no pun intended). The recent research on different species indicates they get more than just calories. Important nutrients are also supplied by the parents. Some of the earlier research indicates that the parental state of discus fish, and with it the changes in their skin to feed the fry, are controlled by the hormone prolactin.[97] That is also the main hormone regulating the production of milk in mammals.[98] Much research is obviously needed on this relatively neglected and fascinating aspect of cichlid biology.

THE SINGLE MOTHER

In this modern era no account of parental care in substrate-brooding cichlids would be complete without mentioning family life sans father. In the dwarf apistos *(Apistogramma),* the female alone cares for the eggs and fry while the male courts other females. Because she is so tiny, she has only about a hundred or fewer young. The father, tending a wide area that contains his harem, is so large and involved with confronting intruding males and seeking females ready to spawn that he plays only a small role in keeping out potential predators on the young. That job befalls the female.

She shepherds the fry in a tangle of leaves and other debris in the stream. Perhaps care by one parent is possible only in a complex environment, such as one rich in rocky shelter, plants, or empty snail shells for hiding.[99] Natural selection has also operated on the fry to help them survive when they have only one parent.

The fry of apistos differ from typical biparental families in being highly cryptic. They are typically yellowish brown with mottled darker markings. They also behave differently. Instead of a schooling cluster of fry above the bottom, they hug the bottom tightly. They are well spaced and so are not really schooling, though they stay near their mother. Much of the time the semi-solitary fry are nearly motionless. They move in quick short darts and independently of one another. That makes them hard to see. The fry of

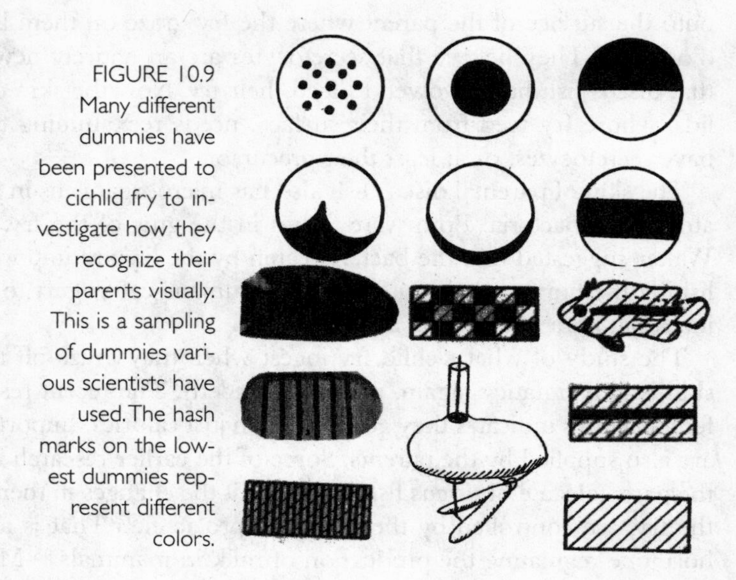

FIGURE 10.9 Many different dummies have been presented to cichlid fry to investigate how they recognize their parents visually. This is a sampling of dummies various scientists have used. The hash marks on the lowest dummies represent different colors.

some biparental species, too, have cryptic fry that hug the bottom, for example, the African cichlids in the genera *Pelvicachromis* and *Julidochromis*.

Many dwarf lamprologines in Lake Tanganyika, Africa, also have exclusively female care of the fry, and the fry have converged on the same kind of behavior as seen in the apistos.[100] For instance, in the haremic cichlid *Lamprologus furcifer,* the fry of the single mothers spread out over the vertical rock face where they feed and are difficult for a predator to spot.[101]

Tetsuo Kuwamura compared the behavior of many species in Lake Tanganyika.[102] He ranked the behavior of seventeen species on a gradient from biparental to solely maternal care. As the care of the fry became more the domain of the female, the fry swam less and less in open water and became more closely associated with the bottom. His account of the behavior of the fry of exclusively maternal species sounds remarkably like that of fry of the apistos, another example of evolutionary convergence.

HOW DO I KNOW THEE?

Newly swimming fry risk not connecting with their parents, and that would be their death knell. Even when the fry are with them, the parents may fight an intruder pair, and that activity can scatter the fry. In the confusion, some of the fry might swim to one of the many predators lurking around the school because they are similar to the parents and the naive fry do not see well.[103] Somehow, a nearby parent serves as a beacon for the fry to assemble under it. The parents might even release a chemical attractant in their urine. Some examples are known, but they are relatively few.[104]

The extent to which the fry can visually distinguish their parents from other fish-like objects has received lots of attention, starting with those early pioneers, G. K. Noble and Brian Curtis. The method they introduced is still used. They made simple dummies of the parents and then altered the appearance of the dummies.[105] To be most effective, the dummy parent should move and be roughly the size of the parent.[106] The color of the dummy also plays a role, though results from several studies have produced different conclusions about the degree to which the color of the dummy should agree with that of the parent.[107] The general conclusion, however, is that fry are not demanding. They even respond to the most unlikely, abstract dummies, but more to some than others.[108]

IS THAT MINE?

Parental cichlids, and especially mothers, will take custody of the most unlikely organisms. Water fleas *(Daphnia)*–small crustaceans about the same size as their young–are a standard part of a cichlid's diet. But some adult cichlids, frustrated by the lack of an opportunity to breed, have been seen to hover over a cloud of water fleas and even to give alarm signals at the right time. Not surprisingly, when motivated to care for eggs or fry, parents can be undiscriminating.

As one example, Peter Weber created dummy clutches of eggs of the convict cichlid using melted drops of wax, colored to approximate real eggs. His first finding was that when he removed the mother's eggs from a wall that he left bare, she continued fanning the nonexistent eggs for some time. However, she does discriminate among eggs to some extent. When he provided faux eggs, the closer they were to the size of the real ones the longer she was willing to invest in them. But when Weber gave a mother a choice between real and dummy eggs of the same size, she cared for both clutches. Oddly, she actually fanned the counterfeit eggs a bit more than the genuine ones.[109]

Early in the history of such studies, cichlids were thought to fixate on the first fry they experienced.[110] This phenomenon of learning one's young from the first experience with them, and having that effect last, is termed parental imprinting. In the case of the jewel fish, if a parent were given the wrigglers of a different species, the next time they spawned, they would eat their own fry but accept those of the alien species. That was soon disproved.[111] In fact, many parental cichlids accept foreign young into their school of fry, irrespective of previous experience.

Given the sizeable investment parents make in their offspring, natural selection should work strongly against parents that accept alien fry, whether of their own or of another species. The costs could be large. That leads to the question, could adopting foreign fry benefit the parents? And by benefit I mean increase the pair's Darwinian fitness. If so, they could evolve behavior favoring adoption. Come to think of it, wouldn't it be nice to evolve more help around the house?

≈ *eleven* ≈

FAMILY LIFE GETS COMPLICATED

To fall in the realm of science, a theory or hypothesis must be falsifiable, and that sets science apart from religion. Organic evolution is clearly falsifiable, as Charles Darwin made clear. He stated that his theory of evolution would be disproved if a species could be shown to have an adaptation that served exclusively the interests of another species. And he didn't mean domesticated animals. That a cichlid parent should invest in the care and protection of the offspring of a different species therefore presents a challenge to the theory of evolution. To get to that challenge, we need first to consider parent-young relationships in more depth.

THE MORE THE MERRIER

That cichlids in nature adopt fry from pairs of the same and even different species has been recognized for some time. The earliest report came from Nigeria, where pairs of *Tilapia melanopleura* and *T. mariae* adopted foreign fry of their own species; some also took in the fry of other species.[1] Two more cichlid species there, *Pelvicachromis pulcher* and *Chromidotilapia guentheri,* also incorporated newcomers of their own species. Svere Sjölander thought adopting might have been in response to the threat of the fry-eating jewel fish, *Hemichromis guttatus*.

In Sri Lanka, two to three pairs of green chromides, *Etroplus suratensis,* pool their fry into one large school,[2] a situation called crèching from the French word for day nursery. In Africa, the substrate-brooding *Tilapia ren-*

dalli sometimes forms large families with four to six parents.[3] Other instances of crèching have been observed in Central America, but not closely studied.[4] In these crèches, the fry were all of the same species.

Why should cichlids protect the fry of other pairs? What is the payoff for the adopters, if any? By observing families of the Midas cichlid, *'Cichlasoma' citrinellum,* in Nicaragua, Ken and Kathy McKaye became the first to muster convincing evidence of not just accepting fry into the family but rather the actual kidnapping of them.[5] That reinforces the proposition that parents somehow benefit from the presence of foreign young. I mentioned earlier that predation on their fry is so intense that the half-life of a guarded school can be as brief as three days. Many pairs lose all the fry under their care to predators.

In the process of documenting this loss, McKaye and McKaye were amazed by the opposite trend in some families: The number of fry actually increased over time, in the most extreme case, from 35 on day 26 to 400 by day 41. Looking closer, they noticed that the fry differed slightly in size, suggesting differences in age; even more surprising, fry of other species were commonly present.

The parents were actively taking fry from other pairs. All pairs of Midas cichlids with young older than five weeks had foreign fry in their midst. And this behavior was not confined to the Midas cichlid. Other pairs of cichlids were adopting one another's young.

How was the acquisition of additional fry accomplished? It turned out that when pairs of Midas cichlids engaged in fights at the boundaries of their territories, their fry were often close by and got mixed up. When the pairs separated, one had captured fry from the other (the same thing happens in aquaria between parental pairs of rainbow cichlids,[6] and those of orange chromides). The McKayes also saw a male slip into a neighbor pair's territory while the parents were fighting yet another neighbor pair; the crafty trespasser then herded about fifty fry back into his own school. When the observers removed parents, their fry were quickly rounded up by neighbor pairs with fry.

The McKayes then netted fry of some pairs of Midas cichlids and released them into the schools of other pairs. The stranger fry were accepted if they were of the same age or younger than the recipient pair's own fry, but not if they were older. Embraced fry of the about the same age survived. But when the fry were much smaller than those of the pair, the pair's fry gobbled up their smaller step-siblings.

So, why did the parents voluntarily incorporate foreign fry into their family? The McKayes contend that increasing the number of fry in the school improves the chances of the pair's own fry surviving to independence. This is called the dilution effect on predation, and it had been proposed as one explanation for adopting by another species, the rainbow cichlid, *Herotilapia multispinosa*.[7] The effect may have an upper limit. In nature, pairs of convict cichlids may not be able to protect and thus raise more than around fifty fry.[8]

The dilution hypothesis presumes that predators take fry at random from the school. By adding foreign fry, the parents lessen the chance that its own young will be consumed; if the predator is more apt to take the alien fry, that is even better. This hypothesis also assumes a larger school will not attract additional predators, which could negate the gain from the dilution effect. Finally, to benefit from adoption, the foreign fry should not compete with the host fry for food.

Another possible advantage arises when smaller fry are taken in. Recall that the bigger fry in the host family devour guests if the size difference is great enough. In the convict cichlid, cannibalism sets in when one fry exceeds the length of another by about 4.5 millimeters.[9] In the Midas cichlid, the fry who eat their schoolmates grow much faster than those who don't; that causes the size disparity to increase, exposing even more of the small siblings to cannibalism.[10]

The McKayes predicted the particular circumstances that should promote adopting between pairs. Colonial breeding is one, merely because families are so close together. More interesting, however, adoption should increase when the number of predators on the fry increase. So far, only one experiment has been done to test the predation hypothesis, and with only suggestive results.[11]

Research on adopting in the convict cichlid was advanced by Brian Wisenden and his mentor Miles Keenleyside in a series of well-conceived field studies.[12] The convict cichlid is ideal for such studies because it is little and is found in small clear streams in Costa Rica where it can be observed closely and captured easily. Over two hundred families were examined, and 29 percent of them contained foreign fry. When present, foreign fry made up on average about 11 percent of the school.

They confirmed many of McKaye's findings: Adoption of the fry of other pairs is common; fry larger than their own are eaten; and smaller fry are accepted. Unlike parental Midas cichlids, alien fry of the same size as those of the adopting pair were not permitted into their schools. A later study revealed the parents actively bite at fry larger than their own when offered in a glass jar.[13] Most important, the smaller adopted fry were more vulnerable to predators than were the host fry.[14] Evidently as a consequence, the fry of families to which alien fry had been added experimentally had higher survivorship than did unaltered families.

In addition, the incorporated fry grew more slowly; the implication is that they suffered when competing for food with the larger host fry.[15] The findings indicated, further, that adoption should be beneficial only when the fry are near the end of care by their parents, which corresponds to a length of 10–12 millimeters in the convict cichlid.

In another experiment, the father was removed. Fewer fry survived.[16] Part of the reason for this was that the fry were more susceptible to being kidnapped by other pairs when the mother was the sole guardian.[17] Wisenden and Keenleyside contended, however, that the unaccompanied mother might have responded to her difficult situation by turning it to her advantage, somehow encouraging other pairs to adopt her fry.

CUTTING ONE'S LOSSES

By giving the fry up for adoption, sometimes called "farming out,"[18] at least some of the parent's fry might benefit from the dilution effect. This would apply especially if the fry were close in size to those of the host family, and if the fry of the host family were young. Apparently, when the fry are young, 7.0 millimeters long or less, they are not effective in evading a predator.

The mother who gives up her family is free from further caretaking. She could spend more time feeding and preparing to enter another reproductive cycle. Accordingly, both adopting and farming out can be adaptive, depending on the circumstances. For instance, she runs the risk that her fry might be eaten before being adopted, so the abundance of predators is a crucial consideration. In general, however, the mother who donates her fry probably does less well than the adopter.

These observations indicate a prudent parent must be constantly assessing the state of its family. If it is losing fry to predators, at what point would it be better for its lifetime Darwinian fitness to stop investing in the current family? At some point, taking care of the young could be a losing proposition. The parent must weigh whether to continue investing or to cut its losses and invest in future offspring.[19] In fact, when the number of eggs is reduced, parental cichlids often eat those remaining and start over.[20] When the number of young fall too low, a sensible tactic is to farm the young out rather than just let them be eaten.

MOUTH-BROODING FARMERS

Although adoption is common among substrate brooders, only the mouth brooders who continue to guard their fry after releasing them are in a position to adopt alien fry. Such mouth brooders are often biparental. In a number of these monogamous species in Lake Tanganyika, the female first takes the brood into her mouth and remains with the male. Later, the female spits out the brood in front of the male, and he then assumes the role of mouth brooder. This is where the term "farming out" originated.[21]

Species that farm out are vulnerable to taking up alien young. For example, among the *Xenotilapia flavipinnis,* one parent of unknown sex was seen to swim about six meters to a neighbor pair who had their own school of fry. The neighbors attacked the intruding bearer of gifts, but the intruder managed to spit its young into the neighbor's school of fry anyway, and they were accepted.[22]

Why should a parent farm out its young? The first explanation offered was that both parents are needed to guard the young. If one member of the pair were to desert, the other would be tempted to fob its young off on a neighbor, and underwater manipulations supported that interpretation.

In Lake Tanganyika, experimenters removed from twenty-six different families of *Perissodus microlepis* the pair member not carrying the young. The remaining parent experienced a steady loss of its fry to predators. Two such mateless females swam up to thirty meters to release their fry into other families, who seemed upset but nonetheless took in the guests. One single male swam off with his young but was disastrously unsuccessful; he released a batch of them in front of a small fish-eating cichlid, *Lamprologus profundicola;* the fry rushed into its mouth, and it proceeded to swallow them without so much as saying grace.[23]

The eagerness of parental mouth brooders makes them vulnerable to farming out. Unfortunately for them, it also exposes them to true aquatic cuckoos.

CAT CUCKOOS AND ALLIES

The cuckoo bird is so famous for parasitizing the nests of other birds that the word cuckoldry has become part of our language. No longer able to raise its own offspring, the cuckoo sneaks into a nest of another bird while the parent is away and quickly lays a large egg.[24] The cuckoo chick is larger than its nest mates, and within hours after hatching it hoists its step-siblings, one by one, onto its back and flings them out of the nest. The foster parents feed the remaining cuckoo chick, even after it fledges and is so large it towers over them as it perches outside of the nest.

Such behavior among vertebrate animals was thought to occur only in birds. The discovery of aquatic cuckoos was made by aquarists trying to recreate a community of Tanganyikan fishes. They added the small catfish, *Synodontis multipunctatus,* to their cichlid aquaria. The "cats" joined spawning pairs of mouth-brooding cichlids. The cats also commenced spawning. But then the female cichlid picked up the catfish eggs and nurtured them as her own. The cats were parasitizing the cichlid's parental nurturing.[25]

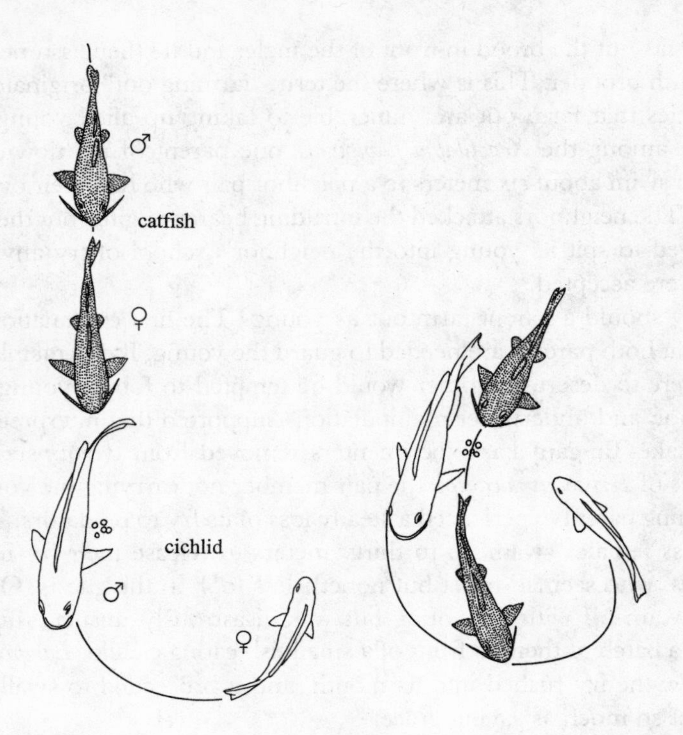

FIGURE 11.1 A pair of mouth-brooding cichlids (in white) is spawning; the female has released some eggs and is circling back to pick them up. However, a pair of catfish, *Synodontus multipunctatus,* follows close on and drops and fertilizes their eggs before the female cichlid can get back to hers. As a result, the mother cichlid takes the catfish eggs into her mouth with her own eggs (after Sato 1987).

Tetsu Sato went to Lake Tanganyika to confirm and expand those observations. In the mouths of thirty-two mouth brooders in six species *(Simochromis diagramma, S. babaulti, Tropheus moorii, Ctenochromis horei, Gnathochromis pfefferi, Pseudosimochromis curviforns)* he found baby parasitic catfish. Sato sought diligently for nonparasitic catfish but found none.[26]

For now, the catfish are thought to be obliged to cede their eggs to maternal cichlids for safe upbringing, just like the cuckoo. Aquarists, however, have induced the cats to reproduce in the absence of spawning cichlids, and populations of catfish elsewhere in Lake Tanganyika might not be cuckoos. (A related Lake Tanganyika catfish, *S. petricola,* may also parasitize maternal cichlids.)

One adaptation by the catfish to this opportunistic mode of reproducing is that females are able to muster and release small numbers of eggs at frequent intervals, as though ever on the ready for yet another maternal cichlid to parasitize.[27] From their side, cichlids are ready dupes. In captivity, many different mouth-brooding females from various lakes happily take up the catfish eggs.

The eggs of the cats are considerably smaller than those of their cichlid host, so they hatch much sooner, and their yolk sacs are used up within three days after hatching. Without sustenance, they dine on their cichlid mouth mates; because they are too large at first for them to engulf, they just bite into the yolk sacs of the tiny cichlids and eat the yolk. The cats grow so fast that within a few days they can swallow an entire cichlid wriggler.

By the time the cats are ready to emerge from their foster mothers' mouths, they have feasted on all or almost all of their mouth mates. Amazingly, after the fostered cats have emerged, the cichlid mother will take them back into her mouth when alarmed. That is just another indication of how uncritical and therefore vulnerable cichlid parents are when it comes to eggs and young.

A conflict of interest between two species sets up an evolutionary dynamic. From the cichlid's side, it ought to evolve countermeasures against the parasitic cuckoo; some birds have done that.[28] However, that may be out of the question for the female cichlid. Unlike a bird, her eggs are laid

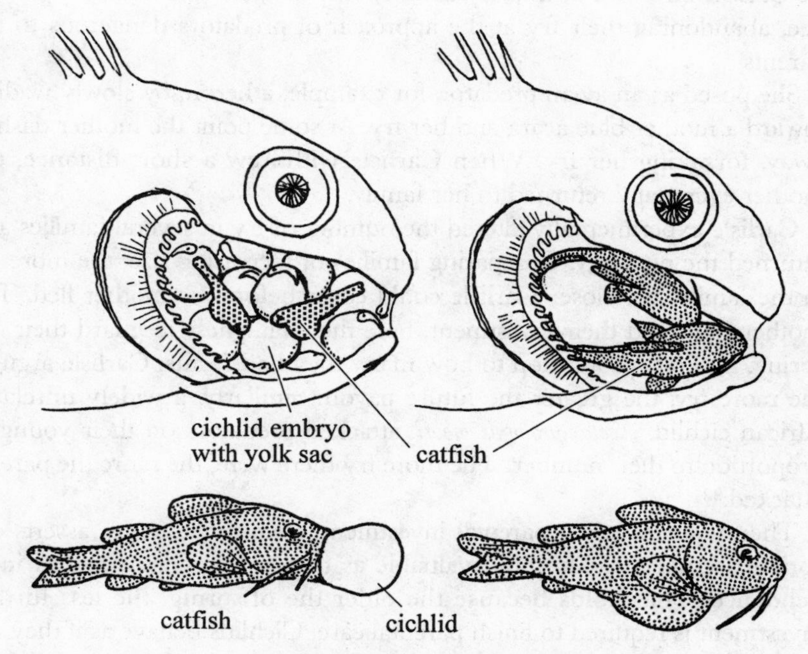

FIGURE 11.2 The smaller catfish eggs hatch quickly, and the tiny catfish suck out the yolk of the larger cichlid larvae. The catfish grow with incredible speed and are soon big enough to engulf the entire cichlid larva (after Sato 1987).

in the open, exposed to near instant predation. That forces her to pick up the eggs so quickly that she cannot take the time to discriminate among them; that is why in many species the female snatches the eggs before they can be fertilized, and why the egg spots on the male are so effective. Perhaps some cichlid is nonetheless evolving countermeasures, but we have only a hint of this possibility in Lake Tanganyika cichlids.[29]

The cat cuckoo has a less severe problem, and perhaps none at all. Its young might develop too rapidly for their foster mother. If the little cats eat all their cichlid mouth mates, but the mother is geared to retain her young until much later, then the baby cats have to induce the mother to release them, or else they will starve. They probably stimulate release through their activity. These issues have not yet been addressed.

QUALITY CONTROL

For animals like cichlids, theory predicts parents should evolve the ability to evaluate the quality of their offspring.[30] A good indicator of brood quality for a substrate-brooding cichlid is the number of fry it is shepherding. This interesting proposition was first tested in a stream in Panama by Tamsie Carlisle on a blue acara, *Aequidens coeruleopunctatu*.[31] She had seen parents flee, abandoning their fry at the approach of predators dangerous to the parents.

She posed as an avian predator, for example, a heron, by slowly wading toward a mother blue acara and her fry. At some point the mother dashed away, forsaking her fry. When Carlisle withdrew a short distance, the mother eventually returned to her family.

Carlisle experimentally altered the number of fry in several families and returned the next day. Comparing families of increasing size, the more fry in the family, the closer Carlisle could come before the mother fled. The mothers adjusted their investment, here the willingness to guard their offspring, in direct proportion to how many they had. Thus, Carlisle argued, the more fry, the greater the future payoff. Similarly, a widely unrelated African cichlid, *Neolamprologus moori,* attacked predators on their young in proportion to their number: The more fry there were, the more the parents attacked.[32]

Theory surrounding parental investment, even for humans, asserts offspring become steadily more valuable as they approach the age of independence. That holds because the older the offspring, the less further investment is required to finish parental care. Cichlids behave as if they understand this, although the prediction about amount of investment for a given age has been tested experimentally only once. Again, the species was the convict cichlid. And sure enough, the older the offspring were, going from eggs to wrigglers to fry, the more active the parents were in defending them from a dummy predator.[33]

Further study, however, proved that it is not age per se but rather the vulnerability of the fry to predation that the parent can assess. In one experiment, the fry were kept at what must have seemed to the mother convict cichlid a constant early stage of development; this was done by regularly substituting younger fry for the current ageing ones. Such mothers attacked at a dummy fry predator not in accordance with how long they had nurtured their young but rather in keeping with the developmental stage of the fry.

The mothers' past investment in rearing young therefore did not determine parental investment, as theory predicts. Rather, the parents adjusted their protective behavior to the fry's developmental stage. Bob Lavery maintains that the behavioral mechanism is the recognition by the mothers of the vulnerability of the fry to predation.[34]

JUST RIGHT

An indication of Darwinian fitness is how many offspring of an individual survive to reproduce in the next generation in relation to other individuals of the species. But just cranking out ever more young might not achieve this objective. Parents may produce more surviving offspring by having fewer of them. Instead of a maximum number of offspring, an optimum number might be better. This concept is known as optimal clutch size, originally studied in birds.[35]

The idea is straightforward. If a bird lays too many eggs in a given clutch, the parents will not be able to feed the nestlings adequately nor to provide enough other types of care. However, if the bird lays too few eggs, then it loses out in the reproductive race because it could have reared more. The way the bird "knows" how many to lay is through its ancestral history, the process of natural selection. Over a great many seasons, the genes of birds who laid too many or too few eggs were gradually eliminated. And so it should be for fishes in general. Those that lay too many eggs reduce their future reproduction. And those who lay too few when they could have produced more also leave too few descendants.

This proposition has so far been tested only on our old friend the convict cichlid.[36] The question can be checked properly solely in the field, under natural conditions of feeding and predation. The scientists knew brood size at the beginning of the fry stage varies enormously, in their experience up to a maximum of about 160 fry, with an average of around 75. In their experiment in a Costa Rican stream, they added fry to several families until they numbered 150 in each. They left the broods of control families unchanged so their broods were variable with a mean number of 70 fry, or about half the number of fry as in the large experimental broods.

The results were dramatic. Soon the average number of fry in the augmented families dropped to just a little more than in the control families. Another ten days later, the difference was even less. In addition, the fry in the large families grew much more slowly than those in the normal ones.

Having more fry than can be well cared for is hence done at a cost to fitness. The parents cannot protect them effectively, and the fry do not get as much to eat in such a crowd. Present and probably future reproduction are reduced. One specific optimal clutch size for all convict cichlids, however, is not knowable. It depends on many factors that vary from time to time in a given stream and between streams. These are availability of food, amount of predation (often depending on where in the stream the family is found), and the size of the female.[37]

HELPING AROUND THE HOUSE

Television specials on the behavior of animals such as wolves and cape hunting dogs have made us aware of the complex social behavior of such mammals. When the dominant female gives birth to pups, for instance, the other pack members help rear them. When the pack hunts, some individual stays behind to baby-sit the pups. And returning individuals, even those who are not the parents, share food with the pups. The nonreproducing members of the group are typically either siblings of the breeding pair or the pair's offspring from a previous litter. (In a few species, however, helpers are not related to the breeding pair.)[38] Helping with offspring is found in a wide variety of mammals and birds.[39]

Those who study helping have sought long and hard to explain why the helpers should forgo their own reproduction, at first sight an apparently severe cost to their Darwinian fitness, and instead help nurture the primary pair's young. The hypothesized benefits are many, and different hypotheses, or combinations of them, may apply to different species. The most commonly proposed benefit is getting one's genes into the next generation by raising siblings, with whom, on average, half of one's genes are shared. Other explanations include learning how to rear young, inheriting a good territory, acquiring one of the parents as a mate, and putting off the dangers of leaving home.

Helping had not been found among cold-blooded vertebrates, the fishes, amphibians, and reptiles.[40] However, if you were told some kind of fish has nest helpers, which would you guess? Right. Cichlids. You might also have forecast, if pressed, that the examples are found in the ancient rift lakes in Africa. And indeed they are. So far, however, they are known only for cichlids in the oldest lake, Tanganyika. But cichlids are not entirely alone. Helping was discovered recently in two fish species in the genus *Betta* (a member of which is the famous Siamese fighting fish).[41]

The study of helping among substrate-brooding cichlids offers advantages over working with mammals and birds. Helping species of cichlids are small and easily monitored in the wild, have short generation time and limited dispersal, and can also be observed readily in naturalistic communities set up in captivity. Compared with birds and mammals, however, cichlid helpers are little known, despite the advantages their study offers.

Helping in cichlids was discovered just two decades ago.[42] The original report was followed up only recently, but considerable progress has been made, considering the short time. Helping has been described in a small number of species from Lake Tanganyika.[43] These include *Neolamprologus savoryi*[44] and *Julidochromis marlieri*.[45] Michael Taborsky and his colleagues are analyzing helping in a small Lake Tanganyikan cichlid, *Neolamprologus pulcher*.[46] (This favorite of the hobbyists is better known as *N. brichardi,* but the two species are now regarded as the same, and *N. pulcher* is the older name.)

The research has been guided by three concepts that have emerged in recent years to explain the evolution of social behavior. One of these is parent-offspring conflict.[47] The theoretical issue is complex but at its simplest it is almost quotidian: Developing offspring demand more care than the parent is willing to give, because the parent needs to withhold some investment for future reproduction. An oft-mentioned example is the weaning conflict between a mammalian mother and her young, though proving that this really is a conflict has not been easy.[48] Closer to home is the money teenagers would like to get from their parents.

The other two concepts could be regarded as one. The core idea is cooperation, here between helper and parent. It posits that both the helper and the parent benefit genetically from any assistance rendered. Because each sibling shares, on average, half of its genes with its other siblings, assuming the mother mates with only one male, the helper gains fitness by assisting its siblings in passing on its genes. The parent and helpers also profit from the help by producing more surviving healthy offspring. For the helper to stay, the prospect of it reproducing successfully ought to be greater than if it leaves and seeks to raise a family on its own.

Another way of looking at this is through the concept of reciprocal altruism, which requires no genetic relationship. Conflicts of interest are best resolved when each participant provides some benefit to the other in excess of the costs. This should work only in a close-knit social unit where the individuals know one another and remain together. If one partner in this bargain cheats, then the other can retaliate.[49] Applying this kind of thinking to simple fish may seem far-fetched at the moment, but you will see how it works after you understand the nature of helping in *Neolamprologus pulcher* and the experiments that were inspired by these concepts. First, we need knowledge of the social system of this tiny cichlid.

Neolamprologus pulcher lives either in tight-knit families or in aggregations of nonbreeders in shallow water along the shores of Lake Tanganyika. Family territories occur in rocky outcrops that are like islands with stretches of sand to the next outcrop, so the fish live in small semi-isolated communities where they feed on plankton just over their territories. Each defended territory is small, only about 50 centimeters (20 inches) in diameter, with the cavity as the focal point. Nest cavities are a limiting resource, so only the strongest adults acquire them. Families take shelter from predators in the cavity, where they also sleep out of harm's way; the eggs and wrigglers are raised there as well. The large dominant male, up to 60 millimeters long, often has two mates, hence two families.[50]

Each family contains a pair of big adults with about seven smaller youngsters. These young are the issue of up to three to four spawnings, so they range in size from small, about 10 millimeters long, to fairly large, occasionally up to about 50 millimeters. The youngest helpers are only about three weeks old.[51] The largest are about one year old; that is the age at which most of them typically leave the family and join the aggregation of nonbreeders.[52]

During the early stage of eggs and wrigglers, most of the care of the offspring is done by the mother.[53] The father provides little direct care, although he is competent to carry out such duties.[54] He maintains his top status in the family through aggression.[55] The helpers, for their part, invest in both direct and indirect care.

Which tasks the helpers provide, and how much, depends on three factors. The first two are obvious: They must live in a territory. Parents must be present. The third factor is the domination by parents and older offspring.[56] The older, larger juveniles are more involved in repelling other members of their species, which is indirect care, than are the younger ones. And dominance of the younger by the older young may stimulate the younger ones to provide direct care of their yet younger siblings.[57] Older helpers do assist in cleaning the eggs. But if one of them becomes the dominant male, as when all the larger members of the family were experimentally removed, he becomes infanticidal, eating the eggs.[58]

Now for a complicating factor: As the helpers age, their relatedness to the new siblings decreases, on average. That follows because of the mortality of their parents, especially the father who is prone to predation as he defends the territory. When a male is lost, a new large male takes over. Some offspring therefore end up helping stepfathers who have replaced their biological fathers. As a consequence, the relatedness of the older helpers to the younger ones tends to decrease with time, from one-half genes shared with younger offspring to as little as one-quarter.[59]

Taborsky asked, what are the costs and the benefits to the helpers?[60] He demonstrated that the main cost to them is slower growth. Later, he and Astrid Grantner actually measured the energy costs of the various activities that make up parental caretaking.[61] Because this is the first time the energetic costs of helping have ever been measured in any kind of animal, we should examine how that was done. The experiment rested on the well-known finding that amount of oxygen consumed is equivalent to energy spent.

First, a family of *N. pulcher* was established in an aquarium. A weighed subject was placed into a flask just large enough to permit it to move around. Helpers in the flask dug and removed sand from the provided shelter and performed submissive displays through the glass wall at the approach of other family members in the aquarium. Imprisoned mothers and fathers responded to their mirror images by attacking them through the glass wall of the chamber.

From their calculations, Taborsky and Grantner estimated the cost of doing various acts of parental behavior by the mother, father, and helper. For all the fish, aggressive interactions raised the metabolic rate three- to fourfold over "normal" metabolism. Surprisingly, the submissive display of helpers, described as a quivering of the tail, was similarly costly, resulting in a 3.5-fold increase. As one might have predicted, digging was even more costly, increasing the rate of oxygen consumption about sixfold. Such activities, however, consumed only about 1.5 percent of the total energy budget. Seen another way, about 98 percent of the fish's energy was devoted to nonparental behavior.

Given the cost to a helper, what are the benefits? The first one is the production of more siblings by the mother. This results from one simple change. With helpers, the mother is free to spend more time feeding. As a consequence, she grows faster. And a larger mother produces more eggs. That increases the fitness of both the helpers and the mother.

The helpers benefit in yet another and more direct way. They are less apt to be eaten by a predator. The most active predator on them is a related cichlid, *Lamprologus elongatus,* but it is not a very large piscivore. When the young reach a length of around 35 to 40 millimeters, they become more difficult for the average *L. elongatus* to capture and to ingest. The parents actively attack this predator, and the large helpers may join in. In a test situation in an aquarium inhabited by a predator, when the young were not afforded the protection of a family, they were soon devoured. Thus, the parents reciprocate for the help by protecting the helpers.

Yet another possible benefit to the young in helping is the possibility of inheriting the territory.[62] Field experiments demonstrated the reality of the benefit for a cichlid fish.[63] But all good things come to an end.

As the young grow, they reach a size where the benefits they provide to the parents might exceed potential costs. For their part, the young are reluctant to leave, even though they are not then so closely related to the new young, or not related at all. Apparently they recognize a good thing. Even though they venture out to feed among the nonbreeding fish in the aggregation, they are reluctant to take on the risk of predation inherent in leaving home for good and finding a place in which to refuge. In fact, remaining longer in the shelter may inhibit their growth. To overcome this, the parents evict helpers when they reach a length of 35–50 millimeters. The timing of the departure therefore presents a conflict between the parents and their offspring, a conflict that doesn't need explaining to many human parents.

The helpers have evolved to prepare for the day they must face the harsh world away from home. Despite not growing as fast as nonhelpers, the helpers lay down more fat.[64] That enables them to call upon reserves and thereby evidently achieve a spurt of growth. That would bring them up to a size better suited to the challenges of their new way of life.

But what might the costs to the parents be of permitting the young to overstay their tenure? Perhaps the most acute cost emanates from the offspring becoming sexually mature while still at home. The young males, especially, have been seen in aquaria to behave in a way that suggests that they may fertilize some eggs as they are being laid.

To test whether helper males are actually fertilizing at least a few eggs, genetic fingerprinting was done on several families. Overall, about 10 percent of the offspring in a family are fathered by the helper male. When the dominant male sees helpers fertilizing eggs, he savagely attacks them and drives them out of the home.[65]

If that were not enough reason to expel the maturing helper males, additional observations detected them making a tasty morsel out of eggs from time to time.[66] That is definitely a loss to the parents. On these grounds, the pair aggressively banish their large sons.

In recounting the behavior of cichlid families with helpers, I have presented only the big picture. The research has revealed yet more sophisticated aspects of their biology, such as the role of individual recognition,[67] so necessary for reciprocal altruism.[68]

ALIEN NANNIES

The observation that one substrate-brooding cichlid protects the fry of an entirely different species might seem to refute Darwin's theory of organic evolution. If not, at the very least it cries out for explanation. Several examples of adopting fry of other species have been reported, such as the cichlids *Neetroplus nematopus, 'Cichlasoma' longimanus,* and *'C.' nicaraguense* in

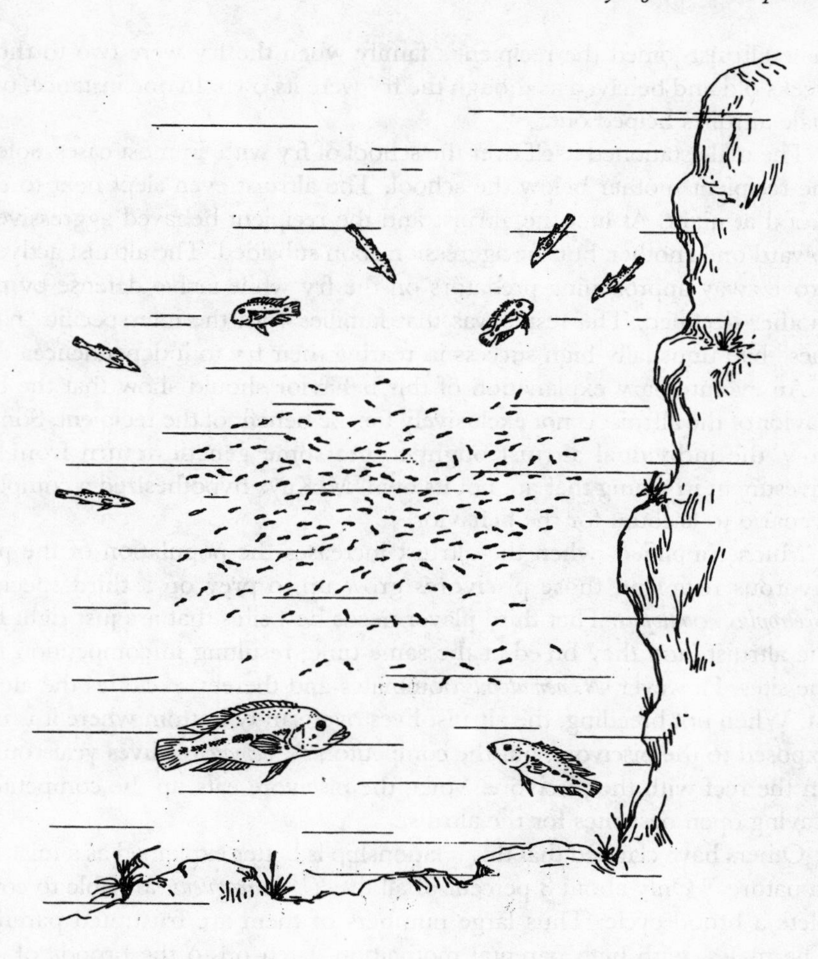

FIGURE 11.3 In Lake Xiloá, Nicaragua, two male *'C.' nicaraguense* hover over the school of fry of the large piscivorous cichlid *'C.' dovii* and protect the fry from predators, which are the slender fish *(Gobiomorus dormitor)* near them. The pair of *'C.' dovii* are below their fry, the larger male stationed to the left (from McKaye 1977).

Central America. But that can be accounted for by the dilution effect I described above.

Ken McKaye discovered a more tantalizing relationship while scuba diving in Lake Xiloá, Nicaragua.[69] He observed and quantified the behavior of two species of cichlids in eight instances of one species helping raise only the offspring of the other species but none of its own. The beneficiary, called here the recipient, was the large piscivorous cichlid *'Cichlasoma' dovii*. The so-called altruist was the much smaller generalist *'C.' nicaraguense*. A male altruist joined the recipient's family when the fry were two to three weeks old and behaved as though the fry were its own. In one instance, two male altruists helped out.

The male stationed itself over the school of fry with, in most cases, solely the recipient mother below the school. The altruist even slept next to the brood at night. At first the altruist and the recipient behaved aggressively toward one another, but the aggression soon subsided. The altruist actively drove away approaching predators on the fry while active defense by the mother declined. The result was that families with the interspecific "nannies" had unusually high success in rearing their fry to independence.

An evolutionary explanation of this behavior should show that the behavior of the altruist is not exclusively for the benefit of the recipient. Somehow, the individual altruist ought to get some genetic return from its investment in young that are not its own. McKaye hypothesized a complex scenario to account for the behavior.

Much simplified, when the altruist increases the population of the piscivorous recipient, those piscivores grow up to prey on a third species, *Neetroplus nematopus*. That third player needs nest sites that are just right for the altruist, and they breed at the same time, resulting in competition for the sites. However, *N. nematopus* dominates and thereby excludes the altruist. When not breeding, the altruist lives over sandy bottom where it is not exposed to the piscivore, but the competitor, *N. nematopus,* lives year-round on the reef with the piscivore. Voilà, the piscivore eats up the competitor, leaving open nest sites for the altruist.

Others have claimed that this relationship is better explained as a mistake in nature.[70] Only about 3 percent of all the *'C.' nicaraguense* are able to complete a brood cycle. Thus large numbers of them are frustrated parents. The males, with high parental motivation, latch on to the broods of *'C.' dovii,* perhaps because the fry of both parents share a prominently striped color pattern, but this is merely an explanation at the level of mechanisms, not that of evolved adaption.

Ken McKaye was perplexed when no more families with interspecific care were found for some years. Had natural selection expunged the genes for such wasteful parental care by males of *'C.' nicaraguense*? Apparently not. McKaye tells me the phenomenon has reemerged and provided several examples of male *'C.' nicaraguense* protecting the fry *'C.' dovii.* One male *'C.' nicaraguense* was even observed shuttling back and forth between two families of *'C.' dovii,* both of which he helped. Moreover, McKaye has captured the behavior on film.

A roughly similar relationship between a fish-eating and a planktivorous cichlid has recently been described taking place in Lake Tanganyika.[71] The piscivore, *Lepidiolamprologus profundicola,* spawns in rock and sand areas, placing the eggs on top of a rock. Just as soon as the eggs hatch, the mother moves the young into a cavity where the wrigglers and fry remain hidden until they disperse during the dark of the moon about two weeks later. The mother alone guards the nest site, vigorously driving away all potential predators on her offspring, and then leaves when her fry disperse.

The planktivore, *Cyprichromis leptosoma,* takes advantage of this free baby-sitting. This species is a mouth brooder, and when the fry are ready to emerge the mother swims to the reef, and if maternal *L. profundicola* are present, she releases the fry nearby. Modestly large schools of the planktivore's fry assemble over the maternal piscivore where they profit from her protection. Fry of *C. leptosoma* are attacked by predators at a rate of about two per hour. When the maternal piscivore leaves, the rate of attacking jumps to about ten per hour, a fivefold increase. In a sense, the fry parasitize the beneficence of the maternal piscivore. I should mention that all fry are not so fortunate. Those that wind up with no baby-sitter at all are attacked about twenty times per hour.

Interspecific brood care among cichlids is remarkable enough, but McKaye has discovered another and more extreme example: consistent and reliable interspecific crèching in some mouth-brooding African cichlids who mix their fry with the those of a catfish.[72] Note that this is feasible only between species that care for their young after the eggs hatch, as this example illustrates. First, I provide a brief, general consideration of the costs and benefits of taking in a nest parasite.

CATS AS ALLIES

Raising a parasite is done at a cost, and at no benefit to the host parent, with one known exception. The Panamanian oropendulas, close relatives of the oriel, raise freeloading cowbirds, who are the North American equivalent of cuckoos. Another parasite, botflies, are even more damaging to the nestling oropendulas. The babies of parents who tolerate a young cowbird survive better because the nestling cowbird pays its way: It eats the injurious botflies.[73] As with so many coevolved adaptations, the behavior has trade-offs. The cost to the parents of raising the parasitic cowbird is outweighed by the benefit of reduced insect parasitism on their young. Thus natural selection has favored tolerant oropendulas.

In like fashion, a bagrid catfish, *Bagrus meridionalis,* and some mouth-brooding cichlids that would appear to parasitize its nest have evolved a relationship to their mutual benefit, although it is the catfish who seem to get the slightly better deal.[74] To reproduce in Lake Malawi, this large catfish forms a monogamous pair and excavates a shallow nest in the sand in relatively deep water. Because of the depth, much of the observation was by means of remote video recording.[75] When the eggs hatch, the male catfish hovers directly over them, and the larger mother takes up a more peripheral position. The parents drive off prospective nest predators, which can be any of a large number of cichlid fishes in the neighborhood.

Two species of cichlids, and possibly a third, have evolved an obligatory reproductive relationship with the catfish. The species are *Ctenopharynx pictus* and *Copadichromis pleurostigmoides* and sometimes an unidentified species of *Rhamphochromis;* most of the following applies to the first two species. When their young are ready to swim, the mouth-brooding mothers bring them to the catfish and spit them into the school of baby cats. Probably as an adaptation to reduce predation on their young (the St. Ignatius effect), up to ten cichlid mothers release their fry simultaneously.

The maternal cichlids approach the much larger catfish parents with care. Female *C. pictus* discharge their fry one to two meters away from the nest and rely on the fry swimming to the cats. Maternal *C. pleurostigmoides,* on the other hand, swim well up over the catfish family and let their fry swim down to the nest. The fry are released only into families of baby cats that are longer than 20 millimeters or shorter than 80 millimeters. Catfish fry 20 millimeters long could be eaten by the cichlid fry, and the parental catfish do not tolerate that. Young catfish longer than 80 millimeters, on the other hand, are big enough to eat the cichlid fry and in fact do, so the cichlid mothers avoid them.

Once the fry are inserted, the show starts. The mother cat tolerates the cichlid fry. But the relationship goes beyond mere tolerance. The cichlid fry eat from her body surface; they even briefly defend miniterritories on her. Father cat, in contrast, won't permit the cichlid fry into the center of the school of baby cats, so they are forced to the periphery of the school where predation is most severe.

The brunt of the predatory attacks are therefore borne by the cichlid fry. When the family is without cichlid fry, the total amount of attacking by predators is about the same, but naturally the predators target only young catfish. Such catfish fry are attacked seven times more than when cichlid fry are present. Those catfish without cichlid associates predictably lose seven times more fry to predators than those with cichlids. Cichlid mothers of one, two, or even three different species hang around and deflect predators. They thereby increase the survivorship of their young but also that of the little catfish.

This is a day-care center. At dusk, the cichlid fry disappear, and around 5:00 A.M. they reappear; where they go has not been determined. Perhaps the mothers who hover at the periphery take them into their mouths overnight.

Whether the catfish or the cichlids benefit more from this coevolution is irrelevant to the cichlids as long as those who place their young in catfish families do better than cichlids of the same species that do not. Because these cichlids do not reproduce without the catfish, one must assume natural selection has favored the behavior. Multispecific parental care is so far unknown for any other kind of animal and is just another example of the evolutionary exploits of cichlid fishes.

In these tantalizing examples of interspecific interactions, one thing remains clear: Darwin's dictum stands. This chapter and the preceding ones have provided many examples of such evolutionary wonders produced by cichlid fishes. But we haven't considered that aspect of their evolution for which they also draw so much attention in the realm of evolutionary biology, the huge number of species they have evolved, some of them in remarkably little time. In the next chapter, I'll describe the most spectacular examples of multiple speciation and discuss the ideas put forth to explain how it happens.

≈ *twelve* ≈

CICHLID FACTORIES

Hobbyists love to collect. The more kinds out there, the better. Cichlid hobbyists, warmly referred to as cichlidiots or cichlidophiles, take naming the species seriously, but they are also attracted by difference for its own sake. When a new cichlid appears in the hobby, even if it is merely a color variant of an already available species, they cherish the fresh cichlid. And when through a genetic accident an attractive mutant pops up, they propagate the sport and spread it through the hobby. In short, hobbyists crave novelty, and that drives a thriving tropical-fish industry.

The font of seemingly endless cichlids are the three Great Lakes of East Africa. There the profusion of cichlid species is mind-boggling.[1,2] In their fascination with that outburst of multiple species, evolutionary biologists have understandably focused their attention almost exclusively on the Great Lakes cichlids. What is so special about cichlids that they can spew forth species by the bushelful? And do the Great Lakes cichlids provide all the answers to that evolutionary riddle? I think not, but the roots of the debate do indeed lie there, so that is where we shall begin.

To approach the larger issue, I start by describing the major features of the three Great Lakes of East Africa and the characteristics of the assemblages of cichlid species in each of them; they are much alike but at the same time differ in revealing ways. To balance the picture, I also describe some assemblages of cichlids that have evolved in other places; they provide a perspective on speciation not afforded by the cichlids in the Great Lakes.

Following that introduction, we'll take a look at the concept of species and define the term in a simplified form and according to the most generally accepted theory of how species originate. With that understanding in

TABLE 12.1 Some Physical Characteristics of the Three Great Lakes of East Africa

Lake	*World Rank in Size*	*Area km² (miles²)*	*Depth m (feet)*	*Clarity m (feet)*	*Age in Years*	*Cichlid Species*
Tanganyika	7th	34,000 (13,100)	1,470 (4,823)	22 (72)	6 million	300
Malawi	9th	31,600 (12,200)	700 (2,310)	17 (56)	1–2 million	500
Victoria	3rd	68,600 (26,500)	95 (305)	1–8 (3–25)	12,400	400

hand, we go on to learn about the role of behavior and trophic morphology in the origin of species. As the theory further unfolds, the logic for, and sequence of, the remainder of the chapter becomes obvious.

Does it make a difference whether other kinds of fishes are already there, and why? How do fish get to a new habitat? Once physically separate populations are established, to what extent are they exposed to genes imported from the other populations? Evolving into separate species is fostered by fish not dispersing but rather staying at home–how is that cultivated by parental care? Finally, the evidence for an alternative hypothesis to explain speciation, sexual selection, is subjected to a detailed, critical review. I do that because the sexual-selection hypothesis of speciation is currently so popular and, in my opinion, overly promoted. Now let's turn to an account of the major theaters of cichlid speciation, the Great Lakes of East Africa.

THE THREE CRUCIBLES

The Great Lakes of East Africa are the only large, enduring lakes in the tropics, and their physical properties are summarized in the accompanying table. The number of species given there is merely a best-guess approximation winnowed from the wide range of estimates by various researchers. Keep in mind that the profusion of species found in each lake is, with a few exceptions, found nowhere else but in that particular lake.

The Great Lakes have often been compared to seas, and the two deep lakes, Tanganyika and Malawi, are indeed moderately salty due to reduced outflow and high evaporation. But the marine analogy can be misleading, especially when the comparison is with tropical marine reefs. There the coral carpet is an extraordinarily rich and complex physical and biological

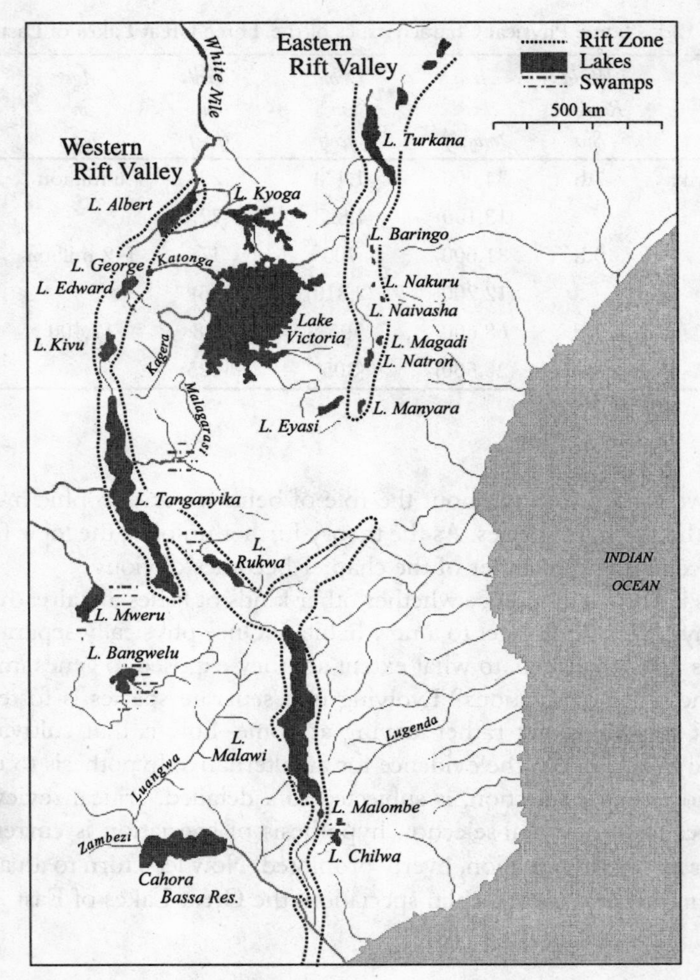

FIGURE 12.1 A map of East Africa with the positions of the three Great Lakes as well as the other large lakes and principal rivers.

environment. In contrast, looking down on the rocky bottom of Lake Tanganyika, one sees only a monotonous stretch of bare rocks.[3] The same holds for the other two Great Lakes. Here and there patches of low, carpeting sponges cling to the reef. Beyond that, snail shells are abundant in some areas. Not to overdo this, the rocky reefs are nonetheless structurally complex, providing many habitats for the cichlids.

Another significant way the Great Lakes differ from the sea is that their surface levels rise and fall greatly over long periods of time. Contemporary visitors to the lakes comment on the fluctuations in water level over mere decades.[4] Between 1390 to 1860, the water surface of Lake Malawi was about 120 to 150 meters below its current level. During the early part of the nineteenth century, the water surface had risen to about 40 meters below current levels. At that time, the Ngonde king Mwangonde could still walk on dry land across the northern part of the present lake to Mwela, there to marry Mapunda. With further rise in the lake level, that walk is no longer possible.[5]

The Rift Lakes

The long, deep, and trough-like shape of both Lakes Tanganyika and Malawi, which makes them rift lakes, has profound implications for them as crucibles of evolution. Tanganyika is the deepest of the two and by far the oldest body of water of the three Great Lakes. Genetic evidence indicates the Tanganyikan cichlids started speciating around 12 to 14 million years ago,[6] well before the lake assumed a shape recognizable today. Lake Malawi is much younger, said to be 1–2 million years old.[7]

Even though in volume the deep Lakes Tanganyika and Malawi are among the largest in the world, their cichlids are able to use only a thin surface layer, so they are restricted mostly to the coasts. The bottom profile of these elongate lakes is steep, descending rapidly to great depths. That, coupled with the relative lack of wind, results in a nearly uniform water temperature to considerable depth. And the consequence is that water lying below a depth of 200 meters (about 610 feet) lacks oxygen at life-sustaining levels.

The steep rocky shores typically alternate with more gently sloping sand or mud bottom.[8] Additionally, the large rivers that mouth into the lake have deposited extensive reaches of soft sedimentary bottom.[9] Numerous pinnacles emerge as small islands rise close to the surface as habitable submerged banks. Reef cichlids are hindered from using deep rocky pathways to move between pinnacles and headlands because of the anoxic water there. Likewise, cichlids specialized for life over open bottom are cut off by steep rocky areas of bottom.[10]

Lake Victoria

Although its surface area falls a bit short of being the largest in the world, Victoria is the most spacious lake in the tropics and is well known as the source of the Nile River. Recently, and stunningly, its age as a continuously filled lake basin was established through the wonders of dating organic material with carbon-14. It has existed only about 12,400 years.[11] The lake was bone dry for some thousands of years before that, wiping out the entire fish fauna of the area. In a geological and evolutionary context, the lake is an infant.

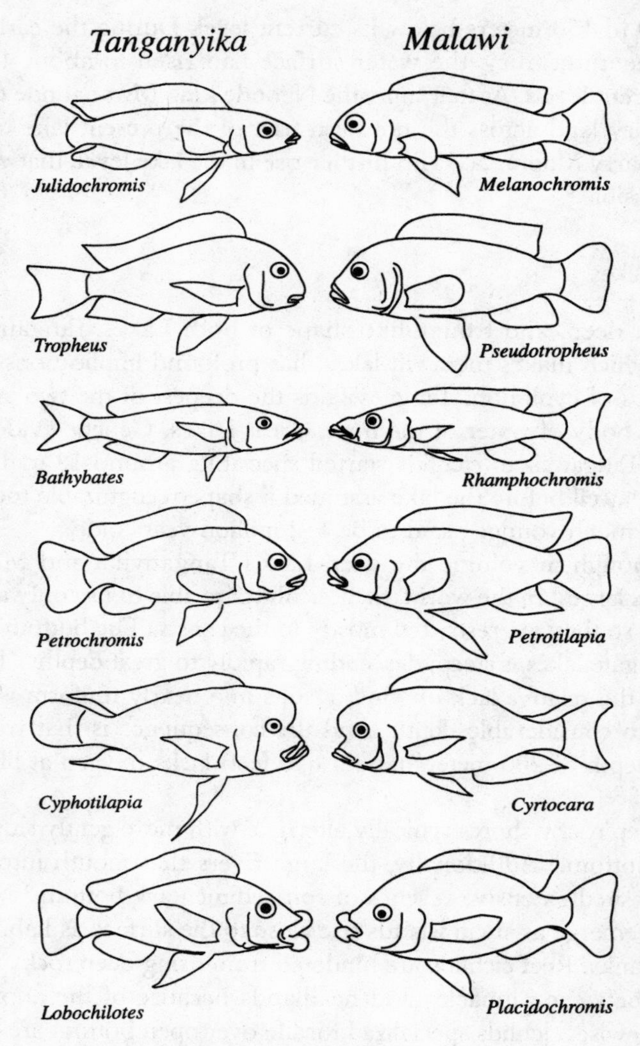

FIGURE 12.2 The independently evolved cichlids in Lake Tanganyika (left) and Lake Malawi (right) have converged on remarkably similar forms (courtesy of Axel Meyer).

Lake Victoria differs from the two rift lakes in other ways. It is a large shallow basin. Most of the water at its depths has, or has had, enough oxygen to sustain fish life and permit extensive movement. The oxygenation results from the shallowness of the lake, coupled with the massaging action of the wind. The shoreline of Lake Victoria is convoluted in some areas where small islands and rocky reefs abound. The environment, therefore, is nicely broken up into different habitats. Because of the complexity of its shoreline, and because of its gradually sloping bottom, relatively minor fluctuations in water level can alternately join or isolate habitats. Consequently, even though Lake Victoria is very young, it presents an exceedingly dynamic environment for the rapid radiation of cichlid fishes.

THREE RADIATIONS

Consider the metaphor of a celebratory rocket in the night sky. The rocket explodes into many starlets that fall in a shower. Imagine the expanding cloud of starlets as cichlids evolving into multiple species and the sky as one lake. Now each starlet bursts into a subset of starlets–more species. That is the way explosive speciation is visualized. Each substarlet's spray of bright embers, moreover, consists of particularly closely related species. And that leads to another metaphor, the species flock, which is a radiation among closely related species.[12]

The explosion metaphor conjures up visions of speedy speciation, and for good reason. That so many species have emerged in so little time is an evolutionary wonder. The radiations in the youthful Lake Victoria, in particular, have been explosive.[13] Ancient Lake Tanganyika, while rich in species, seems slow in comparison. And then we have Lake Malawi, the lake of intermediate age: one thousand species?[14] Perhaps. No one knows for sure how many species we are talking about in these lakes. It depends in part on how one counts species. Nonetheless, the number of species in each lake is remarkably similar, though conventional wisdom puts Lake Malawi first, Victoria second, and Tanganyika last.

Using the Same Mold

One aspect of the radiations at first befuddled biologists. Closely related species, for instance, in the same genus, seemed to be found in all three lakes. Closer examination, however, revealed that it was a deceiving resemblance. In reality, the same cichlids had been, so to speak, reinvented in each lake.[15] This is called convergence by evolutionary biologists.

Species that pursue parallel lifestyles in the different lakes have been shaped by natural selection to the same end.[16] Those that live on rock reefs and scrape algae, such as the mbuna in Lake Malawi and the trophiines in Lake Tanganyika, are ringers for one another. A few species in all three lakes (and in Nicaragua) have evolved great puffy lips that might at first suggest a common heritage, but the shared feature is merely convergence brought about by the same style of feeding. Cichlids in the three lakes that live in exposed situations, such as up in the water, or over open bottom, are elongate, high-speed swimmers and have similar mouths, teeth, and jaws.

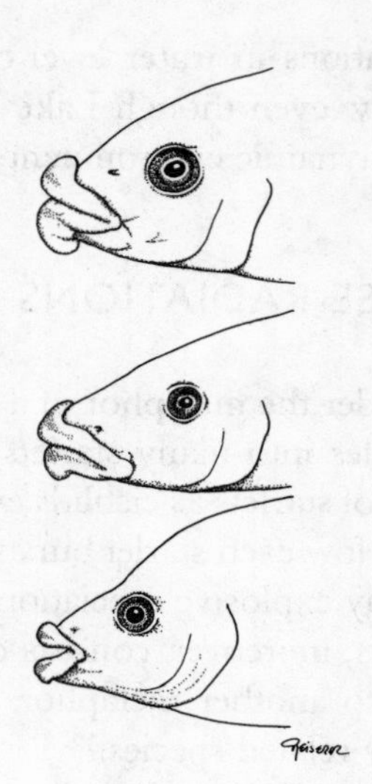

FIGURE 12.3 Puffy lips have evolved several times among cichlids. In the three examples portrayed here, the uppermost is from Lake Malawi, the middle one from Lake Tanganyika, and the bottom one from Lake Nicaragua.

That the radiations have been repeated in the three Great Lakes has fortuitously offered up a slightly fogged window into the process. They were all invaded by cichlids who radiated into hundreds of species. But the radiations are widely different in age. Lake Victoria appeared only "yesterday" in geological time, Lake Malawi is relatively old, and Lake Tanganyika is virtually primordial. As we proceed from youngest to oldest lake, a larger pattern emerges from the three radiations. Let's start with young Lake Victoria.

Youthful Lake Victoria

The situation seemed at first relatively simple in Lake Victoria, with only two tribes of cichlids. One tribe, the tilapiines, has been little more than a footnote because it has produced only two endemic species, *Oreochromis variabilis* and *O. esculentus*. The other is a huge cluster of endemic species. All are haplochromines, and they are the celebrities.[17]

From the original founders, perhaps as many as five hundred species have emerged to fill virtually all possible ecological niches, and in the incredibly brief span of about 12,400 years.[18] Equally astounding, evolving into so many different species has been accomplished with little change in the underlying genetics. When the mitochondrial DNA of fourteen species of nine endemic genera was analyzed, the genetic variation among the species was less than within our own, single species, *Homo sapiens*.[19] By extension, one could imagine that the entire cichlid fauna of Lake Victoria has less variation in mitochondrial DNA than do humans.

Compared with the relative lack of genetic variation, the divergence in feeding behavior is profound and must have happened with remarkable speed. The number and types of trophic specializations in Lake Victoria, as revealed by teeth, jaws, and supportive structures, parallel those found in Lakes Malawi and Tanganyika.[20]

Balancing that, the youthfulness of the Lake Victoria haplochromines is revealed in their greater homogeneity of appearance, compared with the cichlids in the other two Great Lakes.[21] The general body plan is more conservative, and reproductive coloration is bright but limited in scope. The color patterns fall into three main types for females: plain, brown blotches on orange to pink, and black blotches on brown to white. Males are a bit more diverse, as befits polygynous mouth brooders, and sort out into three general types: blue, red below, and red above.[22]

A further lack of diversification in Lake Victoria is seen in the retention of egg spots on the anal fin in all of the haplochromine species.[23] The mbuna of Lake Malawi commonly have egg-spotted anal fins, as do some

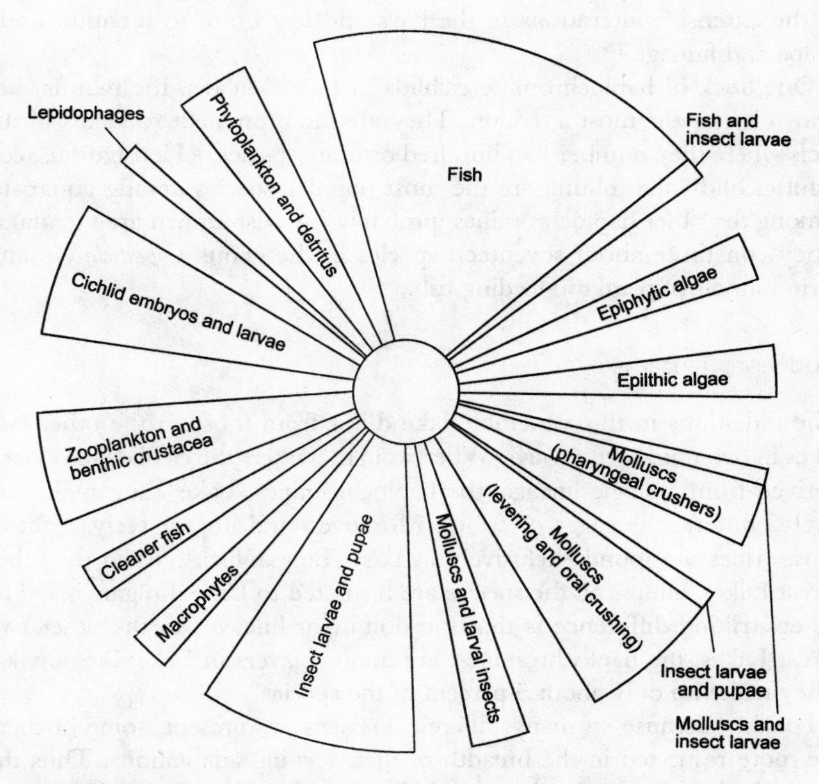

FIGURE 12.4 This food wheel represents the frequency of diets of cichlids in Lake Victoria. The width of each sector is roughly proportional to the number of species sharing that diet (from Coulter et al. 1986).

other types of haplochromines in that lake; however, many species there lack egg spots. Among the Lake Tanganyika mouth-brooding cichlids, egg spots on the anal fin are often either lacking or have been elaborated on the pelvic fins instead of the anal fin, but most of this is a consequence of different lineages of mouth brooders.[24]

Lake Malawi in Middle Age

Despite the physical similarity between Lake Malawi and Lake Tanganyika, the cichlids of Malawi are more allied in their haplochromine heritage to the cichlids of Lake Victoria. A genetic study revealed that a single lineage of haplochromines has produced almost all the species of cichlids in the lake, the other few being tilapiines.[25] Other scientists have suggested, however, that the Malawi haplochromines may consist of more than one lineage, though still haplochromines.[26]

The haplochromines have evolved to fill all the feeding niches and then, within feeding adaptations, have produced species flocks. This is reflected in the extensive alterations in the jaws and teeth, not to mention body shape and finnage.[27]

One flock of haplochromine cichlids in Lake Malawi, the mbuna, has drawn by far the most attention. They are the prominent residents of the reefs where they number two hundred or more species.[28] Herbivorous, colorful cichlids, the mbuna are the most popular species among aquarists. Among the other haplochromines, probably the best known are the utaka. They constitute about seventeen species in the genus *Copadichromis* and form schools of plankton-feeding fish.

Doddering Tanganyika

The radiations in this venerable lake differ from those in the other two lakes in two noteworthy ways. Whereas in the others the cichlids have been derived from a single lineage, the haplochromines, Lake Tanganyika has twelve different lineages or tribes. Moreover, and impressively, eight of those tribes are found exclusively in Lake Tanganyika.[29] As in the other Great Lakes, almost all the species are restricted to Lake Tanganyika. The other striking difference is that the dominant lineages in the other two Great Lakes, the haplochromines, are minor players in Lake Tanganyika. They make up only about 5 percent of the species.

Possibly because so many different lineages are present, some of them are more restricted in the breadth of their feeding adaptations. Thus the lamprologines are predominantly predators, although different genera and species have specialized on particular types of prey, and at least one, *Neolamprologus moorii,* is herbivorous.[30] Other tribes, predominantly mouth brooders, have taken over the herbivore feeding niche, most prominently the subtly colored *Tropheini,* which are distant relatives of the haplochromines.

The substrate-spawning lamprologines are the major players in Lake Tanganyika, with more than eighty species.[31] A more recent estimate reckons the lamprologines make up nearly 50 percent of all the species in the lake[32] and an impressive 75 percent of all the species inhabiting the littoral zone.[33] But as one aquarist friend puts it, they are just a bunch of dull brown fish. I find their behavior fascinating nonetheless.

One last cichlid bears mentioning, the huge bass-like piscivore that lurks near reefs over open bottom. This substrate-brooding cichlid, *Boulengerochromis microlepis,* is the only representative of a primitive tribe that is distantly allied to the tilapias. Although it is the only species of its tribe, this prima donna of a recent National Geographic film is an unforgettable and imposing inhabitant of the lake.

When instead of the total number species just the number of reef-dwelling species in Lake Tanganyika is compared with those in Lake Malawi, the count is similar. Lake Tanganyika has fewer total species of cichlids because of its smaller number of open-bottom species; the open-bottom species are also more generalized and less differentiated than those in Lake Malawi.[34] Apparently, the depth of the anoxic water fluctuates more in Lake Tanganyika, often rising up into relatively shallow depths. That makes it difficult to evolve specializations for life at different depths, although some such specialization is apparent.[35]

Age and Speciation

A reasonable starting assumption about the pattern revealed by comparing the lakes would be that the older the body of water, the more species it should have evolved and that the number of lineages, or tribes, should likewise be greater. That generalization holds for the number of lineages, but not for the number of species, though even that is debatable.[36] The obvious trend with passing time lies not with the number of species but rather with the degree of differentiation.[37]

Lake Tanganyika has the fewest cichlid species, but they are by far the most heterogeneous, which is in part attributable to its numerous lineages. According to the anatomist Melany Stiassny, just one of its tribes has "done it all": "In terms of general divergence of appearance, body form, and size range, lamprologine diversity rivals that found within the entire African cichlid radiation."[38] Another perspective is provided by Christian Sturmbauer and Axel Meyer who are now molecular biologists but got their initial training in behavior and ecology. They assert that the mouth-brooding Ectodini, with about half the number of species as the Lamprologini, "are probably the ecologically, morphologically, and behaviorally most diverse tribe of Tanganyikan cichlids."[39] Not to worry. Both tribes are wonderfully diverse, though in different ways.

Considering the immense differences in age, the three Great Lakes in East Africa have an intriguingly similar number of species. That suggests speciation within each lake was rapid in the beginning but later leveled off, hitting a ceiling or even declining as the community of fishes matured.[40] That the oldest lake has the fewest species raises the possibility that Lake Tanganyika may once have had many more species but that over time some species disappeared.

Progress Through Extinction

Reflect on what might happen when what was originally one species has evolved in isolation into two neighboring but reproductively divorced species. If they come into contact again, chances are high that they would compete for some limiting resource.[41] It might be for a refuge or breeding site,[42] but most of the research has been directed to competition for food.[43]

As competition for food becomes more intense, deeper differentiation leads to refinement of specializations. Eventually, the possibilities for feeding in different ways are exhausted. With the passage of time, the number of species declines as the less-efficient ones become extinct.[44] Over time, the most-able species gradually disperse in stepping-stone fashion to more and more distant habitats, where they either eliminate competitors or are themselves effaced. As this process approaches culmination, the number of species is reduced to some relatively unchanging or slowed state as the most successful species become widely distributed.[45] In some circles this is referred to as evolutionary stasis.[46]

This pattern becomes apparent when comparing the three Great Lakes. Speciation is rapid at first, but eventually the number of species stabilizes or even decreases with the passage of time. However, that conclusion depends in part on how species are counted. The younger the lake, the more difficulty in deciding whether differently colored populations of a kind of cichlid are simply geographical variants of one species or are full species. I'll discuss this in more detail a bit later on.

Crypto-Radiations in the Great Lakes

Radiations are not confined to reef dwellers. Cichlids living over the open bottom, or at considerable depths, in both Lakes Malawi and Tanganyika have their own radiations and along trophic lines.[47] In Lake Malawi, for instance, the genera *Diplotaxodon* and *Rhamphochromis* have evolved a rich assemblage of barracuda-like species that feed variously on other fish and on plankton. Some of the species have become truly pelagic.

I have shamelessly ignored those fishes because so little is known about them, but they do present special challenges to the question of how radiations are produced inasmuch as they are so mobile and have the potential, perhaps unrealized, to disperse widely throughout the lake.

THE SPECIES PROBLEM

Evolutionary biologists spiritedly but cordially debate the definition of a species.[48] The roots of this controversy lie in history. The first species concept was essentially biblical: A horse is a horse, a pigeon is a pigeon; if they can interbreed with another species, they are not separate species. That approach carried over to the way scientists first handled species. Whenever a new species was described, a type specimen was deposited in some museum, and that practice persists.

That led to the typological conception of a species. It infers that each species is sharply defined and fully separate from other species. The more modern and dynamic view is that the species connotes a genetic assemblage of individuals.

The contemporary theoretical definition of a species, more properly the postulational definition, stipulates which conditions must be fulfilled before one can conclude that a group of animals constitutes a valid species. For our purposes, a species is (1) a population of individuals who (2) ordinarily mate only with one another and thus (3) share a common reservoir of genes, called a gene pool.[49] This is a slimmed-down presentation of the biological-species concept.

In application, a collection of specimens from, say, one reef in a lake, is sorted into alike and not alike, taking into consideration recognizable sexual and age differences. The different lots are regarded as separate species because even if they can potentially interbreed, they do not. Since they occur in the same place, they are said to be sympatric.

Problems arise when two similar but somewhat different populations of what seem to be the same species exist at different places and are physically separated. Barriers to movement between populations of a cichlid fish, for instance, might be deep water or long stretches of open bottom between reefs at opposite ends of Lake Malawi. Because they do not occur together, the populations are said to be allopatric.

Two cutoff populations on neighboring reefs present a problem. Would fish of the two populations readily interbreed if the reefs were joined? If they would, then they are only geographical variants of one species. If not, they are different species. In practice, this is difficult to determine, although one approach is to test experimentally whether, given a choice, one species responds selectively to fish of its own population or not.

The term species is an abstraction that we impose on nature. In reality, aggregations exist in myriad states, ranging from separate populations of the same species to discretely different species. Speciation is an ongoing process, and one that takes time; we study snapshots in the progress of the process.

HOW SPECIES GET STARTED

In the classical model for how new species arise, some members of the species become cut off from the main population, as when a few individual cichlids somehow disperse to a different island or lake. This founder population takes root there and, in the fullness of time, becomes different, most likely because the environment to which they adapt differs in some way.[50] Natural selection favors perfecting the behavior associated with mating, so obstacles to mating with other species are expected to evolve.[51] Such behavioral obstacles to interbreeding appear to be strengthened when the two species come together. This is known as the allopatric model of speciation.

Behavior as the Entering Wedge

Behavior both leads the way and refines the process of speciation. Dispersing to new habitats is a behavioral act. Even if we assume that the next step is specializing on a type of food, that is first and foremost a behavioral innovation. Changes in morphology follow.

To develop but one of the many possible examples, imagine the course of natural selection when some deep-bodied, darkly colored reef cichlids, with small down-turned mouths for scraping *Aufwuchs,* disperse to a reef where the current brings an abundance of plankton, but *Aufwuchs* is scarce. If they are to survive, the fish must swim up into the water to capture the nutritious plankton. Snaring mobile prey favors suction feeding and strong swimming. That selects for more elongate individuals with streamlined finnage and sharp forward vision.

Camouflage becomes important because they become more exposed to predation away from the reef. Bright hues are shed, and the color pattern converges on a silvery blue. The sheen results from the retention of crystal-like pigments in the scales, transforming the fish into a mirror. That makes them difficult to see because they reflect the color of the water around them.[52]

This evolutionary path can take place independently at different places within a lake, producing multiple species of planktivores. What, then, happens when they meet? Competition favors a partitioning of the trophic environment. In Lake Victoria, several species of planktivores have behaviorally partitioned even the open water where they capture their prey. They have specialized in the depths where they feed, distance from shore and type of bottom under them, and even in the time when they feed.

The trophic morphology of planktivores departs from that of bottom-dwelling cichlids in a major and general way. Cichlids that forage on the bottom take a great variety of foodstuffs, and this is reflected in the profoundly different trophic equipment among their species.[53] In contrast, the trophic morphology of planktivores is much alike from species to species, as it must because they all feed on plankton.[54] They have divided up the environment through behavioral specializations, not through morphological differences.

As others have noted about speciation, two different things seem to be going on. An early stage is behavioral separation into clear differences in type of food eaten, and that probably happens soon after the invasion of a lake. That step is accompanied by profound changes in trophic morphology and further changes in behavior. The second step is radiation within each type of feeding specialization. That results in a flock of closely related species that feed in much the same way. This may be accompanied by changes in behavior for more refined partitioning of the environment, while retaining the same basic trophic morphology.[55]

Jaws Promote Speciation

Much has been made of the highly adaptable teeth and jaws of cichlids, which enable them to invade new surroundings, feed on whatever is available, and then evolve closer specialization. First, a reminder of what is so special about cichlid pharyngeal jaws.

In most teleost fishes, these jaws are limited in their movement because the upper and lower pharyngeal jaws are bound together by ligaments. They work as a unit, and so are said to be coupled. In cichlids, these jaws have become decoupled.[56] The pharyngeal jaws, together with changes in the outer jaws and their teeth, afford cichlids great flexibility in the food taken.[57] Thus different cichlids can manipulate and process a great variety of food items ranging from crushing tough snails to shredding plant material and packeting scales to macerating fish prey; the list is in actuality much longer.[58]

So when a generalized cichlid, such as a typical *Haplochromis,* enters a new environment, it is equipped to exploit whatever kind of food it encounters. However, a trophic jack-of-all-trades pays a cost, because when a wide variety of foodstuffs is taken, no one fare is processed with high efficiency. If a single food resource is both nutritious and locally abundant through time, individuals who can handle that manage better than their competitors. Often, members of one's own species are the chief competitors, especially when food becomes scarce. Therefore, natural selection favors individuals best equipped to deal with the accessible food.

For instance, where snails are available, individuals with sturdy pharyngeal jaws that bear pavement-like teeth, called molariform, are favored because they can efficiently crush and process hard-shelled snails.[59] But when the most accessible food consists of relatively soft insect larvae, then the more delicate papilliform pharyngeal jaws are more effective.[60] In reality, the pharyngeal jaws of different species are complexly different, often carrying a combination of molariform and differently shaped papilliform teeth. Moreover, jaw structure varies slightly but continuously among individuals within a population providing the raw material on which natural selection works.

Some species go a step further. Different individuals have discretely different pharyngeal jaws. And the differences are not continuous, so they are said to be polymorphic. One individual may have molariform jaws with more robust supporting bones, whereas a second may have papilliform jaws with more delicate supporting architecture. In some instances, the difference is genetic.[61] In others, the difference is induced: An individual may develop either the robust or the delicate structure depending on diet.[62]

Despite increasing specialization, the diets of different species commonly overlap.[63] Cichlids retain sufficient trophic flexibility to track the changing abundance of type of food and to take advantage of sudden outbreaks of a highly desirable food for which the species is not ideally adapted,[64] as when they invade a new habitat.

In all three of the Great Lakes, the range of trophic specializations is remarkably alike, albeit with some differences.[65] They have converged on the same trophic adaptations. As in the other two Great Lakes, in the young Lake Victoria the number of different feeding specializations packed into a given reef is spectacular.[66] The head end of the fish seems to evolve first. The teeth and jaws are highly differentiated, but the body plan is more conservative than seen in the other two lakes. Compare this with the oldest lake, Tanganyika, which displays the most diverse array of body plans.[67]

I have dwelt at some length on the pharyngeal jaws of cichlids because of the information about them that is available from intensive research. The findings illustrate how speciation works. The fish split up the environment behaviorally. The ability of their jaws to handle diverse foodstuffs lets them exploit new feeding niches. Selection then produces specialization.

This section started with a brief account of the allopatric model of speciation and progressed to a discussion of how the process of speciation gets started. The actual process of speciation is difficult to observe, but the results are not.

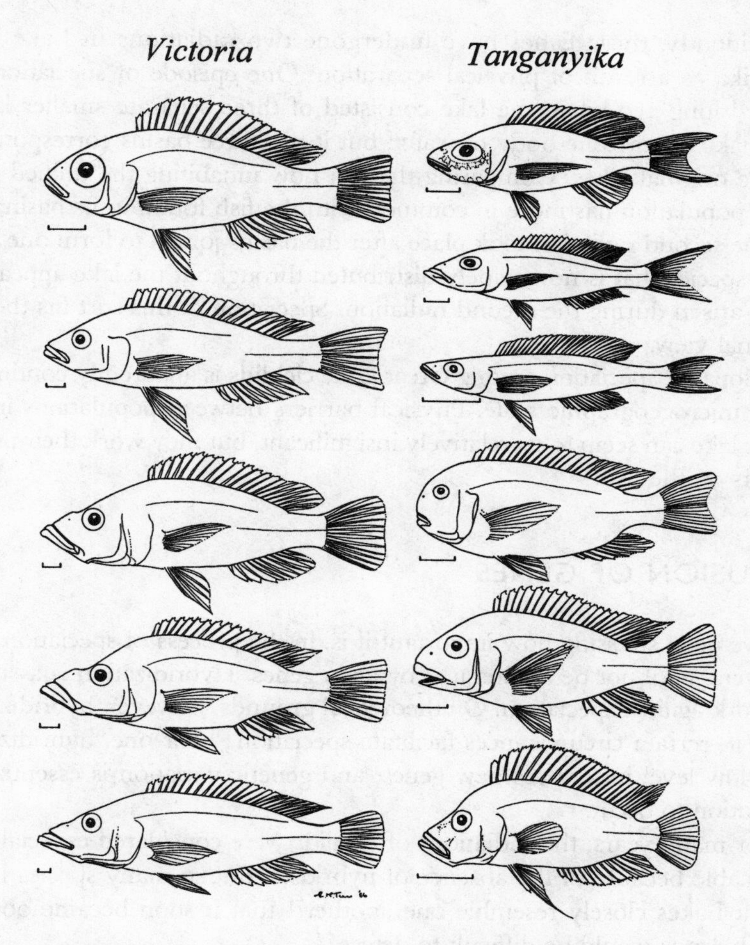

FIGURE 12.5 The cichlids in Lake Tanganyika (right) have evolved great diversity of shape in both body and head. In contrast, the cichlids of Lake Victoria have evolved comparably different heads, but their bodies have remained relatively unaltered (after Coulter et al. 1986).

Does the Model Fit?

Do populations of cichlids in the Great Lakes fit the allopatric model? Let's look at the results of some investigations using powerful genetic techniques. A large-scale analysis of some cichlids in Lake Tanganyika provides a good case history, one fully congruent with the allopatric model of speciation.

Mitochondrial DNA was used to characterize many separate populations of cichlids inhabiting the wave-washed rocky shores of Lake Tanganyika. Six species of *Tropheus* were analyzed plus their close relatives, two species of *Simochromis,* and four species of the *Eretmodini,* the goby cichlids.[68]

Evidently, these fishes have undergone two radiations in Lake Tanganyika as a result of physical separation. One episode of speciation occurred long ago when the lake consisted of three separate smaller lakes. The lake is now one body of water, but it has three basins corresponding to the original lakes. Comparing the fish now inhabiting the unified lake, each population has more in common with the fish found in its basin.

The second radiation took place after the basins joined to form one lake. One species that is now widely distributed throughout the lake appears to have arisen during the second radiation. Speciation of this sort fits the traditional view.

Allopatric speciation among Great Lake cichlids is apparently continuing on a microgeographic scale. Physical barriers between populations in the same lake can seem to us relatively insignificant, but they work their potent effects on the fish.

INFUSION OF GENES

I have been stressing how important it is, in the process of speciation, that the gene pool not be adulterated by alien genes. Hybridization is assumed to work against speciation. On theoretical grounds, however, hybridization may in certain circumstances facilitate speciation.[69] For one, hybridization at a low level introduces new genes, and genetic variation is essential for evolution to occur.

For many years, the radiations of cichlids were considered especially remarkable because of the absence of hybrids.[70] But so many species in the Great Lakes closely resemble one another[71] that it soon became obvious that hybrids would be difficult to detect.[72]

We know aquarists regularly hybridize cichlids in captivity, though under disturbed conditions, and sometimes hybridize radically unrelated species.[73] If not too unrelated, the hybrids are as fertile as their parent species. Because of this, some cichlid specialists have suggested that hybridization may be involved in speciation,[74] though this issue remains little studied.

When the cichlid *Cynotilapia afra* was transplanted from northern Lake Malawi to an island at its southern end, some of them interbred with the existing *Metriaclima zebra,* producing intermediate hybrids. The two species occur together naturally at the northern end of the lake but do not interbreed there,[75] which may be telling us something about the evolution of barriers to mating. Examples of naturally occurring hybrids are also known for cichlids in places other than the Great Lakes.[76]

A close analysis of the genetic structure of nine species of mbuna from Lake Malawi, however, showed that "hybridization has not been rampant during the speciation of the flock." The researchers did allow that hybridization may have played a role in the early evolution of the flock, though it is rare today.[77]

FINDERS KEEPERS

Cichlids do not automatically take over lakes when they gain access to them. They are relatively minor players in some other prominent but lesser lakes in East Africa. Lake Turkana, for example, has thirty-nine fish species, only seven of which (18 percent) are cichlids. For Lake Albert, the numbers are nine of forty-four species (20 percent).[78]

The difference may lie in the extent to which the other inhabitants of the lake got there before the cichlids and adapted well to the lake. They are competitors and conceivably could preempt ecological space. Cichlids, therefore, may enjoy much of their success in radiating in the Great Lakes simply because they got there early in the game. Those lakes contain many other species, and some of them have produced minor radiations in other kinds of fishes, but nothing approaching the cichlid radiations.

Lake Tanganyika, for instance, is home to about seventy-five species of noncichlid fishes, in twenty-one families, that dwell in the lake proper.[79] The pelagic realm is dominated by two species of herring relatives and four species of Nile perches. Mini-radiations have evolved in three families, two of them catfishes. The spiny eels, the Mastacembelids, have produced a mini-radiation. Their feeding behavior is worth noting because of their interactions with cichlids: At least four of the most common species of mastacembelids specialize on eating the eggs of cichlid fishes.[80]

The general conclusion about the noncichlid fishes in Lake Tanganyika, in stunning contrast to the cichlids, is that absolutely none has developed local forms.[81] Most of the species are found in all parts of the lake, though some are confined to just one region. That suggests those species are so mobile that the opportunity for local speciation is nil. Being so mobile, many of them, especially the cyprinids, migrate out of the lake to spawn.[82] But beyond that, the cichlids differ in ways that suggest they are preadapted, and better adapted, to life in a lake.

MADE FOR LAKES

The ostariophysan fishes are attuned to the riverine situation in which they evolved, and they dominate flowing water throughout the world.[83] Something about their basic body plan gives them an advantage in rivers. They have relegated riverine cichlids to a minor role. On the other hand, the ostariophysans are less well prepared to exploit lakes.

Moreover, seasons have a greater impact on riverine than on lake-dwelling fishes. The ostariophysan fishes are adapted to spawning large numbers of eggs in one burst at the height of the breeding season, which is usually associated with the annual pattern of rainfall; they tend to retain this life-history trait even after colonizing a lake. Likewise, breeding by cichlids in rivers tends to be confined to the dry season because the fry are vulnerable to strong flow. When cichlids gain access to a lake, in contrast, they are liberated from rapidly flowing water and can reproduce year-round and do so (though some seasonal effect is present). As a result, they can quickly increase their numbers.

The body plan of cichlids is a marine heritage. Physically, lakeshores are more like marine coasts than they are like rivers. The general body plan of cichlids appears well suited to lakes.[84] Consider just their fins. The elevated pectoral fins, together with the forward position of the pelvic fins, provide superior maneuverability in tight quarters. Therefore, when cichlids gain a foothold in a new lake, they are well equipped to compete with any ostariophysan inhabitants, especially if the cichlids arrive before the ostariophysans, or other kinds of fishes, have become well established.

Some may wonder why I do not mention here the advanced jaws of cichlids. The reason is that as important as the jaw structure is to cichlid speciation, that does not confer dominance in rivers. Now that we have the cichlids in a lake, we need to consider the issue of their vagility, because the tendency to stay in one place, once there, promotes separation into unique gene pools.

THE PUSH-PULL OF GETTING AROUND

If cichlids are to speciate, they need to colonize new habitats. The haplochromines may be specialists in doing this. Much is made of the singularity of the radiations of haplochromines.[85] About two-thirds of the species of cichlids in the Great Lakes are haplochromines, whereas only a paltry fifty species or so of haplochromines are found in African rivers. Perhaps they are simply specially adept at finding and invading new lakes. They are the primary colonizers of other East African lakes as well.

To evolve into different species, on the other hand, cichlids shouldn't be too vagile, or else they could never adapt to a different habitat. First, let's look at the cichlid's ability to disperse and how that has been assessed:

Housing Projects

Rock-dwelling species need a structured habitat where they can seek refuge and reproduce. In Lake Malawi, this was demonstrated for juveniles by setting out rocks that offered small hiding places on rocky reefs in spots where shelter was otherwise not available. Juvenile cichlids quickly settled in and became territorial, excluding their brethren. Mortality was high, however. Less than 10 percent of the juveniles survived three weeks after being released from their mothers' mouths.[86] Evidently, cichlids can disperse as juveniles within a reef habitat, but they risk being eaten when they settle in places with inadequate shelter.

Some cichlids travel yet further. In Lake Malawi, a cluster of man-made reefs was set out on sandy bottom a full kilometer from the nearest inhabited rocky area. Eighteen of the many species that moved in over a period of five years had never been seen venturing out over sand bottom. Of eighteen colonists, ten were rock-dwelling mbuna, and nine of those set up house and reproduced in the new dwellings.[87] A prominent newcomer was *Metriaclima zebra,* which is one of the most widely distributed mbunas in Lake Malawi.[88]

Evidence of another sort revealed that dispersal has its limits. A storm in Lake Tanganyika in the early 1970s exposed a new shallow rocky reef habitat that is isolated by a 15-kilometer (9-mile) stretch of sandy bottom. Years later, no appropriate cichlid inhabitants, such as *Tropheus* or goby cichlids, had managed to populate the new reef.[89] Remarkably, no one has reported visiting the islands in recent years to survey which cichlids, if any, have managed to get there.

Home Bodies

Parental behavior may play a crucial role in speciation because of the site fidelity associated with drawn-out family life.[90] The fry spend much of their early days with their parents at one place before setting out on their own. That provides both a prolonged opportunity to learn the site and the time to become large and sedentary. Despite documented examples of limited dispersal, young fish probably do not move far from their "home,"[91] and that tends to preserve the local gene pool.

More problematical is the situation with those species adapted to an existence over open bottom or up in the water feeding on plankton.[92] Although they possess great mobility, they are generally not widely distributed in any one of the Great Lakes. Stretches of inhospitable habitat could be barriers, but parental behavior may hold the answer.

When they reproduce, open-water cichlids evidently return to their natal habitat.[93] Those open-water cichlids that drop and catch their eggs in midwater would seem fully emancipated from the bottom and free to roam. Yet, to mate they gather in embayments or along the faces of underwater cliffs. Further, open-water mouth-brooding females bring their fry to the reef when they release them. The fry require a reef environment even when the adults do not.

Thus even the most seemingly liberated cichlids retain the ancestral trait of breeding at a certain type of location and perhaps return to the specific spawning area whence they came. The fry might imprint on the habitat they first experience, as do salmon.[94] That would provide a behavioral mechanism maintaining a degree of separation between populations, even of open-water cichlids.

Oddly, the one example showing that marked fish return to the same place to spawn in successive seasons is a tilapia, *Oreochromis variabilis,* in Lake Victoria.[95]. I say oddly because the tilapias are exceptional in showing little or no tendency to speciate in the Great Lakes,[96] although some local differentiation can be found among tilapias in Lake Victoria.[97]

GENES FLOW, BUT NOT MUCH

Reef cichlids can disperse over open bottom. But how many of them successfully do so? Genetic analysis of neighboring populations of reef cichlids in Lake Malawi suggests that successful movement between them is too low to matter.

A team of scientists concentrated on four species of mbuna in Lake Malawi inhabiting four rocky reefs. The reefs were close together, ranging from 700 to 1,400 meters apart, or roughly one-half to one mile, with open sand bottom in between. The populations were large in each instance. A sophisticated method of genetic analysis was used, microsatellite loci (these are numerous highly variable elements of DNA and are thus sensitive indicators of small genetic differences).[98]

Within the four species of mbuna, each population differed genetically from nearby populations of its own species. The differences were small but of the same order as those found between other, clearly distinct species of mbuna living together, that is, sympatrically. The scientists detected a trickle of genes flowing from one population to the other; a few mbunas had managed to cross the open bottom to reach a neighboring reef, but that had precious little genetic impact.

The populations of reef-dwelling mbuna separated by small stretches of sand bottom are therefore differentiating into populations as genetically different as indisputably distinct species of mbuna. The scientists suggest that the thousands of different isolated populations of each species conceivably have the potential of becoming, or even of being, distinct species.

The reefs where the four unique species of mbuna currently live were high and dry two hundred to three hundred years ago. Those mbunas apparently developed their genetic differences within that short span of time, equivalent to about sixty-seven generations. The cichlids in the Great Lakes of Africa are therefore continuing to rapidly evolve, and they have been doing so since the creation of the lakes, as other evidence suggests.

COLORFUL GENE FLOW

Populations of the many species in the different Great Lakes vary in color pattern from place to place,[99] but morphological differences are slight to nonexistent. The gnawing question, then, is do the color differences reliably indicate genetic differences? The answer is mostly yes and a bit no. At least, that is what one penetrating, investigation revealed.[100]

The widely distributed mbuna *Labeotropheus fuelleborni* resides in Lake Malawi in the shallowest, waved-washed areas on rocks where it scrapes algae for food. Where analyzed at the southern end of the lake, separated populations displayed a continuous gradient of coloration running north and south.

Microsatellite evidence showed a corresponding genetic gradient. For instance, *L. fuelleborni* inhabiting long continuous rocky shores were genetically the same at opposite ends of a large reef, and the color was the same everywhere, as well. But between reefs, gaps greater than two kilometers or water deeper than 50 meters corresponded to abrupt, though small, changes in genes and coloration. Further, isolated neighbor populations were most alike in coloration and genes, and the further apart the populations, the more different they became. The genetic study also disclosed that a few fish must be moving between populations.

The counterevidence, that against congruence of color and genes, sprung up when researchers examined populations of *L. fuelleborni* inhabiting much more isolated and remote insular reefs. One would anticipate marked color differences. Puzzlingly, the coloration of the isolated populations was much like that of the coastal populations although they differed genetically. Genes and color in this case are not congruent.

The preponderance of evidence therefore supports the proposition that, first, cichlids are sufficiently vagile to settle new habitats. Second, despite their vagility, they tend to say put and become genetically distinct from other local populations. So far, all the evidence is consistent with the well-established allopatric hypothesis of speciation.

CHOOSE A MATE AND SPECIATE

Some scientists attribute the explosive radiations of cichlids to the part played by coloration in mating behavior in speeding up the process of speciation.[101] That sexual selection might accelerate speciation is not in conflict with the allopatric model.[102] The problem arises with this hypothesis when it is proposed as a means of producing sympatric speciation, that is, in the absence of spatial barriers or timing of activities. But first things first. What role does sexual selection play in speciation? This line of thought stems historically from the early work of Humphrey Greenwood on Lake Nabugabo, a satellite of Lake Victoria.

Lake Nabugabo

Lake Nabugabo is a small, narrowly separated lake that budded off from Lake Victoria less than 4,000 years ago. Each of the five endemic haplochromine species in Lake Nabugabo have obvious counterparts in Lake Victoria, which are most likely their progenitors. What makes this situation so special is that Greenwood contended that the five new species did not differ in any adaptive way from their parent species in Lake Victoria. The only difference lay in the breeding colors of the males; female coloration was unchanged.[103] That no adaptive differences exist, however, is a tenuous assumption.

If the two lakes were again joined, the daughter species would encounter their parent species. Then the look-alike females from the two lakes would maintain the difference by choosing as mates only those males having their species' coloration.[104] Thus the daughter species in Lake Nabugabo would have evolved behavioral obstacles to mating while in isolation.

Published in 1965, this study has become a repeatedly cited and hence classic example of nonadaptive speciation brought about by a change in male coloration and female mate choice. Regrettably, in the thirty-five years since its proclamation, no one has done the simple experiment of testing whether females would actually discriminate between their own males and those of the parent species.

The main point so often drawn from this example is that interbreeding has been blocked through chance changes in male coloration. In some situations, that would accelerate speciation.

Cranking Up Speciation

Mate choice in highly polygynous species in general has been suggested as an accelerator of speciation.[105] To illustrate, consider the consequences of one exceptionally colorful male mating with several females. His genes for color would be rapidly propagated and might thereby hasten evolution. The proponents of this view for cichlids have in mind polygynous mouth brooders.[106] But as I have shown, those species are in reality typically polygynandrous. Females consequently dilute the polygyny effect by mating with multiple males themselves.

Turning back to the role of sexual selection in speciation, we lack clear answers. That mate choice is important, indeed crucial, to maintaining the integrity of a species has never been disputed. Research on cichlids has shown repeatedly that sister species properly choose mates of their own species when given a choice.[107]

The effect of male color as a barrier to interbreeding was shown by experiments with some Lake Victoria cichlids. Under normal lighting, females chose the appropriately colored males for spawning.[108] But when color cues were erased by manipulating the illumination, species discrimination disappeared,[109] proving that color provides indispensable information for species recognition.

"One Fish, Two Fish, Red Fish, Blue Fish"

That sexual selection is the driving force proposed by Ole Seehausen to explain speciation in polychromatic cichlids in Lake Victoria. This case study centered on three closely related, sister species. But first a general comment about the coloration of cichlids in Lake Victoria. Polychromatism in Lake Victoria differs somewhat from that in the other Great Lakes in that it is limited to just a few types (perhaps another indication of the recency of its radiation). All the Lake Victoria species can be categorized into three general types of morph for each sex: Females have dark blotches on orange, or black blotches on white, or are plain. Males are predominantly blue, or red dorsally, or red ventrally, according to species. Some species are polychromatic, manifesting various color morphs within the same population, whereas other species are limited to one or another coloration.[110]

Pairs of species that differ by color are common. The predominant color of the males in one species may be blue while in the other it may be some variant of red through yellow.[111] That suggests one polychromatic species may have given rise to two species, each of them having just one color of male. In support of this, some species are polychromatic at one location but exist at different places as separate sister species bearing one or the other coloration.

Such color morphs have been said to be vestiges of an "ancient" polychromatism that persists, providing the raw material for repeated episodes of speciation through female choice.[112] Further, goes the argument, because of this disposition to produce what is deemed a different type of polychromatism, the haplochromine cichlids in Lake Victoria are singularly equipped to speciate the fastest, hence radiate explosively.

In an experiment done by Seehausen, female choice verified the validity of three species, each with males differing in coloration. The cichlids were taken from separate reefs in the lake, so they were allopatric. To simplify, assume the polychromatic species has males of two of the color types, A and B. In a second species, male coloration is only of type A, and in the third, only type B. In the A and the B populations, females chose only the male color corresponding to their species' coloration. "This may explain rapid speciation by sexual selection on male coloration."[113]

Unfortunately, mate-choice tests were not done on females from species AB. If they do not discriminate between, say, type A males from population A, and A males from their own AB population, then where is the barrier to interbreeding? Further, in the AB population, how do females choose among males A and B?

Splitting Up but Staying Together

The debate becomes intense when sexual selection is proposed as a mechanism for speciation in sympatry. Most evolutionary biologists are intensely skeptical of claims of sympatric speciation and insist that the allopatric model must first be shown to be inadequate before trying to demonstrate how sympatric speciation could happen.

With regard to the cichlids in the Great Lakes, this controversy was obscured in the past by inconsistent usage of the term sympatry. That so many species of cichlids sprung up within one lake was accepted in some circles as prima facie evidence of sympatric speciation. But as we've seen, the vast Great Lakes are rich in different habitats, providing the conditions for within-lake allopatric speciation.[114]

The crux of the debate surrounding sympatric speciation is whether fish of the same species, who regularly encounter one another in their daily affairs, can split into two sister species. That can happen only if different natural-selection forces favor each type of fish and gene flow between the two types is cut off. These strictures are difficult to meet, especially in a sizeable population.[115]

What has been offered as evidence for sympatric speciation is also consistent with allopatric speciation. For instance, in Lake Victoria, closely related species living apart resemble one another more than where they overlap.[116] That is precisely what one would predict from an allopatric model: Differences may be enhanced when the two new species again come into contact.[117]

So far, only one piscine example of sympatric speciation has been convincingly demonstrated, the evolution of the kokanee from the sockeye salmon.[118] That situation is completely different from the one in which the Great Lake cichlids find themselves, and hence it is an inappropriate model. No one has demonstrated that female choice of mate has produced, or is producing, separate species of cichlids from within an interacting population. Sexual selection, pure and simple, has nonetheless been proposed and repeatedly endorsed, though often with some hedging, as the ultimate behavioral mechanism driving or refining sympatric speciation among cichlids.[119] Opposition to this proposal has also been vigorous.[120]

Theoretical Sympatric Speciation

Recent mathematical models of sympatric speciation have established that it is theoretically possible, given the assumptions of the models.[121, 122] Some of the models produce two sister species even in the absence of discrete polymorphism in males, simply requiring that the males vary widely in coloration.[123] Another model makes the interesting prediction that sympatric speciation will be accelerated in new environments where predators are scarce.[124]

The seductive feature of a model is that its starting assumptions tend to predetermine the outcome. When the models support the hypothesis, that does not mean the answer has been found. Other models, with other premises, might produce the same or even better outcomes. As only one example, the models mentioned here assume equal Darwinian fitness for the two-color morphs, a questionable assumption.[125]

Where does this leave us? Perchance mate choice could drive sympatric speciation, especially when some aspect of habitat partitioning is involved.[126] That conclusion, however, is not widely accepted and remains in the realm of controversy.

This entire line of thought suffers further from a provincial outlook on cichlid speciation. The debate, if we can call it that, deals entirely with mouth-brooding haplochromine cichlids, mostly in Lake Victoria, but also in Lake Malawi. Those species fit well into the hypotheses that are based on sexual selection in highly dimorphic species. Does the sexual-selection hypothesis apply only to those cichlids, or is it more general? Radiations have occurred in other lakes, and these reveal a different answer.

THE NEGLECTED RADIATIONS

Our story would be incomplete if we did not consider other radiations of cichlids. The others do not fit well the proposed pattern of speciation through sexual selection. The first one is an instance of radiation propelled by a special ecological situation that excludes other kinds of fishes but suits the lamprologine cichlids. The second radiation has us returning to Lake Tanganyika. The third and fourth examples are found in two small crater lakes.

Life in the Fast Lane

Cichlids are characteristically minor players among the fishes inhabiting rivers. The prevalent fishes there are the ostariophysans. Cichlids are, as we've noted, a minor though important element in rivers.

A stunning reversal of the subordinate role of cichlids in tropical rivers has evolved in the cataracts of the lower Zaire River.[127] Twenty-five fish families are found in that river, and in most places the ostariophysans predominate. In the extensive rapids of the Zaire, however, cichlids have twice as many species as all the other families combined. Likewise, they contribute more than a third of the individual fish there. These cichlids are lamprologines, here *Lamprologus* and the somewhat related *Teleogramma* and *Steatocranus*. All are elongate, negatively buoyant, and live in crevices and burrows under rocks, all of which are adaptations to resist being swept away by strong currents. These fishes are dully colored and are relatively isomorphic substrate brooders.

The other cichlid lineages inhabiting the Zaire are haplochromines and hemichromines. They are absent from the rapids, and they make up only a fraction of the fish fauna.[128] Apparently, their more compressed and relatively high body ill equips them to cope with torrential flow. In contrast, the elongate morphology of the lamprologines and their typical close association with the bottom preadapted them to refuging in crevices among rocks, the only way to survive in the cataracts. Thus although lakes seem propitious for cichlid speciation, a riverine environment that closes out competitors also seems favorable.

Lake Tanganyika

Recall that in ancient Lake Tanganyika, lineages other than the haplochromines, minor players there, dominate the radiations. The most prominent tribe, prevailing on the reefs and over shallow sand bottom, is the Lamprologini. Yet they are substrate brooders, many of them monogamous. Sexual dichromatism is muted within the group, even among species that are haremic.[129] Female choice is mostly a matter of nest site, not appearance of the male. Radiations are hence not restricted to dichromatic polygynandrous mouth brooders. This radiation cannot be accounted for by the sexual-selection hypothesis of speciation, but keep in mind that this radiation has had millions of years to evolve.

Diminutive Crater Lakes

The West African nation of Cameroon is pierced by several volcanoes, some of them active. Lakes form in dormant ones, and fishes enter through outflowing streams and colonize the lakes. Cichlid radiations in two small Cameroon crater lakes, Barombi Mbo and Bermin, do not easily fit the allopatric model of speciation.[130] That model calls for a patchy environment, but the crater lakes are impressively uniform. On the other hand, speciation in those lakes also cannot be explained through sympatric speciation driven by sexual selection.

The lakes are tiny. The larger one, Barombi Mbo, has a surface area of only 4.15 square kilometers. Smaller lake Bermin is almost too diminutive to dignify the appellation lake; its surface is only 0.6 square kilometers.[131] Each bowl-shaped lake is ecologically monotonous with a steeply sloping bottom consisting of sand and some rocks, logs and forest debris. Fluctuations in water level would not materially change the situation as it would in a shallow lake with a complex coast. The lakes are said to lack, and have always lacked, physical barriers to the movements of cichlids.

The age of Barombi Mbo has been judged to be about 350,000 years,[132] thus much older than Lake Victoria but also much younger that Lake Malawi. That does not mean the cichlid radiation there necessarily started so far back in time. Crater lakes are dangerous places for fishes because any given lake might suddenly become a deadly soup, either cooking the fish through volcanic activity or poisoning it with gas.[133] Consequently, the time required for the radiation in Barombi Mbo remains uncertain.

A species of mouth-brooding cichlid of the genus *Sarotherodon* got into Barombi Mbo some time in the dim past. The biparental mouth brooder, St. Peter's cichlid *(S. galilaeus)*, is a likely candidate for the founder species.[134] The tilapiine cichlids in this genus are only distantly related to the mouth-brooding haplochromines.

Once in Barombi Mbo, the founder radiated into eleven distinct species that are regularly in contact with one another without interbreeding and are hence reliably different species. The eleven species can be arrayed in miniflocks that have the following feeding adaptations: Three species of predators take large prey, one species eats sponge, one takes insects from the bottom and another from open water, yet another feeds on phytoplankton, and four are herbivores or feed on detritus on the bottom.[135] A degree of habitat partitioning has taken place. For instance, some species with similar feeding behavior are found at different depths. This is a remarkable radiation along trophic lines that had to have taken place within the crater.

Eleven distinct species in Barombi Mbo, some twenty to fifty times fewer than those in the largest Great Lake, may not seem impressive. However, when the numbers are adjusted for the minuscule size of the lake, they have originated in an area about 13,000 times smaller.[136] Seen another way round, if that Great Lake had proportionally as many species as Lake Barombi Mbo, they would number somewhere around 145,000! This radiation is therefore more spectacular than any seen in the Great Lakes.

The cichlid radiation in nearby Lake Bermin may be even more stunning, given its lesser size. And its radiation had an entirely different origin from that in Barombi Mbo and those in the Great Lakes. Their nearest existing relatives are *Tilapia guineensis* and the well-known species *T. zilli.*[137] The nine species of tilapias in Lake Bermin are, as in Barombi Mbo, monophyletic and have radiated into two types according to feeding behavior.

Now for the compelling point: These cichlids are all monogamous biparental substrate spawners. To underscore, males and females look much alike and are not colorful.[138] Also of importance, one species, *Tilapia gutturosa,* exists in three distinct color morphs. But the fish mate without regard to type of color. Infrequent yellow females are also present in another species, *T. flava,* and they pair with normally colored males.

Barombi Mbo and Lake Bermin are so tiny and ecologically uniform that they appear neither to lack any physical barrier to the movement of cichlids within the lake nor the opportunity to speciate by occupying different habitats. Note well that this does not preclude evolving differential usage of the habitat, which could arise as a consequence of speciation.

As an alternative, sympatric speciation has been proposed to account for the radiations.[139] Sympatric speciation was also suggested for a pair of sister species of monogamous cichlids in a crater lake in Nicaragua.[140] We are in the dark, however, for an explanation of how this could happen. Ironically, the least touted candidates for sympatric speciation, the monogamous monomorphic cichlids, may be the ones who have achieved it. A caveat, however. Although the assumption of ecological uniformity within the crater lakes is reasonable, it may wrong. If the fish somehow divided up the habitat prior to speciation, that could lead to within-lake speciation.

The Sinkhole Called Guinas

The recent discovery of a single endemic species, *Tilapia guinasana,* in a tiny sinkhole in South Africa,[141] may be a window into the process of speciation,[142] though one that may not generalize. Guinas is little more than a cavern without a roof in a dolomitic formation; its surface area is only about 2,880 square meters, making it the smallest body of water to have an endemic species of tilapia. The entire fish population consists of between 250 to 400 individuals, and it is endangered.

Tilapia guinasana exists in five distinct color morphs, and genetic analysis has proved they are slightly but significantly different. All morphs occur in both sexes. To some degree, the different color forms of this monogamous species mate with like-colored individuals. Because of that, the scientists who analyzed this situation suggested this may represent an early stage of speciation, based on reciprocal mate choice according to color.[143] If so, it would be radically different from the condition of the tilapia that have radiated in crater lakes, because they do not mate assortatively according to color.

Is Polychromatism the Key?

The sexual-selection scenario for sympatric speciation in cichlids has the males of one species present in at least two color morphs, the common normal form and the less usual gold one. Females are of two genetic types: one chooses normal males, and the other gold ones.

Because the Midas cichlid has two color morphs, two of my graduate students, being young and bold, published papers touting the Midas cichlid as an exemplar for how sympatric speciation could occur as a consequence of polychromatism.[144,145] Those papers are frequently cited as supporting the reality of sympatric speciation among polychromatic cichlids in Africa,[146] although the evidence was only suggestive and the mating system so different as to be questionably relevant.

Nonetheless, I was also taken at first by the possibility that the color morphs of the Midas cichlid might presage sympatric speciation. If it could be shown that females really are of two types, one preferring gold and the other normal males, that would be strong supportive evidence. An experimental test of this revealed that regardless of their own color, the color of their parents, or of their schoolmates as young fish, females have a weak but consistent bias for the normal color pattern.[147] That indicates selection has led to a stabilization of preference for the normal male and not to incipient speciation along color lines.

Later, in an unpublished analysis, Axel Meyer looked for genetic differences between the two morphs that should be there if they are speciating. He found no difference, confirming that polychromatism here is not involved in speciation.

In Central America, distinct species of cichlids have seemed separable by differences in color pattern. And that is generally true but misleading. The three species in the Midas-cichlid complex, in contrast, cannot be told apart by coloration.[148] According to Ken McKaye, some newly discovered species in that complex share the same coloration. Yet all of these species are polychromatic. If polychromatism were the entering wedge, then speciation should have produced sets of one normal and one piebald/orange species, but it has not.[149] Evidently, coloration plays little or no role in speciation among these substrate-brooding species. Indeed, instead of facilitating speciation, species-specific differences in color pattern may not emerge until well after speciation.

THE SECRET REVEALED?

Not really. Just how cichlids have produced so many species is by no means entirely understood. Nonetheless, a pretty good outline now exists. A number of individuals of a species manage to enter a body of water in which little competition from other fishes is present. If it is a lake, all the better, because cichlids bring with them a body plan, physiology, and mode of reproduction, better suited to lakes than exists among the ostariophysan fishes. Within that lake, they are vagile enough to get from one suitable habitat to another. Further, suitable habitats should be patchy, separated by hostile conditions such as deep water. In addition, they are basically homebodies, staying on a suitable patch, and that is promoted by prolonged parental care. During that time, the offspring may "lock in" on the local habitat. Even open-water species return to a certain place in the lake to nurture their offspring.

The result is local differentiation into unique gene pools. Whether these are called species or not is often difficult to say because the various populations are in different stages of becoming valid species. That speciation is accelerated by female choice of mates remains an interesting and open question, but it applies only to highly dimorphic species. Sympatric speciation among them remains unproven and is unlikely.

Ah, those hundreds of species, one of the most spectacular examples of adaptive radiation known for any kind of vertebrate animals. Fishes of great beauty and exquisite adaptations. Will it last?

≈ *thirteen* ≈

FISH AT RISK

A few decades ago, evolutionary biologists tended to ask just about the advantage of a particular trait, say, the vermiform appendix. Or, if it disappeared, such as the rear limbs of a whale, its disadvantage. That way of posing the question is slightly misleading. Now biologists rightly ask, what are the trade-offs between the costs and benefits of any given trait? That more dynamic view recognizes that evolution is never done. The environment, physical and biological, is in constant flux. Consequently, the cost-benefit ratio of any trait fluctuates through time for a given population, or at the same time across different populations of a species.

Everything we humans do likewise has its costs and benefits, pros and cons, ups and downs, good news and bad news. Where humans judge the balance to lie, however, depends enormously on their personal stake in the equation, perceived or real. As the population of the world increases and more and more food is needed, the environmental conflict becomes more acute when we try to evaluate the trade-offs between harvesting and polluting, on the one hand, and conserving, on the other. Some would calculate the cost-benefit ratio strictly in terms of trophic yield and how it is obtained. For others, issues such as stewardship of the domain, or aesthetics, loom large.

In this the final chapter I present some of the stresses imposed on nature by the earth's growing population of people. In keeping with this book, my focus is on cichlids, but I also want to illustrate how they fit into a larger, more general pattern of diminishing resources.

DEPLETING OUR PISCINE HERITAGE

Just a few decades back, oceans seemed a limitless larder. No matter how many fishes were harvested, more remained. As for pollutants, the volume of water is so great that human contamination is just a drop in the ocean, or so it was believed.

When the sardine fishery in California collapsed in the mid-twentieth century, conflicting positions were taken. The fault lay in overexploitation with the incredibly efficient purse seines, said one side. Opposed to that, the fishing industry ascribed the crash merely to natural fluctuation. Evidence for large swings in sardine abundance does indeed exist in the sedimentary record over the past two millennia,[1] but the ubiquitous slump in the world's oceanic fisheries points to overfishing.

In 1989, the global catch of fishes peaked at 89 million tons and since then has seesawed around 85 million tons. Had fishing effort remained the same, the numbers caught after that would have dropped. Fishers, however, worked harder. Since 1970, the fishing fleet of the world has doubled in size, often subsidized by the local governing body. Fishers, moreover, use ever more effective equipment, including electronic fish finders and smaller-meshed nets to harvest younger, smaller fishes. The take does not grow, and the catch-per-unit effort steadily falls. Clearly, this is an unfavorable ratio of benefits compared with costs.

Almost 70 percent of the world's major fisheries are harvested up to or beyond their natural limits.[2] Crucial species, such as cod, hake, and haddock, are so threatened that severe limits have been placed on their exploitation as several seafaring nations, for example, Canada, Iceland, the United Kingdom, and Portugal, squabble over access to the fish, sometimes violently.

Years ago fish was much less expensive than beef, but in the past twenty years the price of fish has risen much faster than that of beef, pork, and chicken. The good news is that fish is nutritious and provides a healthier diet than red meat. The bad news is that about one-third of the world's fish catch is used to feed animals, an inefficient squandering of this resource. At the same time, because of the declining catch, the per capita consumption of fish is necessarily decreasing and is projected to slump further, particularly in the developing world.

Even as the demand increases, the world population swells. On the very day I wrote this passage, the local newspaper reported that India had reached the population milestone of one billion people. The world is now home to six billion people. United Nations demographers predict world population will level off in the middle of the twenty-first century at eight to ten billion inhabitants and may even decline a little after that. Pressure on fisheries is already intense but will inevitably increase as a consequence.

This is not just a matter of overexploitation. As human numbers increase, the amount of pollutants from homes, agriculture, and industries, including aerial contaminants, entering our rivers and the ocean climbs higher and higher. Environmental modification, such as deforestation, dams, and dredging, further alters the aquatic environment to the detriment of fishes. Nets trawled along the bottom to catch shrimp and fish such as halibut and flounder do enormous damage and drastically alter the ecosystem. During my life I have seen the rich runs of salmon in the rivers of California dwindle to almost nothing. Diverted and damaged waterways and overfishing have done them in. That this was going to happen was forecast decades ago, but the conflict among powerful interests prevented meaningful action.

FARMING FISHES

Predictably, we have turned to farming fishes to compensate for declining fisheries. Production from aquaculture, including shellfish and shrimp, more than doubled both in value and weight in the decade preceding 1996. More than one-quarter of the fish eaten by humans in the world has been farmed.[3]

The art of aquaculture was pioneered in Southeast Asia. Over centuries, farming fishes there became more and more refined, capitalizing on various native cyprinid fishes, the minnows, in this case better known as carp. These fishes are well suited to aquaculture because they feed on plant matter. Plants capture energy from the sun, and this energy is converted to consumable meat at an efficiency of roughly 20 percent by herbivores such as cattle and carp. Thus it makes little sense to cultivate carnivores on cattle and carp because of the roughly 80 percent further loss of energy in adding another step to the food chain.

The benefits of farming fishes are demonstrably positive, provided the right species are used, and that means herbivorous fishes. They provide nutritious food at relatively low cost with high productivity per acre. Because fishes are "cold blooded," they don't waste energy keeping warm as must the "warm blooded" birds and mammals, so fishes can divert more of their energy to growth. Fish (in the United States, catfish) and rice can be cultivated on the same land, increasing the yield and the farmers' profits.

However, fish farming has drawbacks, and these are intensified when carnivorous fishes are cultivated.[4] Farming fishes such as Atlantic salmon and trout is a losing game because the cost exceeds the gain. Growing one pound of salmon takes up to four pounds of fish as food. Salmon require nutrient-rich food, mostly fish meal and fish oil, that has been obtained from wild-caught fishes. Consequently, salmon farming does not relieve pressure on natural populations of fishes but further depletes them.[5] As one biologist wryly put it, "farm-raising salmon is like farm-raising tigers." In contrast, farming herbivorous fishes lessens the pressure on wild fishes.

The way that Atlantic salmon are cultured at high densities in huge pens brings with it a host of problems.[6] Under these conditions, epidemics break out, and then the salmon are treated with antibiotics. Many of them escape, infecting wild salmon, not to mention adulterating the adapted gene pool of the natural salmon. Escapees may even become established where they are not native and supplant the locally well-adapted populations, as in Norway and possibly the Pacific Northwest of America. Biopollution is impressive, too, as feces and excess food build up on the bottom under the pens. But not all fish farming is for food.

THE AQUARIUM REEF

Tropical fish are cultivated for the aquarium trade. I have heard estimates as high as one billion dollars spent yearly in the United States at various levels of that trade. One part of that economy lies in the production of the more common tropical fishes, mostly in the state of Florida.

The warm climate, abundant water, and flat terrain there are suited to pond culture. In Florida, 203 tropical fish farmers accounted for $57 million in sales in 1997.[7] Frequently, some of the fish slip out of the ponds and into the local waterways where they survive.[8] Many species of escapees have become well established in Florida, including cichlids that now outnumber the species of native sunfishes.

As a consequence of releases of tropical fish by aquarists in Hawaii, fifteen species of cichlids inhabit the freshwaters there. Because of their popularity and market value, cichlids get moved all over the world.

The most popular cichlids in the aquarium trade are those from Lake Malawi, and understandably so, because they are the most colorful. As you have read, coloration of a given species varies from location to location in Lake Malawi, and many species are found only at certain locations. Advanced aquarists are eager to get the newest importation from the lake, or the most colorful variety, so they command high prices.

The exporters have a home base to which they bring captured fish and process them for export. What a nuisance when a particularly desirable cichlid lives at the other end of the lake! The solution: Bring the cichlid to where you operate and establish it on the nearest convenient reef. That is just what has been done.

One reef at Thumbi West Island is now known as "the aquarium" by the people living in the nearest village because it has so many alien species.[9] Although the reef is little studied, some research has been done on the extent to which the different strangers have succeeded, and it varies.[10] The critical issue is whether the successful new species displace those already there.

Some interloper species seem better at securing territories on the reef.[11] And three species, *Cynotilapia afra, Pseudotropheus callainos,* and *P. tropheops* 'red cheek,' grow faster and are more fecund than their counterparts "back home."[12] These findings, while not conclusive, suggest that some of the translocated species are a threat to the original inhabitants.

I have been hearing recently that collectors in Lake Tanganyika have also been moving cichlids from their home areas to more convenient reefs where they do not normally occur. These translocations have the potential to disrupt the local assemblages of fishes, even potentially extirpating local races.

ANGLING FOR CICHLIDS

Most cichlids are too small for the table, but many reach one-half to one pound in weight, and even smaller ones are cherished as food. They have firm tasty flesh, and unlike carp, they have only a few rib bones and they are thick. Throughout the tropics, fishing for cichlids is chiefly artisanal. In various countries in Central America, I often saw boys or men heading home with strings of a few cichlids, called mojarras. They use traditional techniques such as angling with hook and line, trapping with baskets, and throwing cast nets. Unfortunately, unconscionable wasteful methods have appeared, such as nonselective poisoning and dynamiting of fishes.

In Nicaragua, fishers located on the shores of large Lakes Managua and Nicaragua accumulate live cichlids and hold them in pens until they have enough fish to make worthwhile a trip to the market in Managua or Granada. Species of about one pound or less, such as the Midas cichlid, are served intact in a bowl of soup. Larger cichlids, known as guapote or laguneros, if very big, are baked; then the vertebral column is gently pulled out taking with it the stout rib bones, leaving an almost boneless, delicious dinner.

Such utilization is sustainable because the rate of capture does not exceed the replacement capacity of the cichlids. The development of more efficient means of catching the fishes, on the other hand, can result in an unsustainable rate of removal.

Subsistence fishing in Africa is widespread in the basins in which the Great Lakes lie and in other areas. Many kinds of fishes, such as minnows and catfishes, are eaten, but cichlids are the mainstay of the fisheries. In Malawi, 75 percent of the animal protein consumed by the people comes from fishes captured in and around Lake Malawi. Because of the lack of refrigeration, the fish are usually dried in the sun to preserve them for transport to markets. About 35,000 fishers provide the fish. The complete processing of them, from capture to market, including making and maintaining boats and fishing gear, is estimated to involve about two million people.[13]

The highly productive Lake Malombe, near Lake Malawi, has come under increasingly heavy artisanal fishing pressure from the use of larger nets with finer mesh. Total catch has accordingly dropped severalfold. The captured cichlids have progressively shrunk to smaller sizes. Of the haplochromine cichlids recorded from the lake, nine larger species appear to have disappeared locally.[14]

The Great Lakes are unusual in their size. The presence of extended areas of open bottom in accessibly shallow water in some of their parts permits the use of large lift nets, trawls, and purse seines deployed from motorized boats. Previously unknown to the fishers, such boats were introduced fairly recently by Westerners.

The fisheries in Lake Malawi are following the usual path. Catches were good at first, then began to diminish. With the decline, the fishers compensated by employing larger trawls cutting wider swaths with finer mesh and for longer times in the water. Ever smaller fishes are harvested, protected areas are poached, and practices in general are destructive. In some parts of Lake Malawi, larger species are in decline, and some have disappeared.[15] Similar changes are taking place in the other two Great Lakes, particularly in Lake Victoria.

Lake Tanganyika has suffered less, but trouble looms on the time horizon. Commercial fishing began in the 1960s and within twenty years the catch began shrinking. The large-scale destruction of forests has released silt into the lake such that the delta of the Ruzizi River in the lake has seen a tenfold increase in sedimentation within the past two decades. Several kinds of fishes and invertebrates there have vanished.[16]

AQUATIC CHICKENS

Those ancient but progressive Egyptians farmed fish, and the fish of choice was our old friend the Nile tilapia.[17] Other cichlids have been tried in aquaculture, such as *Etroplus suratensis* in India and Sri Lanka, but it grows too slowly.[18] In Central America, indigenous species such as the guapote *'Cichlasoma' managuense,* a predator, and other species are problematical because they rapidly overpopulate with consequent stunting.[19]

The cichlids of choice are therefore the mouth-brooding tilapias in the genus *Oreochromis*. They are herbivorous, thriving on planktonic algae, although they can be omnivorous. This kind of tilapia grows faster than other cichlids, and so they more quickly reach a good size for market. They are such superior fishes for aquaculture that they have been spread all over

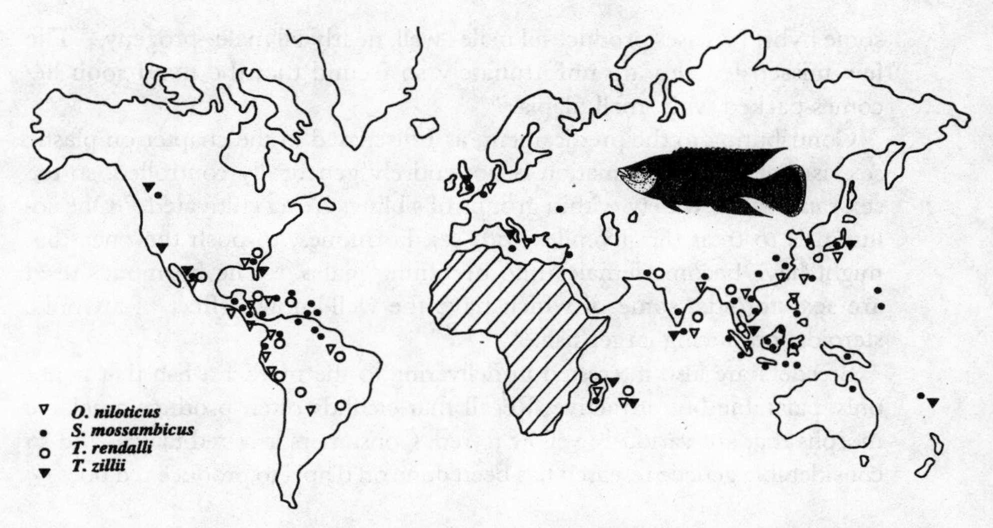

FIGURE 13.1 Four species of tilapias have been introduced throughout the world where warm freshwater is present. No symbols are placed on Africa because the cichlids are indigenous to that continent (after Leveque 1977).

the tropical world, and into temperate zones where thermal springs provide warm water, as in the state of Idaho.

Tilapias are now the most widely cultivated fish in the world and are dubbed "aquatic chickens" by people in the trade. As one measure of the tilapia's success in aquaculture, within the brief span of 1985 to 1988 their worldwide production totaled 260,000 tons, mostly produced in tropical Asia.[20] Ironically, they are not farmed much in Africa, the continent of their birth, apparently for cultural reasons.

Another irony is that the most widely spread tilapia, *O. mossambicus,* is not the best one for farming. The top species is the Nile tilapia, *O. niloticus.*[21] It grows the fastest, and its reproduction is the easiest to manage. In practice, however, all the species leave something to be desired. Consequently, an enormous amount of research has been invested in trying to create the ideal tilapia through hybridization between various species and their backcrosses. Currently, genetic engineering is being applied in which desired genes are lifted from one tilapia and inserted into the eggs of another.[22]

The goals of genetic alteration are several. First, the farmer would like his fish to grow larger faster while more efficiently converting food to meat.[23] The fish should also resist disease and crowding.

Given the chance, tilapia will divert their energy to breeding instead of growing, which results in small mature fish. Obviously, that is undesirable. Tilapia can also quickly overpopulate a pond, which further contributes to stunting. Thus, another goal is creating a tilapia population of only males, because they are the sex that grows faster, or of sterile fish.[24] Remarkably, some hybrid crosses produce all male–well, nearly all male–progeny.[25] The few missed females are unfortunately so fecund that the pond soon becomes packed with small tilapia.[26]

Contributing to the predicament, as I discussed in the chapter on plastic sex, is that sex determination is not entirely genetically controlled, so the sex ratio varies widely within groups of siblings being cultivated.[27] One solution is to treat the juveniles with sex hormones, to push the ones that might have become females into becoming males.[28] The hormones used are sex steroids, some of which have the well-known effect of anabolic steroids, producing larger fish.[29]

Breeders are also interested in delivering to the market a fish that is not only palatable but attractive. Recall that cichlids often produce amelanic morphs that are variously yellow to red. Consumers love red tilapia, and so considerable genetic research has been done on tilapia to produce red ones.[30]

FREE-RANGE TILAPIA

More than one person has had the bright idea that if tilapia are such wonderful fish to farm, then let's follow the model of range cattle and just release them into warm-water lakes and ponds. Unlike range cattle, once tilapia are stocked they can't be removed, that is, not entirely. Because they are highly adaptable to different ecological situations, they regularly thrive where stocked, and they spread from there. One population of tilapia has even taken root in the seawater lagoon of a remote coral atoll in the Pacific Ocean.[31] They have been planted around the tropical world. Inevitably, tilapia have an impact on the native species, and often those species are other cichlids. Consider the following example.

In the 1980s, tilapia were released into Lake Nicaragua, the largest lake in the tropics outside of Africa,[32] and one that had a rich and productive fauna of cichlids.[33] The introduced tilapia were of three species, *Oreochromis aureus, O. mossambicus,* and *O. niloticus*. Within a few years, fishers started catching them in their nets. At the same time, they noticed a drop in the local species of cichlids as the tilapia started to replace them. Sadly, the total catch of cichlids, including the tilapia, ebbed. Moreover, the people prefer the flesh of their native cichlids. Nonetheless, tilapia have been and continue to be released all over Central America and other parts of the tropical world.

ANGLER'S DELIGHT

Anglers perpetually yearn for a yet bigger fish, one that fights on the line and that satisfies the stomach (though fishing for tarpon and bonefish prove the latter consideration is inconsequential for anglers in it just for the sport). Cichlids have played a role here, both as perpetrators and as casualties of ill-advised translocations of large predaceous fishes.

Building the Panama Canal required creating a long lake across the isthmus for the passage of oceanic ships. The result was Lake Gatun. From the rivers that fed the lake, it was quickly colonized by various South American fishes, most notably silversides, characins, cichlids, sleepers, and poeciliids, as well as tarpon from the oceans.

The peacock cichlid, *Cichla ocellaris,* also called the peacock bass, was soon to make an appearance.[34] This is a huge piscivorous fish. I say fish, but it consists of a number of species of *Cichla* in South America. In fact, the species that we are about to consider is, according to Paul Loiselle, surely not *C. ocellaris* and most likely *C. monoculus*. This cichlid commonly grows to a weight of two kilograms (4.5 pounds) and a total length of 50 centimeters (20 inches). The name derives from the big ocellus at the base of its tail. Such a grand predator caught the attention of anglers.

In 1965, about one hundred juveniles were put into a pool in a dammed portion of a Panamanian creek. During the rainy season of 1966 the water overflowed, the fish escaped, and the rest is, as they say, history. The peacock cichlid rapidly advanced as a wave into Lake Gatun, multiplying and growing large, and that elated the anglers. As they appeared, they cut a swath of destruction through the rich fish fauna ahead of them and left a piscine desert behind them. Like an army of vacuum cleaners, they ate every fish in their path.

At Barro Colorado Island, where their impact was studied most closely, six of the eight very common species were extirpated and the seventh was severely reduced. The last species, the modestly large black-belt cichlid, *'Cichlasoma' maculicauda,* actually increased in number, apparently because the peacock cichlid annihilated the small predators that normally prey on the black-belt's young.[35]

The most worrisome consequence of the invasion is that peacock-cichlid predation compressed the complex food chain into a simple one. That may be desirable for fish farming, but in nature it means a less productive ecosystem and one that is also less stable, more vulnerable to perturbations.

Some animals disappeared not because they were eaten but because the peacock cichlid beat them to their normal prey. To mention one visible example, tarpon had previously chased schooling silversides to the surface where terns joined in the feeding. Peacock cichlids ate all the silversides. Without their customary food the tarpon vanished, and so did the terns and other fish-eating birds, such as kingfishers.

Perhaps more troubling (though not well proved) is that as the mosquito-eating fishes vanished, mosquitoes became more abundant. Malaria victims became infected mainly with *Plasmodium falciperum,*[36] which is often fatal.

Previously, the malarial protozoan was primarily *P. vivax,* which is more benign and rarely fatal but is nonetheless difficult to treat.

At the time of that report, more than twenty-five years ago, the peacock cichlid was still expanding its range into Lake Gatun and associated rivers. No further report has been issued, so we are left to guess the final outcome, such as how the ecosystem has adjusted to the predator. Also at the time, peacock cichlids were being transplanted from the lake to Panamanian rivers unconnected to Lake Gatun, and one can only wonder about the consequences.

Peacock cichlids of more than one species have also made their way into North America. Where the water is warm enough, as in Florida, fishers have counted its introduction a success.[37]

The example of the peacock cichlid demonstrates a fundamental and indisputable point because of the simplicity of the situation. Before its introduction, the fish fauna of Lake Gatun was stable, threatened neither by overfishing nor by pollution. The decline was unarguably and solely the result of the onslaught of a large efficient predator, itself free of predators and without significant competitors. The local fishes lacked appropriate adaptive escape behavior and were consequently eaten up.

VIOLATING VICTORIA

Extinction is an integral part of evolution. Since the beginning of life, species have continuously arisen and then disappeared, and nearly all species that ever existed have become extinct.[38] Some species simply metamorphosed through geological history, but most were displaced by new ones. In addition, now and then physical forces, such as volcanism, meteors, and hurricanes, caused widespread extinctions.

The characteristics of those species most likely to become extinct should get your attention because they share so many attributes with species most apt to speciate. Vulnerable species have the following features: They are specialized and patchily distributed; they exist in small or fluctuating populations; and they have limited ability to disperse.[39] Note how much these traits apply to cichlids in the Great Lakes of Africa.

Cichlids are at risk. As Les Kaufman put it, "What we're seeing in Lake Victoria is the greatest vertebrate mass extinction in the modern era, and one of the most extreme disruptions ever observed in a body of water this large."[40]

In *Darwin's Dreampond,* Tijs Goldschmidt chronicles with disarming charm and cynical humor the demise of Lake Victoria. His book is but one culmination of what has become a large literature on the tragedy of that

FIGURE 13.2
The Nile perch of Lake Victoria grows to a huge size and is prized by fishermen. Here we see a specimen next to a person to add scale.

lake. Most of the attention has focused on the introduction of the huge piscivorous fish, *Lates niloticus,* the Nile perch.

An immense detailed overview appeared in 1995, titled *The Impact of Species Changes in African Lakes.*[41] The larger part of this edited collection of articles is devoted to Lake Victoria and the Nile perch. The authors present different perspectives, some of which strongly conflict. At one extreme, a fisheries biologist intimated extinctions may not have occurred and that they probably will not happen.[42] I can't imagine how he reached that conclusion.

Fish Gangsters or Saviors?

No one is sure just why, when, or by whom the Nile perch was planted in Lake Victoria, from which lake it came, and to what extent the Nile perch was repeatedly planted, but the caper appears to have been done initially on the sly around 1950 to 1960.[43] Some say the goal was to promote angling to attract tourists, others maintain it was to create a commercial fishery. It doesn't matter now.

After its introduction, the presence of the Nile perch wasn't felt until they irrupted around 1980.[44] Then things changed rapidly. Prior to that, roughly 80 percent of the fisheries consisted of cichlids. After the upsurge of big Nile perch, however, around 80 percent of the catch became Nile perch and only 1 percent cichlids (these figures and those that follow are "soft" because accurate statistics are difficult to get in that area). As the fishery matured, juvenile Nile perch became an increasingly important part of the catch, especially in shallow water.

By 1992 Lake Victoria had achieved the status of the most productive lake on earth with a harvest of Nile perch estimated at 200,000 to 300,000 tons per year.[45] Its increase was closely paralleled by the upsurge of two other species, a native cyprinid fish known as dagaa or omena, *Rastrineobola argentea,* but more easily remembered as the lake sardine, and the introduced Nile tilapia.[46] Parallel changes in species occurred in Lakes Kyoga and Nabugabo following the introduction into them of Nile perch and Nile tilapia.[47]

Up to that time, the preferred fish for market was the endemic tilapia, *Oreochromis esculentus,* and a large minnow, the migratory *Labeo victorianus.* The haplochromine cichlids were taken in goodly numbers and eaten, but they are smaller fishes and not as desirable.[48] With the advent of the Nile perch, the main fishing effort turned to them, in part because the easily caught fishes vanished. Huge specimens of Nile perch, some up to two meters (six feet) long and weighing 135 pounds, came to dominate the fisheries.

Unlike the small cichlids, Nile perch are too large and the flesh too oily to dry in the sun, so they are smoked. The result: considerable destruction of the local forest to provide firewood.[49] Increasingly, insulated trucks transport Nile perch to central processing facilities. Some of the frozen flesh, as well as traditionally sun-dried fishes, are distributed within East Africa. More and more, however, the freshly refrigerated fillets of Nile perch are flown to Europe and Israel, and even to North America. This has driven the price up, and that has stimulated over exploitation of the Nile perch.[50]

In some areas, the catch of Nile perch has fallen sharply, by as much as 50 percent. The local fishers blame the downturn on overexploitation. Fine-mesh seines are used to take juvenile Nile perch only four to twelve centimeters long.[51]

Nonetheless, the planting of Nile perch is heralded by some as a great success. It has created jobs and attracted foreign currency. The Nile perch is also said to have brought so much protein to the interior of Tanzania that its name in the language spoken there means savior.[52]

Against that, the export trade priced the fillets out of the reach of the indigenous people.[53] Cichlids and other fishes previously eaten became so scarce that the local fishmongers went to the processing plants and purchased carcasses of Nile perch. They fried and sold head, tails, and other scraps; although at first resisted by the local people, they are now accepted and are said to be nutritious.[54] Protein in their diets, typically consumed in the form of fish, has nonetheless dwindled. Hence the fundamental economic progression has been from providing protein to the indigenous people to exporting protein out of Africa.[55]

The other significant introduction into Lake Victoria was the Nile tilapia; it proved remarkably effective in exploiting the abundant phytoplankton. In the process, the Nile tilapia appears to have replaced the prized native tilapia, *Oreochromis esculentus*.[56] Nonetheless, the new source of protein was welcomed by the fishers.

To throw in another bit of bad news, the ornamental water hyacinth, *Eichhornia crassipes,* of South America has infested warm waters throughout the world, starting back in the nineteenth century. It has done enormous damage in almost all instances, as boaters in Florida can attest. It has been proclaimed the world's worst aquatic weed, a botanical gangster.

Hyacinth multiplies with alarming speed, covering calm bays and plugging up boat landings. It harbors mosquitoes and snails, thereby spreading disease. This weed also lowers the oxygen in the water beneath it. It has recently been seen in Lake Malawi, and many fear it will invade Lake Tanganyika. In poor Lake Victoria, water hyacinth is now abundant.[57]

Hyacinth does have an up side in Lake Victoria. Nile perch do not penetrate under its dense mats, perhaps because they are sensitive to reduced levels of oxygen,[58] but some haplochromines do survive there. Consequently, in a minor way, water hyacinth has created a refuge for some cichlids.[59]

Mugged Fishes

Prior to the advent of the Nile perch, the great diversity of cichlids in Lake Victoria fed on every imaginable kind of food and in all habitats, and the fishes flourished. Many of the species were found offshore, and a small motorized fishery for them began in the 1970s. At one point, the boasts trawled about 1,300 individual fish of thirty-seven species per ten minutes. The haul was dominated by cichlids, although it included several other families of fishes. The catch supported local communities.[60]

During the rapid ascent of the Nile perch in a period of roughly five years, about two-thirds of the cichlid species disappeared along with other kinds of fishes.[61] The Nile perch seeks its prey near the bottom in areas away from reefs, and down to depths of 60 meters. The most vulnerable cichlids, therefore, were those that lived over the open bottom and some species that fed up in the water column near shore.

Those species of catfishes inhabiting the more exposed areas of the bottom, where they overlapped with Nile perch, were the first to be decimated.

More than one species is thought to have become extinct, for example, the endemic deep-water cat, *Xenoclarias eupogon.* Those catfishes dwelling in the shallows around reefs, such as the little species of *Synodontis* and others, are holding on.[62]

Ironically, several kinds of invertebrate animals have increased in numbers, such as the lake fly.[63] The most widely discussed case is that of a shrimp, *Caridina nilotica,* which has taken the place of the bounteous haplochromine cichlids that fed on detritus in exposed soft-bottom areas. With the irruption of Nile perch, the abundance of the shrimp-eating cichlids decreased a thousandfold, or more. The numbers of planktivorous cichlids also plummeted, probably releasing more phytoplankton to drop to the bottom and thereby increase the amount of detrital food. Those haplochromine shrimp eaters were among the first to be eliminated by the Nile perch, but even the nonspecialist cichlids commonly devoured the juvenile stage of the shrimp.

The numbers of shrimp were few prior to the Nile perch, but they experienced competitive release with the removal of the cichlids, and their numbers grew rapidly. The Nile perch did not at first consume many of the shrimp, and that may have allowed them to reach a critical mass. Then the Nile perch fed voraciously on the shrimp. As the Nile perch increased in size, they became too large to consume the small shrimp. The smaller young Nile perch, however, did eat them, and they in turn were cannibalized by the larger Nile perch.[64] A parallel appears to have occurred with the removal of the planktivorous cichlids. That loss may have provided an opportunity for the lake sardines to take over, as they have.[65]

The overall effect was a simplification of the food chain–some say an oversimplification. As many as three hundred species of cichlids, by one account, and several species of other fish families have become, for all practical purposes, extinct.[66] Removed from the open-bottom ecosystem were the cichlids that ate molluscs, other fishes, insects, algae, phytoplankton, zooplankton, and detritus. They have been replaced, in essence, by the detritivorous shrimp, which in turn are consumed by the Nile perch.

The food chain has therefore been greatly simplified. The Nile tilapia eats algae and the likes, the lake sardine feeds on plankton, the young Nile perch eat the shrimp, and the adult Nile perch eat all three species. And those three species form the fishery. The simpler the food chain, the more unstable it becomes. For instance, the removal of the major phytoplanktivores may have contributed to the increasing layer of persistent anoxic bottom water and its vertical pulsations.[67]

As if that were not bad enough, the use of chemicals has raised a life-threatening situation. According to *The Nation,* a Kenyan newspaper, a chemical used to curb water hyacinths resulted in widespread death of fish and was stopped. Others dispute that this is how the use of toxic chemicals got started, but it is now indeed a problem. It struck unscrupulous persons that poisoning fish is an excellent way to capture them.

Now at least four different chemicals are being used illegally. Fish killed by the chemicals poison people who eat them, and so they are not sold locally because people have learned to recognize contaminated fish. The rogue fishers are armed and prone to violence, and some people have been killed. Although the penalties for this ugly form of fishing are high, including life in prison, the local authorities lack the resources to control it.

Finding the Cause

The cause of the demise of cichlids in Lake Victoria is not as clear as it could be, compared with the case of the peacock cichlid in the Panama Canal. The problem goes deeper than the Nile perch. Comparing the three Great Lakes, Victoria's catchment basin has by far the highest density of people. And their numbers around the lake climb steadily. In Uganda, for instance, the population increased seventeenfold in seventy years reaching 17 million people by 1991. The overall population growth has been 3–4 percent annually and is accelerating.[68]

With that expansion has come rising urban and agricultural pollution, and deforestation that bleeds more silt into the lake. Cyanobacteria bloomed as nutrients increased, making the water increasingly opaque. A disturbing correlate is that the anoxic layer of water on the bottom has crept up, sometimes in sudden spurts.[69] A remotely operated submarine, for instance, sent back video images of scattered dead fishes and invertebrates lying on the bottom at a depth of 40 meters.[70] Anoxic water must have moved up so fast it overtook the fish living there. "Much of Lake Victoria may now be incapable of sustaining aerobic life."[71] Remarkably, the shrimp *Caridina nilotica* can tolerate such low levels of oxygen for some time.[72]

The growing number of people created a relentless demand for more fish on the table. Signs of overfishing began to appear before the Nile perch became abundant.[73] The introduction of motorized trawlers can share the blame.[74] Moreover, even though the numerous species of rock-dwelling cichlids are seldom taken by Nile perch, their numbers have seen a notable decrease.[75] I'll return to a plausible reason for this in a moment.

If overfishing and altered water chemistry were the entirety of the problem, we might have expected further decline in the cichlid fishes of Lake Victoria, but not the precipitous extinction or near extinction of an estimated three hundred species.[76] The Nile perch seems to be the prime culprit.

Consider this: Cichlids were the chief diet of the Nile perch until they disappeared, and cichlids also vanished in areas that were not heavily exploited by the fishers. The Nile perch advanced as a wave throughout the lake, and each time they became abundant the cichlid populations slumped.

Locally, the cichlid species most susceptible to the Nile perch are the first to go. Nile perch differ in where they spend their time. At the depths where they are most abundant, cichlids are at their lowest. Similarly, cichlids that inhabit reefs and areas of submerged vegetation are least effected.[77] With the cichlids of Lake Victoria already suffering from overfishing and altered water chemistry, the Nile perch, with a little help from the Nile tilapia, put another nail in the coffin.

Changed Survivors

The first cichlids wiped out were the big (mostly piscivores) and rare species. Those that survived have adapted by becoming much smaller,[78] a consequence often resulting from natural selection for reproducing while still young–the individual fish may not live long enough to reproduce, otherwise. Another such response selected for is increasing fecundity to produce more young, and some cichlids have done that. Also, species with the same feeding behavior, such as the planktivores, have proved differently susceptible to the Nile perch. Those that feed more in the open water have fared the best.[79] Clearly, evolution continues, driven by the Nile perch.

A remarkable reduction in number of species has taken place where the water has become very murky in recent years.[80] Early research on hybrid haplochromines, and the importance of mate color in species recognition, led to the prediction that if the water were turbid enough, cichlids in Lake Victoria might hybridize.[81] In accordance with this prediction is a suggestive correlation:[82] The clearer the water, the more species are found, and with different colors. Where reefs in Lake Victoria have very turbid water and therefore a limited spectrum of wavelengths, typically only one drab species per genus is seen. The coloration of those species is intermediate to the red through blue species found in clear water.

Females, unable to distinguish their species by color in such water, are thought to mate willy-nilly with any closely related males and thereby produce intermediate decedents. The putative hybrids would be amalgamations of species, which means the loss of species.[83] This process may have been building up since the lake started becoming more turbid, some eighty years ago, and may now be gaining speed with further loss of visibility.

The Future?

The collapse of the cichlids naturally drew attention to those species most easily captured. These were mainly the detritivores and planktivores because they could be harvested en masse with large nets. They were also the species whose taxonomy had been the best studied. Then came the discovery that Lake Victoria actually has a rich fauna of rock-reef cichlids, some hundreds of species, in fact.[84] They feed mainly by scraping *Aufwuchs* or capturing insect larvae. More importantly, they can hide among crevices when Nile perch approach.[85] Accordingly, they have suffered less than the other cichlids since the irruption of the Nile perch.[86]

A recent report, on only one region of the lake, indicates a reduction of species of reef-dwelling cichlids from thirty-three down to twenty-four, a 27 percent loss.[87] One would like to see a smaller loss, but still it is much better than what has happened to the open-water cichlids. Signs are, too, that the underinvestigated reef regions of Lake Victoria may harbor a rich fauna of haplochromine cichlids still unknown.[88] Grasping at straws, perhaps, are the few anecdotal reports of unfamiliar haplochromines. Are new species already evolving, say, from the odd hybridization between two nearly extinct species?[89] Perhaps.

Lake Tanganyika might seem to offer some hope, but to believe that is to engage in self-delusion. The species-rich and diverse cichlids there coexist with four species of Nile perch. Given just a few thousand years or so, perhaps the surviving cichlids of Lake Victoria would evolve effective anti-perch behavior and gradually move back into the parts of the lake from which they have been liquidated.[90] Aside from the fact that none of us will be there to test this hypothesis, such a scenario would have to presume that the damage to the lake resulting directly from adulteration and overfishing by humans would abate or even reverse. Hardly. But if the degradation is not ameliorated soon, the lake will become uninhabitable even for the Nile perch.

MESSAGE FROM THE CICHLIDS

I am by nature a cheerful and optimistic person. Growing up in the age of atomic weapons, I learned to practice denial, which was essential for mental health. I find it increasingly difficult, however, to avoid mulling over the progressive deterioration of our world. What I have just written about cichlids in Lake Victoria is but one instance of the growing conflict between the demands on the environment made by ever more people and the complexly interwoven ecosystem it has been our good fortune to inhabit in the recent past. But as our planet amasses ever more people we are compelled to ask, what are the benefits, what are the costs?

The problem is acute, even when we limit the question to the destiny of freshwater. That precious fluid constitutes only 2.5 percent of all the water on the earth, and most of it is unavailable as polar ice caps, deep groundwater, and other forms. Only 0.01 percent of all the earth's water is found in rivers and lakes.[91] Yet we dump our excrement and industrial wastes into that essential liquid and extract water at an ever growing rate for drinking and bathing, agriculture and commercial processing. Given this, I have lost my optimism for the future of freshwater fishes (or humans, for that matter).

Perhaps you have shared the following kind of experience when falling asleep. The other night, after spending so much time researching and writing this, the last chapter, my mind kept ruminating on the Lake Victoria situation. I dimly imagined the Nile perch as an icon for humans, and then the Nile perch as mankind. There it was, the dominant force expanding into all habitable places, devouring resources, growing rapidly, and reproducing without limit. The food chain became simpler and simpler and less and less stable. The Nile perch turned on one another in their desperation, cannibalizing their young, as Nile perch are want to do. Eventually, the population of Nile perch foundered.

Unthinkable and silly? Perhaps. But in their models demographers seriously consider a catastrophic crash of human populations. And then, if we recover, further cycles of growth and collapse. As Lake Tanganyika teaches us, Nile perch and cichlids can coexist in a balanced state. We need not destroy our planet to achieve a balance, to optimize the cost-benefit ratio. Maybe that is the message cichlids send to us.

Cichlid fishes are nature's grand experiment. In Lake Victoria, we are terminating one of the grandest experiments of them all. Are the other lakes next? Is Lake Victoria a model for us?

GLOSSARY

Acanthopterygii A subset of "true" teleost fishes, characterized by spiny fins and ctenoid scales, among many other traits; the percomorph fishes, such as cichlids, are placed in this taxon.

allopatric speciation The evolution of new species as a result of geographic separation of populations of the same original species.

allopatry The state of a population or species occurring in a geographic region different from another species or population.

anisogamy The reproductive products, the gametes, of an animal exist in two unequal sizes, sperm and eggs.

Aufwuchs Assemblage of organisms that encrust rocky surfaces in rivers and especially in lakes, consisting of algae and other plants and their inhabitants, for example, tiny animals and diatoms.

Baldwin effect The process whereby environmentally induced variation becomes genetically reinforced. With subtle differences, also called genetic assimilation.

bar The prominent dark vertical markings on a fish.

basal The condition or species regarded as the starting point in the evolution of a trait or of other species.

basipharyngeal joint A joint unique to cichlids, consisting of a bony protuberance on the top of the upper pharyngeal jaw meeting the bottom of the skull.

benthic Living in water in close association with the bottom.

binomen The formal scientific name of an organism, consisting of two parts, the genus and species.

biparental Two parents together raise the offspring.

branchiostegals The thin elongate rays that support the throat region of a fish.

buccal Refers to the mouth region.

carotene Any of several orange or red crystalline pigments that are synthesized by plants. Vitamin A consists of a mixture of three such pigments.

carotenoid pigments Consisting of up to several pigments such as carotene and xanthophyl, and usually perceived as yellow, orange, or red. Often referred to simply as carotenoids.

carouselling When two fish circle one another rapidly, head to tail, reminiscent of a merry-go-round.

chromatophore A cell, usually found in skin but also elsewhere, that has intricate spider-web-like extensions. When the pigments spread out into the web, the skin darkens, and when the pigments withdraw into the center of the cell, the skin lightens. Most commonly, the pigments are brown (see melanophore, below), but they also occur in several other colors.

cladogram A branching diagram that portrays relationships among taxa, for example, species, families, orders.

conspecific Belonging to the same species.

contacting When a baby fish, a fry, swims to a parent and seemingly bumps into its side. Done to take food from the surface of the parent.

convergent evolution Evolution of similar traits, occurring independently in different evolutionary lineages, often from different previous traits.

crèche From the French term for day nursery. Applied to animals when unrelated offspring congregate, usually under the care of a few parents. The best-known example is found in penguins.

crepuscular An adjective applied to animals that are active during periods of twilight, thus around sunrise and sunset.

ctenoid scales Scales that are typically squared off at one end and bear tiny tooth-like protuberances.

cycloid scales Roundish scales that lack any protuberance.

daphnia Small animals that live up in the water in ponds and lakes. These cladocerans are called water fleas and are eagerly eaten by fish.

diatoms Small unicellular organisms that are not easily classified as either animal or plant but rather are in their own taxon. They contain chlorophyll and nutritious lipids and are found almost everywhere in the oceans and in freshwater.

dichromatism The state of having two discontinuously different colors of individuals. The term is mostly applied to differences between males and females, although it can also occur within one sex.

dimorphic The state of having two discontinuously different types of individuals. Most generally applied, it refers to size, shape, or color, but when narrowly applied it refers only to size and shape. Mostly seen in males and females, although it can also occur within one sex.

diurnal Active during daylight.

ecology The science of how organisms interact with their environment that seeks to explain the distribution and abundance of organisms.

ectoparasites Parasites that dwell on the surface of an animal. A conspicuous case is the fish louse.

facultative The ability to give appropriate alternative responses under different conditions.

fitness A measure of the success of an individual in leaving progeny in the next generation, typically with regard to other individuals in the species and thus as relative fitness. In a fuller form, called Darwinian fitness.

fry The early swimming stage of "baby" cichlids.

gamete A specialized sex cell such as a sperm or egg. It carries only half the genetic complement (chromosomes) of the adult individual.

genetic drift Changes by chance in the gene pool of a population that result from the small number of individuals in that population.

genital papilla A small fleshy pad just behind the abdominal vent of a cichlid and many other kinds of fishes. More pointed in males and more disk-shaped in females.

genome The total complement of genes in a single cell.

glancing When one fish touches its mate or parent with its "shoulder" region in a quick glancing motion.

gonochorist A species in which the ovaries and testes are carried in separate individuals who do not change sex. Humans are a well-known example.

gravid State of the ovary when it is full of eggs ready to be laid.

guanine A crystalline purine chemical compound found in the inner layers of fish skin. There it produces iridescent blues and greens called "structural" coloration. (Common in bird excrement, hence the word guano.)

habituation When with repeated stimulation the organism no longer responds even though it a capable of doing so.

handicap principle The male evolves some physical handicap that indicates to a prospective mate his genetic superiority, because despite the handicap, he can out-compete other, less-handicapped males.

harem An enduring group having more than one female but just one male who fertilizes their eggs.
herbivore An animal whose diet consists mainly of plants and algae.
hermaphrodite, sequential One individual starts life as one sex but later changes into a different one.
hermaphrodite, simultaneous One individual has both functional ovaries and testes.
heterogametic When one sex has two different types of sex-determining chromosomes.
homogametic When one sex has only one type of sex chromosome.
hue The attribute of colors that leads to their seriation according to wavelength of light.
hybrid An offspring from the mating of two different species, but the term can be used in a wider sense to speak of the interbreeding of genetically different populations.
hypothesis A proposition that is tentatively assumed in order to bring out its logical consequences to be tested against the facts, present or to be determined.
intersexual Involving both sexes.
intrasexual Involving only one of the sexes.
introgressive hybridization After the initial act of hybridization, the progeny of successive generations continue to mate with members of the parent species or population and thereby spread the genes of one species or population into the other.
isogamous When the gametes are of equal size.
isolating mechanism Any behavioral or genetic difference that prevents one species from mating with the other.
isomorphic Individuals look alike.
jolting A movement shown by alarmed parental cichlids. Typically, the head snaps to one side and pelvic fins flick downward. Jolting attracts fry to the parent.
kleptoparasitism One animal steals food from another.
larva In cichlids, a stage of development that follows hatching. The larva is incompletely developed, in many ways much like an embryo and is therefore helpless. The plural is larvae.
lek An aggregation of males with small closely packed territories on which they display. Females approach males, spawn, and then depart.
lepidophagous The diet consists chiefly of scales and their covering skin that are rasped from the surface of other fishes.
limnologist One who investigates the physical or biological properties of freshwater systems.
lineage A loosely used term inferring a group of species, sometimes populations, that share a common ancestry. Sometimes interchangeable with tribe.
littoral Along the shore. When broadly defined, includes the shores of lakes as well as of the sea. Usage with regard to depth is inconsistent, but the littoral extends below the intertidal in the sea and at least as deep as plants grow in lakes.
lumen The passage or cavity of a tubular organ, or simply the space occupied by the membrane of a cell. Often applied more generally to mean an opening or "window."
mbuna A cluster of colorful species of herbivorous haplochromine cichlids in Lake Malawi, most of whom dwell on reefs where they busily scrape at the rocky substrate to garner food.
median fins The fins projecting from the midline of a fish, thus the dorsal, caudal, and anal fins.
melanophore A chromatophore (see above) whose pigment is melanin and thus produces dark to black on a fish.
micropyle The tiny opening in an egg through which a sperm enters to effect fertilization.
microsatellite loci The location on the chromosome of tandem-repeat elements called microsatellites.
milt An everyday expression for fish testes and for their excretion.
molariform teeth Teeth that resemble cobblestones–blunt and rounded on top and closely packed. In cichlids, found on the pharyngeal jaws.
molecular biology The field of molecular genetics, but taken in a broader sense to include application of molecular techniques to issues such as speciation and phylogenetic relationships among organisms.
molecular genetics The study of genetics at the level of the molecules that make up the genes and associated elements on the chromosomes.

monogamy A male and female form an enduring pair that remains together at least until one parental cycle has been completed.
monomorphy Existing in only one form, as when a male and female are indistinguishable or nearly so.
monophyletic A group of related taxa that have been produced by one ancestor. Most often used in discussion of large groups, such as the haplochromines in Lake Malawi, but also applied to small species flocks.
morph A distinctively different individual, as opposed to the extremes of continuously varying difference.
morphology Study of the form and structure of animals and plants. Also used when referring to the structure of parts of an animal or plant.
mouth brooding One or both parent cichlids carry the eggs or larvae in the mouth cavity.
natural selection The increased survival and reproduction of individuals better adapted to the environment. Often defined in a much more intricate way.
nuchal hump A fleshy lump on the forehead of a fish. Common among cichlids, especially on males.
obligatory The animal has no option. Contrast with facultative.
ocellus Part of the color pattern. A dark disk with a thin light-colored ring around it. In many cases, it is thought to mimic an eye. The plural is ocelli.
olfaction Sensing an odor.
oophagy Feeding on the eggs of others.
operational sex ratio The ratio of males and females available to mate with one another. This may vary greatly from the actual sex ratio.
opercle The bony plate that covers the gills of a fish.
Ostariophysi The dominant order of freshwater fishes. It includes the minnows, suckers, catfishes, and weakly electric fishes.
osteology A branch of biology dedicated to the study of bony structures.
ostracophilous Literally, loving snails. Refers to the tiny cichlids that inhabit empty snail shells.
otoliths The three "ear bones" of fishes that enable balancing and hearing.
ovigerous Carrying eggs. Customarily used in reference to a mouth-brooding cichlid whose mouth is full of eggs or larvae.
paedophagous Among cichlids, those who eat larvae and fry, but often used broadly to include the eating of eggs (oophagy).
papilliform teeth Relatively slender teeth that are typically elongate and pointed.
parasitic spawning While the female and the dominant male are spawning, a subordinate male sneaks in and tries to fertilize some of the eggs.
pedicle A small, short, and slender stalk, regularly with a larger base.
pelagic Adapted to a life in open water, away from the bottom.
pendle A behavior in which two fighting cichlids face one another and one advances while the other retreats, and then they exchange roles, resulting in ebbing back and forth like a pendulum.
Perciformes The largest and most diverse order of fishes. Spiny-rayed fishes.
Percomorpha Used here almost interchangeably with Perciformes, but it is one level higher in the classification of fishes. The Perciformes are one of several orders within this division.
pharyngeal teeth Teeth that cover the two plates that make up the pharyngeal jaws.
pheromones Chemical signals.
phylogeny The history of the evolution of a group of taxa such as species from their common ancestors. This includes the order from which they branched off and, when possible, when that happened.
phylogeography An analysis of the phylogeny of a group of taxa in relation to their geographic distribution, for example, the mbuna of Lake Malawi.
phytoplankter One who eats phytoplankton.
phytoplankton Microscopic to barely visible plants such as algae that live suspended in the water.
piscivore One who eats fishes.
planktivore One who eats plankton.
polyandry One female forms enduring bonds with more than one male.

polychromatic Occurring in more than one discontinuous color type or morph.
polygynandry A male mates with multiple females, and a female mates with multiple males.
polygyny A male mates with multiple females.
polymorphism Occurring in more than one discontinuous type or morph. Commonly refers just to shape or size but may also be used for coloration.
polyphyletic A group of similar and possibly related taxa that have actually been produced by more than one ancestor. Most often used in discussions of large groups, such as the cichlid fishes. Compare with monophyletic.
premaxilla In perciform fishes such as cichlids, the bone that forms the outer, upper jaw. The other element of that jaw, the maxilla, has been excluded from the gape and sits as a tiny bone atop the maxilla at the angle of the mouth.
primitive A loosely used term that connotes an early state in the evolution of a trait of a group of species. Often employed as synonym of basal.
process In morphology, a part of a structure, for example, a protrusion.
propagule A stage of development specialized for dispersal. In fishes, this takes the form of a pelagic egg or larva.
protandrous A sequential hermaphrodite that starts life as a male and later becomes a female.
protogynous A sequential hermaphrodite that starts life as a female and later becomes a male.
radiation Here meaning evolutionary radiation. A pattern of rapid diversification into many species without assuming the differences are adaptive (though they most likely are).
reproductive barrier Same as isolation mechanism (see above).
resource Any element of the environment that, by controlling it, an animal can increase its fitness. This includes both biological and physical components, such as place of refuge, food, or mate. If the resource is in short supply and critical, it becomes a limiting resource.
riverine Of or dwelling in rivers and streams; inflowing water.
saggita The largest otolith (see above) in the fish ear.
secretocytes Highly specialized cells in the skin of the discus fish that secrete nutrients for their fry.
sexual selection, intersexual The evolutionary changes in male traits that result from females choosing males based on those traits.
sexual selection, intrasexual The evolutionary changes in male traits that result from competition among males for access to females.
sigmoid position A posture taken by fish in social interactions. Seen from above, the body is bent into an S-shape.
speciation Evolution of reproductive isolation that produces two or more descendent species.
species A group of populations whose members interbreed or potentially interbreed with each other in nature. They share a common reservoir of genes, called a gene pool.
spermatozeugma A mucus packet of sperm.
stasis The absence of change in one or more characters for a long period, on an evolutionary timescale.
stripe Prominent dark markings on a fish that run longitudinally along the body or head.
substrate brooder A cichlid that lays its eggs on the substrate, for example, a rock, leaf, or hair roots, and nurtures the eggs and subsequently the larvae and fry.
symbiosis A close association between two or more species.
sympatry Two species or populations occupy the same geographic locality and have the opportunity to interbreed.
systematics The branch of biology that deals with the evolutionary relationships of animals and plants.
taxon With reference to a particular grouping of species at any level. Thus a species is a taxon, but so is a genus, a family, and so on. Plural is taxa.
taxonomy Similar to systematics but with more attention to the process of naming species and higher taxa.
teleost A bony fish. Often treated at different levels by taxonomists, but here used to mean the Euteleostei, or "true" bony fishes.
territoriality The behavior of defending a place that contains some limiting resource. That may be food, a nesting site, or a refuge.

tetrapod Any of the many vertebrate animals that have four limbs, including snakes because they originally had four legs.
trophic Pertaining to gathering and eating food.
utaka A cluster of species of planktivorous haplochromine cichlids in Lake Malawi that feed in open water; typically they are silvery and elongate.
vagility The relative ability of an animal to move about, as when dispersing.
zooplankter An animal that captures and eats animal plankton.
zooplankton Small to microscopic animals that live suspended in water.
zygote The organism that results from the fusion of two gametes, here the egg and sperm.

NUMERICAL GUIDE TO CITATIONS

Introduction

1. Springer, V.G. (1982)
2. Turner, G.F. (1994b)
3. Turner, G.F. (1997)
4. Kullander, S.O. (1998)
5. Maitland, P.S. (1977)
6. Helfman, G.S., Collette, B.B., et al. (1997)
7. Myers, G.S. (1960)
8. Grant, P.R. & Grant, B.R. (1998)
9. Brooks, J.L. (1950)
10. Poll, M. (1986)
11. Ribbink, A.J., Marsh, B.A., et al. (1983)
12. McKaye, K.R. (1983)
13. Stauffer, J.R., Jr., Bowers, N.J., et al. (1996)
14. Greenwood, P.H. (1974)
15. Barel, C.D.N., Ligtvoet, W., et al. (1991)
16. Kornfield, I. & Parker, A. (1997)
17. Johnson, T.C. Scholz, C. A., et al. (1996)
18. Nielsen, J.L. (1994)

Chapter 1

1. Stiassny, M.L.J. & Raminosoa, N. 1994)
2. McCune, A.R. (1987)
3. Long, J.A. (1995)
4. Forey, P.L. (1995)
5. Hoogland, R.D., Morris, D., et al. (1957)
6. Myers, G.S. (1966)
7. Kocher, T. & Stepien, C. (1997)

8. Helfman, G.S., Collette, B.B., et al. (1997)
9. Barlow, G.W. & Munsey, J.W. (1976)
10. Miller, R.R. (1966)
11. Stiassny, M.L.J. (1991)
12. Nelson, J.S. (1994)
13. Queiroz, K.d. & Gauthier, J. (1994)
14. Kocher, T. & Stepien, C. (1997)
15. Meyer, A. (1993)
16. Queller, D.C., Strassmann, J.E., et al. (1993)
17. Strauss, E. (1999)
18. Stepien, C.A. & Kocher, T.D. (1997)
19. Stiassny, M.L.J. (1991)
20. Stiassny, M.L.J. (1994)
21. Nelson, J.S. (1994)
22. Streelman, J.T. & Karl, S.A. (1997)
23. Farias, I.P., Orti, G., Sampaio, I., et al. (1999)
24. Roberts, T.R. & Stewart, D.J. (1976)
25. Gaemers, P.A.M. (1986)
26. Zihler (1982)
27. Martinez-Palacios, C.A., Ross, L.G., et al. (1996)
28. Whitfield, A.K. & Blaber, S.J.M. (1979)
29. Coe, M.J. (1966)
30. Stiassny, M.L.J. (1991)
31. Stiassny, M.L.J. (1994)
32. Geerts, M. (1996)
33. Stiassny, M.L.J. (1994)
34. Kullander, S.O. (1998)
35. Stiassny, M.L.J. (1994)

Chapter 2

1. Rossiter, A. (1995)
2. Coulter, G.W. (1991)
3. McKaye, K.R. & Marsh, A. (1983)
4. Taylor, F.J.R. (1999)
5. Yamaoka, K. (1991)
6. Marsh, B.A., Marsh, A.C. et al. (1986)
7. Barlow, G.W. (1974b)
8. Winemiller, K.O., Kelso-Winemiller, L.C., et al. (1995)
9. Randall, J.E. & Hartman, W.D. (1968)
10. Trewavas, E., Green, J. & Corbet, S.A. (1972)
11. Trewavas, E. (1962)
12. Stiassny, M.L.J., Schliewen, U.K. et al. (1992)
13. Kohda, M., Hori, M. & Snhombo, M. (1997)
14. Yamaoka, K. (1991)
15. Montgomery, W.L. (1975)
16. Yamaoka, K. (1991)
17. Barlow, G.W. & Munsey, J.W. (1976)
18. Yamaoka, K. (1991)
19. Losey, G.S., Jr. (1978)
20. Kocher, T. & McKaye, K.R. (1983)
21. Dominey, W.J. & Snyder, A.M. (1988)
22. Yuma, M., Narita, T., et al. (1998)
23. Bergmann, H.-H. (1968)

24. Wyman, R.L. & Ward, J.A. (1972)
25. Ribbink, A.J. & Lewis, D.S.C. (1982)
26. Stauffer, J.R., Jr. (1991)
27. Eccles, D.H. & Lewis, D.S.C. (1976)
28. Roberts, T.R. (1970)
29. Springer, V.G. & Smith-Vaniz, W. (1972)
30. Trewavas, E. (1947)
31. Loiselle, P.V. (1985)
32. Marlier, G. & LeLeup, N. (1954)
33. Nshombo, M. (1991)
34. Liem, K.F. & Stewart, D.J. (1976)
35. Hori, M. (1993)
36. Mboko, S.K., Kohda, M. et al. (1998)
37. Stauffer, J.R., Jr. (1991)
38. Ribbink, A.J. (1984)
39. McKaye, K.R. (1981)
40. Lucanus, O. (1998)
41. Ribbink, A.J. & Ribbink, A.C. (1997)
42. Ribbink, A.J. & Lewis, D.S.C. (1982)
43. Ribbink, A.J. & Ribbink, A.C. (1997)
44. Barel, C.D.N., Ligtvoet, W., et al. (1991)
45. Nshombo, M. (1991)
46. McKaye, K.R. & Marsh, A. (1983)
47. Gottfried, M.D. (1986)
48. Montgomery, W.L. (1977)
49. Norris, K.S. & Prescott, J.H. (1959)
50. Barel, C.D.N. (1983)
51. Liem, K.F. (1979)
52., 53., 54., 55. Liem, K.F. (1991)

Chapter 3

1. Edwards, A.W.F. (1998)
2. Warner, R.R. (1984)
3. Ghiselin, M.T. (1969)
4. Warner, R.R. (1975)
5. Buxton, C.D. (1993)
6. Moyer, J.T. & Nakazono, A. (1978)
7. Harrington, R.W. (1971)
8. Williams, G.C. (1975)
9. Williams, G.C. (1988)
10. Fischer, E.A. (1981)
11. Barlow, G.W. (1975)
12. Kuwamura, T., Nakashima, Y., et al. (1994)
13. St Mary, C.M. (1998)
14. Barlow, G.W. (1981)
15. Francis, R.C. (1992)
16. Bull, J.J. (1980)
17. Struessmann, C.A., Cota, J.C.C., et al. (1996)
18. Conover, D.O. (1984)
19. Heiligenberg, W. (1965)
20. Römer, U. & Beisenherz, W. (1996)
21. Baroiller, J., Chourrout, D., et al. (1995)
22. Desprez, D. & Melard, C. (1998)

23. Baroiller, J., Chourrout, D., et al. (1995)
24. Tuan, P.A., Little, D.C., et al. (1998)
25. Crapon de Caprona, M.D. & Fritzsch, B. (1984)
26. Lovshin, L.L. (1982)
27. Peters, H.M. (1975)
28. Ohm, D. (1978)
29. Barlow, G.W. (1970)
30. Francis, R.C. & Barlow, G.W. (1993)
31. Brown, M.E. (1957)
32. Short, R.V. & Balaban, E. (1994)
33. Francis, R.C., Soma, K., et al. (1993)
34. Fernandes, M.D.O. & Volpato, G.L. (1993)
35. Fox, H.E., White, S.A., et al. (1997)
36. Francis, R.C. (1992)
37. Yamamoto, T. (1975)37.
38. Bull, J.J. (1980)

Chapter 4

1. Kleiman, D.G. (1977)
2. Westneat, D.F., Sherman, P.W., et al. (1990)
3. Blumer, L.S. (1979)
4. Goodwin, N.B., Balshine-Earn, S., et al.(1998)
5. Williams, G.C. (1975)
6. Trivers, R.L. (1972)
7. Barlow, G.W. (1981)
8. Maynard Smith, J. (1977)
9. Vincent, A., Ahnesjö, I., et al. (1992)
10. Ward, J.A. & Samarakoon, J.I. (1981)
11. Barlow, G.W. (1991)
12. McKaye, K.R. & Barlow, G.W. (1976)
13. Rogers, W. (1987)
14. McKaye, K.R. (1986a)
15. Barlow, G.W. & Siri, P. (1997)
16. Keenleyside, M.H.A. (1985)
17. Mertz, J.C. (1967)
18. Meral, G.H. (1973)
19. Beeching, S.C., Gross, S.H., et al. (1998)
20. Meral, G.H. (1973)
21. Barlow, G.W. (1974a)
22. Nuttall, D.B. & Keenleyside, M.H.A. (1993)
23. Heiligenberg, W. (1964)
24. Warzel, F. (1996)
25. Oehlert, B. (1958)
26. Stewart, D.J. & Roberts, T.R. (1984)
27. Goodwin, N.B., Balshine-Earn, S., et al.(1998)
28. Leibel, W.S. (1998a)
29. Yanagisawa, Y. & Nshombo, M. (1983)
30. Kuwamura, T. (1988)
31. Yanagisawa, Y. (1986)
32. Seehausen, O. (1997)
33. Kuwamura, T. (1997)
34. Kondo, T. (1987)
35. Matsumoto, K. & Kohda, M. (1998)
36. Loiselle, P.V. (1985)

37. Newman, L. (1993)
38. Trewavas, E. (1983)
39. Schwanck, E. & Rana, K. (1991)
40., 41. Trewavas, E. (1983)
42. Aronson, L.R. (1949)
43. Oppenheimer, J.R. & Barlow, G.W. (1968)
44. Barlow, G.W. (1991)
45. Balshine-Earn, S. & McAndrew, B.J. (1995)
46. Barlow, G.W. (1993)
47. Kawanabe, H., Hori, M., et al. (1997)
48. Yanagisawa, Y. & Nishida, M. (1991)
49. Ochi, H. (1993)
50. Ribbink, A.J., Marsh, B.A., et al. (1983)
51. Balmford, A. (1991)
52. Kruijt, J.P., Vos, G.J.d., et al. (1972)
53. Andersson, M. (1994)
54. Kodric-Brown, A. & Brown, J.H. (1984)
55. Vehrencamp, S.L. & Bradbury, J.W. (1984)
56. Loiselle, P.V. & Barlow, G.W. (1978)
57. Andersson, M. (1994)
58. Högland, J. & Alatalo, R.V. (1995)
59. McKaye, K.R. (1983)
60. Kellogg, K.A., Markert, J.A., et al. (1995)
61. Ochi, H. (1996)
62. Ochi, H. (1993)
63. Karino, K. (1996)
64. Faaborg, J. (1981)
65. Borgerhoff Mulder, M. (1991)
66. Yamagishi, S. & Kohda, M. (1996)
67. Martin, E. & Taborsky, M. (1997)
68. Haussknecht, T. & Kuenzer, P. (1991)
69. Rossiter, A. (1995)
70. Walter, B. & Trillmich (1994)
71. Konings, A. (1988)
72., 73. Schütz, D., Taborsky, M., et al. (1997)
74. Sato, T. (1994)
75. Grinnell, J. & McComb, K. (1996)
76. Hiraiwa-Hasegawa, M. (1988)
77. Emlen, S.T. (1997)
78. Emlen, S.T. (1995)
79. Taborsky, M. (1994)

Chapter 5

1. Jakobsson, S., Brick, O., et al.(1995)
2. Baylis, J.R. (1981)
3. Barlow, G.W. (1993)
4. Barlow, G.W. (1987)
5. Sikkel, P.C.(1995)
6. Barlow, G.W. (1974)
7. Myrberg, A.A.J., Montgomery, W.L., et al. (1988)
8. Li, H.W. & Brocksen, R.W. (1977)
9. Barlow, G.W. (1976)
10. Power, M.E. (1990)
11.Barlow, G.W. (1993)

12. Ribbink, A.J., Marsh, B.A., et al. (1983)
13. Jaeger, R.G. (1981)
14. Yuma, M. & Kondo, T. (1997)
15. Jaeger, R.G. (1981)
16. Barlow, G.W. (1974c)
17. Apfelbach, R. (1967)
18. Genner, M.J., Turner, G.F., et al. (1999)
19. Barlow, G.W. (1974)
20. Barlow, G.W. (1974b)
21. Williams, G.C. (1989)
22. Marsh, A.C. & Ribbink, A.J. (1986)
23. Kohda, M. & Takemon, Y. (1996)
24. Ribbink, A.J. (1991)
25. Baerends, G.P. & Baerends-van Roon, J.M. (1950)
26. Gray, J.A.B. & Denton, E.J. (1991)
27. Hurd, P.L. (1997)
28. Barlow, G.W., Rogers, W. & Fraley, N. (1986)
29. Maynard Smith, J. (1974)
30. Simpson, M.J.A. (1968)
31. Maynard Smith, J. (1979)
32. Zahavi, A. (1979)
33. Maynard Smith, J. (1974)
34. Enquist, M., Leimar, O., et al. (1990)
35. McKaye, K.R. & Barlow, G.W. (1976)
36. Enquist, M., Leimar, O., et al. (1990
37. Keeley, E.R. & Grant, J.W.A. (1993)
38. Braddock, J.C. (1949)
39. Davies, N.B. (1978)
40. Barlow, G.W. (1983)
41. Barlow, G.W., Rogers, W. & Fraley, N. (1986)
42. McKaye, K.R. (1977a)
43. Barlow, G.W. (1983)
44. Holder, J.L., Barlow, G.W. et al. (1991)
45. Neat, F.C., Huntingford, F.A., et al. (1998)
46. Neat, F.C. (1998)

Chapter 6

1. Dickman, M.C., Annett, C., et al. (1990)
2. Fox, H.M. & Vevers, G. (1960)
3. Baerends, G.P. & Baerends-van Roon, J.M. (1950)
4. Lanzing, W.J.R. & Bower, C.C. (1974)
5. Seehausen, O., Mayhew, P.J., et al. (1999)
6. Barlow, G.W. (1972)
7. Barlow, G.W. & Munsey, J.W. (1976)
8. Baerends, G.P., Wanders, J.B.W., et al. (1986)
9. Hurd, P.L. (1997)
10. Baldaccini, N.E. (1973)
11. Beeching, S.C. (1995)
12. Heiligenberg, W., Kramer, U., et al. (1972)
13. Beeching, S.C. (1993)
14. Radesäter, T. & Fernö, A. (1979)
15. Evans, M.R. & Norris, K. (1996)
16. Barlow, G.W. (1976)
17. Webber, R., Barlow, G.W., et al. (1973)

18. Barlow, G.W. (1983)
19. Barlow, G.W. & McKaye, K.R. (1982)
20. Barlow, G.W. (1983)
21. Barlow, G.W. & Siri, P. (1994)
22. Marler, P. (1955)
23. Barlow, G.W. & Siri, P. (1994)
24. Lobel, P.S. (1998)
25. Brown, D.H. & Marshall, J.A. (1978)
26. Baylis, J.R. (1974)
27. Lanzing, W.J.R. (1974)
28. Schwarz, A. (1974)

Chapter 7

1. Thornhill, R. & Gangestad, S.W. (1996)
2. Langlois, J.H. & Roggman, L.A. (1990)
3. Møller, A.P. & Thornhill, R. (1998)
4. Clarke, G.M. (1998)
5. Symons, D. (1995)
6. Wedekind, C. & Furi, S. (1997)
7. Darwin, C. (1871)
8. Williams, G.C. (1992)
9. West-Eberhard, M.J. (1983)
10. Zahavi, A. (1975)
11. Ryan, M.J. (1985)
12. Andersson, M. (1994)
13. Fisher, R.A. (1930)
14. Kirkpatrick, M. & Ryan, M.J. (1991)
15. Kodric-Brown, A. & Brown, J.H. (1984)
16. Zahavi, A. (1975)
17. Alatalo, R.V. & Ratti, O. (1995)
18. Ryan, M.J. (1994)
19. Endler, J.A. (1992)
20. Andersson, M. (1994)
21. Trewavas, E. (1983)
22. Barlow, G.W. (1974a)
23. McKaye, K.R. (1991)
24. Ribbink, A.J., Marsh, B.A., et al. (1983)
25. Seehausen, O., Mayhew, P.J., et al. (1999)
26. Knight, M.E. & Turner, G.F. (1999)
27. Baerends, G.P. & Baerends-van Roon, J.M. (1950)
28. Lowe (McConnell), R.H. (1956)
29. Stauffer, J.R.J., LoVullo, T.J., et al. (1993)
30. Lowe (McConnell), R.H. (1956)
31. Burchard, J.E. (1965)
32. Greenwood, P.H. (1991)
33. Lowe (McConnell), R.H. (1956)
34. Lowe (McConnell), R.H. (1959)
35. McKaye, K.R. (1991)
36. Karino, K. (1997a)
37. Krüter, R. (1991)
38. Falter, U. & Charlier, M. (1989)
39. Holzberg, S. (1978)
40. Stauffer, J.R., Jr., Bowers, N.J., et al. (1997)
41. Seehausen, O. & Alphen, J.J.M.v. (1998)

42.Ribbink, A.J., Marsh, B.A., et al. (1983)
43.Knight, M.E. & Turner, G.F. (1999)
44. Crapon de Caprona, M.-D. (1980)
45. Falter, U. & Dolisy, D. (1989)
46. Hert, E. (1986b)
47. Hert, E. (1991)
48. Kellogg, K.A., Markert, J.A., et al. (1998)
49. Stauffer, J.R., Jr. & Kellogg, K.A. (1996)
50. McKaye, K.R., Louda, S.M. et al. (1990)
51. Taylor, M.I., Turner, G.F., et al. (1998)
52. Karino, K. (1997a)
53. Nelson, C.M. (1995)
54. Clarke, G.M. (1998)
55. Thornhill, R. & Gangestad, S.W. (1996)
56. Brooks, R. & Caithness, N. (1995)
57. Sheridan, L. & Pomiankowski, A. (1997)
58. Karino, K. (1997a)
59. Lobel, P.S. (1998)

Chapter 8

1. Goodwin, N.B., Balshine-Earn, S., et al. (1998)
2. Bereczkei, T., Voros, S., et al. (1997)
3. Francis, R.C. (1990)
4. Baerends, G.P. (1986)
5. Barlow, G.W., Rogers, W., et al. (1977)
6. Barlow, G.W. (1976)
7. Barlow, G.W. (1968)
8. Coleman, R.M. (1993)
9. Barlow, G.W. (1991)
10. Seehausen, O., Mayhew, P.J., et al. (1999)
11. Barlow, G.W. & Siri, P. (1994)
12. Heiligenberg, W., Kramer, U., et al. (1972)
13. Crapon de Caprona, M.-D. (1986)
14. Barlow, G.W., Francis, R.C., et al. (1990)
15. Barlow, G.W. & Rogers, W. (1978)
16. Kosswig, G. (1947)
17. Barlow, G.W. & Munsey, J.W. (1976)
18., 19. Baylis, J.R. (1976)
20. Holder, J.L. (1991)
21. Hale, E.A., Stauffer, J.R., Jr., et al. (1998)
22. Holder, J.L., Barlow, G.W., et al. (1991)
23. Warner, R.R. (1978)
24. Noble, G.K. & Curtis, B. (1939)
25. Loiselle, P.V. (1985)
26. Bleick, C.R. (1975)
27. Meek, S.E. (1907)
28. Barlow, G.W. & Siri, P. (1997)
29. Holder, J.L. (1991)
30. Schwarz, A. (1974)
31. Balshine-Earn, S. & Lotem, A. (1998)
32. Noble, G.K. & Curtis, B. (1939)
33. Reebs, S.G. (1994)
34. Barlow, G.W. (1970)
35. Kano, T. (1989)

36. Hunt, G.L., Jr. & Hunt, M.W. (1977)
37. Greenberg, B. (1961)
38. Lamprecht, J. (1973)
39. Trivers, R.L. (1972)
40. Holder, J.L. (1991)
41. Greenberg, B. (1961)
42. Rogers, W. (1988)
43. Rogers, W. & Barlow, G.W. (1991)
45. Peters, H.M. (1957)
46. Barlow, G.W. (1968)
47. Barlow, G.W. (1970)
48. Noonan, K.C. (1983)
49. Rogers, W. & Barlow, G.W. (1991)
50. Rogers, W. (1995)
51. Barlow, G.W., Francis, R.C., et al. (1990)
52. Holder, J.L., Barlow, G.W., et al. (1991)
53. Schwanck, E. (1987)
54. Barlow, G.W. (1992)
55. Nuttall, D.B. & Keenleyside, M.H.A. (1993)
56. Lorenz, K.Z. (1966)
57. Barlow, G.W. & Siri, P. (1994)
58. Barlow, G.W. (1983)
59. Barlow, G.W., Rogers, W., et al. (1977)
60. Baylis, J.R. (1976)
61. Westneat, D.F., Sherman, P.W., et al. (1990)
62. Thornhill, R. & Gangestad, S.W. (1996)
63. Baker, R.R. & Bellis, M.A. (1995)

Chapter 9

1. Marconato, A., Rasotto, M.B. et al. (1996)
2. Billard, R. (1986)
3. De Silva, S.S., Maitipe, P., et al. (1984)
4. Albrecht, H., Apfelbach, R., et al. (1968)
5. McKaye, K.R. (1977)
6. Lavery, R.J. (1991)
7. Barlow, G.W. (1964)
8. Weber, P.G. & Weber, S.P. (1971)
9. Konings, A. (1998)
10. Ward, J.A. & Wyman, R.A. (1975)
11. Barlow, G.W. (1974a)
12. Keenleyside, M.H.A. & Bietz, B. (1981)
13. Timms, A.M. & Keenleyside, M.H.A. (1975)
14. Keenleyside, M.H.A. & Prince, C.E. (1976)
15. Nakai, K., Yanagisawa, Y., et al. (1990)
16. Knowlton, N. (1979)
17. Aronson, L.R. (1951)
18., 19. Kraft, A.v. & Peters, H.M. (1963)
20. Ostrander, G.K. & Ward, J.A. (1985)
21, 22. Stiassny, M.L.J. & Mezey, J.G. (1993)
23. Wickler, W. (1956)
24., 25. Bern, O. & Avtalion, R.R. (1990)
26. Peters, H.M. (1971)
27. Sato, T. (1994)
28. Kellogg, K.A., Markert, J.A., et al. (1998)

29., 30. Baerends, G.P. & Baerends-van Roon, J.M. (1950)
31. Lowe (McConnell), R.H. (1956)
32. Trewavas, E. (1983)
33. Wickler, W. (1997)
34. Lowe (McConnell), R.H. (1956)
35. Wickler, W. (1997)
36. Wickler, W. (1965)
37. Grier, H.J. & Fishelson, L. (1995)
38. Taborsky, M. (1998)
39. Baerends, G.P. & Baerends-van Roon, J.M. (1950)
40. Kirchshofer, R. (1953)
41. Wickler, W. (1962a)
42. Wickler, W. (1962b)
43. Wickler, W. (1962a)
44. Loiselle, P.V. (1985)
45. Jackson, P.B.N. & Ribbink, T. (1975)
46. Hert, E. (1989)
47. Wickler, W. (1962a)
48. Mrowka, W. (1987)
49. Hert, E. (1986b)
50. Hert, E. (1986a)
51. Hert, E. (1989)
52. Goldschmidt, T. (1991)
53., 54. Ochi, H. (1996)
55. Nelissen, M. (1976)
56. Wickler, W. (1962a)
57. Konings, A. (1999)
58. Mrowka, W. (1987)
59. Goldschmidt, T. (1991)
60., 61. Hert, E. (1986a)
62. Newman, L. (1993)
63., 64 Birkhead, T.R. & Møller, A.P. (1992)
65. Taborsky, M. (1998)
66. Quinn, T.P., Adkison, M.D., et al. (1996)
67. Taborsky, M. (1994)
68. Albrecht, H. (1968)
69. Ribbink, A.J. & Chan, T.-Y. (1989)
70. Crapon de Caprona, M.-D. (1986)
71. Parker, A. & Kornfield, I. (1996)
72. Ochi, H. (1996)
73. Sato, T. (1994)

Chapter 10

1. Williams, G.C. (1966)
2. Nagoshi, M. & Yanagisawa, Y. (1997)
3. Williams, G.C. (1966)
4. Trivers, R.L. (1972)
5. Taborsky, M. (1985)
6. Mrowka, W. (1987a)
7. Rangeley, R.W. & Godin, J.-G.J. (1992)
8. Barlow, G.W. (1976)
9. Greenberg, B., Zijlstra, J.J., et al. (1965)
10. Lamprecht, J. (1973)
11. Nakano, S. & Nagoshi, M. (1990)

12. Ward, J.A. & Wyman, R.L. (1977)
13. De Silva, S.S., Maitipe, P., et al. (1984)
14. Ward, J.A. & Samarakoon, J.I. (1981)
15. Hale, E. & Stauffer, J.R., Jr. (1997)
16. McKaye, K.R. (1977)
17. Coleman, R.M. & Galvani, A.P. (1998)
18. Hale, E. & Stauffer, J.R., Jr. (1997)
19. McKaye, K.R. (1977)
20. Bergmann, H.-H. (1968)
21. Chien, A.K. & M.S., S. (1972)
22. Zoran, M.J. & Ward, J.A. (1983)
23. Rechten, C. (1980)
24. Lamprecht, J. (1973)
25. Reebs, S.G. & Colgan, P.W. (1991)
26. Baylis, J.R. (1974)
27. Heijns, W. (1996)
28., 29. Chien, A.K. & M.S., S. (1972)
30. Praetorius, W. (1932)
31. Baylis, J.R. (1974)
32. Courtenay, S.C. & Keenleyside, M.H.A. (1983)
33. Jones, A.J. (1972)
34. Kühme, W. (1963)
35. Loiselle, P.V. (1985)
36. Leibel, W.S. (1998b)
37. Leibel, W.S. (1992)
38., 39. Dambach, M. & Wallert, I. (1966)
40. Oppenheimer, J.R. (1970)
41. Tyler, M.J. & Carter, D.B. (1981)
42. Peters, H.M. (1965)
43. Oppenheimer, J.R. & Barlow, G.W. (1968)
44. Nagoshi, M. & Yanagisawa, Y. (1997)
45. Balon, E.K. (1981)
46. Williams, N.J. (1972)
47. Dambach, M. (1963)
48. Neil, E.H. (1964)
49. Wickler, W. (1959)
50. Paulo, J. (1995)
51. Hale, E. & Stauffer, J.R., Jr. (1997)
52. Sjölander, S. (1972)
53. Hale, E. & Stauffer, J.R., Jr. (1997)
54. Barlow, G.W. (1976)
55. Ward, J.A. & Wyman, R.L. (1977)
56. Ward, J.A. & Samarakoon, J.I. (1981)
57. Magurran, A.E. (1990)
58. Williams, G.C. (1989)
59. Baerends, G.P. & Baerends-van Roon, J.M. (1950)
60. Perrone, M. (1978)
61., 62., 63. Baerends, G.P. & Baerends-van Roon, J.M. (1950)
64. Shennan, M.G.C., Waas, J.R., et al. (1994)
65. Chien, A.K. & M.S., S. (1972)
66. Cole, J.E. & Ward, J.A. (1969)
67. Ward, J.A. & Wyman, R.A. (1975)
68. Gaines, S.D. & Lubchenco, J. (1982)
69. Kramer, D.L. & Bryant, M.J. (1995)
70. Yamaoka, K. (1985)
71. Martin, E. & Taborsky, M. (1997)
72. Williams, N.J. (1972)

73. Wisenden, B.D., Lanfranconi-Izawa, L., et al. (1995)
74. Meral, G.H. (1973)
75. Balon, E.K. (1985)
76. Mrowka, W. (1987a)
77. Yanagisawa, Y. & Ochi, H. (1991)
78., 79. Yanagisawa, Y. & Sato, T. (1990)
80. Yanagisawa, Y., Ochi, H., et al. (1996)
81., 82. Noakes, D.L.G. & Barlow, G.W. (1973)
83. Schütz, M. (1995)
84. Noakes, D.L.G. (1973)
85. Oosten, J.v. (1957)
86. Noakes, D.L.G. (1973)
87. Schütz, M. & Barlow, G.W. (1997)
88., 89 Ward, J.A. & Barlow, G.W. (1967)
90. Wyman, R.L. & Ward, J.A. (1973)
91. Ward, J.A. & Barlow, G.W. (1967)
92. Quertermus, C.J., Jr. & Ward, J.A. (1969)
93. Ward, J.A. & Barlow, G.W. (1967)
94. Blüm, V. & Fiedler, K. (1964)
95. Hildemann, W.H. (1959)
96. Bremer, H. & Walter, U. (1986)
97. Blüm, V. (1966)
98. Nicoll, C.S. & Bern, H.A. (1971)
99. Loiselle, P.V. (1985)
100. Nagoshi, M. (1983)
101. Yanagisawa, Y. (1987)
102. Kuwamura, T. (1986)
103. Baerends, G.P., Bennema, B.E., et al. (1960)
104. Barnett, C. (1981)
105. Noble, G.K. & Curtis, B. (1939)
106. Brestowsky, M. (1968)
107. Baerends, G.P. (1993)
108. Kuenzer, E. and P. Kuenzer (1962)
109. Weber, P.G. (1970)
110. Noble, G.K. & Curtis, B. (1939)
111. Greenberg, B. (1963)

Chapter 11

1. Sjölander, S. (1972)
2. Ward, J.A. & Wyman, R.L. (1977)
3. Ribbink, A.J., Marsh, A.C., et al. (1981)
4., 5. McKaye, K.R. & McKaye, N.M. (1977)
6., 7. Baylis, J.R. (1974)
8. Wisenden, B.D. (1994)
9. Fraser, S.A., Wisenden, B.D., et al. (1993)
10. Valerio, M. & Barlow, G.W. (1986)
11. Fraser, S.A. (1996)
12. Wisenden, B.D. & Keenleyside, M.H.A. (1995)
13. Fraser, S.A. (1996)
14. Wisenden, B.D. & Keenleyside, M.H.A. (1992)
15. Wisenden, B.D. & Keenleyside, M.H.A. (1994)
16., 17. Wisenden, B.D. & Keenleyside, M.H.A. (1992)

18. Yanagisawa, Y. (1985)
19. Sargent, R.C. & Gross, M.R. (1986)
20. Schwanck, E. (1986)
21. Yanagisawa, Y. (1985)
22. Yanagisawa, Y. (1986)
23. Yanagisawa, Y. (1985)
24. Davies, N.B. & Brooke, M. (1991)
25. Finley, L. (1983)
26. Sato, T. (1986)
27. Wisenden, B.D. (1999)
28. Payne, R.B. (1977)
29. Wisenden, B.D. (1999)
30. Sargent, R.C. & Gross, M.R. (1986)
31. Carlisle, T.R. (1985)
32. Karino, K. (1997b)
33. Lavery, R.J. & Colgan, P.W. (1991)
34. Lavery, R.J. (1995)
35. Wisenden, B.D. & Keenleyside, M.H.A. (1995)
36. Fraser, S.A., Wisenden, B.D., et al. (1993)
37. Lack, D. (1954)
38. Reyer, H.U. (1986)
39. Emlen, S.T. (1991)
40. Wilson, E.O. (1975)
41. Witte, K.E. & Schmidt, J. (1992)
42. Kalas, K. (1976)
43. Taborsky, M. (1984)
44. Taborsky, M. (1994)
45. Yamagishi, S. & Kohda, M. (1996)
46. Taborsky, M. (1985a)
47. Trivers, R.L. (1974)
48. Bateson, P. (1994)
49. Axelrod, R. & Hamilton, W.D. (1981)
50. Limberger, D. (1983)
51. Kalas, K. (1976)
52. Taborsky, M. & Limberger, D. (1981)
53. Taborsky, M. & Grantner, A. (1998)
54. Kalas, K. (1976)
55. Taborsky, M. & Grantner, A. (1998)
56. Siemens, M.v. (1990)
57. Balshine-Earn, S., Neat, F.C., et al. (1998)
58. Siemens, M.v. (1990)
59. Taborsky, M. & Limberger, D. (1981)
60. Taborsky, M. (1985a)
61. Grantner, A. & Taborsky, M. (1998)
62. Woolfenden, G.E. & Fitzpatrick, J.W. (1984)
63. Balshine-Earn, S., Neat, F.C., et al. (1998)
64. Taborsky, M. & Grantner, A. (1998)
65. Dierkes, P., Taborsky, M., et al. (1999)
66. Siemens, M.v. (1990)
67. Hert, E. (1985)
68. Taborsky, M. (1985a)
69. McKaye, K.R. (1977)
70. Coyne, J.A. & Sohn, J.J. (1978)
71. Watanabe, T. (2000)
72. McKaye, K.R. (1986)

73. Smith, N.G. (1968)
74. McKaye, K.R., Mughogho, D.E., et al. (1992)
75. McKaye, K.R. & Oliver, M.K. (1980)

Chapter 12

1. Fryer, G. & Iles, T.D. (1972)
2. Seehausen, O. (1996)
3. Yanagisawa, Y., Ochi, H., et al. (1997)
4. Konings, A. (1998)
5. Owen, R.B. Crossley, R., et al. (1990)
6. Nishida, M. (1997)
7. Banister, K. & Clarke, M.A. (1980)
8. Fryer, G. & Iles, T.D. (1972)
9. Konings, A. (1998)
10. Eccles, D.H. (1986)
11. Johnson, T.C., Scholz, C.A., et al. (1996)
12. Echelle, A.A. & Kornfield, I. (1984)
13. Seehausen, O. (1996)
14. Turner, G.F. (1994)
15. Greenwood, P.H. (1974)
16. Kocher, T.D., Conroy, J.A., et al. (1993)
17. Meyer, A., Kocher, T.D., et al. (1990)
18. Johnson, T.C., Scholz, C.A., et al. (1996)
19. Meyer, A. (1993a)
20. Witte, F. & Oijen, M.J.P.v. (1990)
21. Seehausen, O. (1996)
22. Seehausen, O., Alphen, J.J.M.v., et al. (1999)
23. Seehausen, O. (1996)
24. Seehausen, O. & Bouton, N. (1996)
25. Moran, P., Kornfield, I., (1994)
26. Meyer, A. (1993a)
27. Fryer, G. & Iles, T.D. (1972)
28. Ribbink, A.J., Marsh, B.A., et al. (1983)
29. Nishida, M. (1997)
30. Brichard, P. (1978)
31. Stiassny, M.L.J. (1997)
32. Loiselle, P.V. (1985)
33. Sturmbauer, C., Verheyen, E., et al. (1994)
34. Fryer, G. (1996)
35. Coulter, G.W. (1991)
36. Sturmbauer, C. (1998)
37. Greenwood, P.H. (1984)
38. Stiassny, M.L.J. (1997)
39. Sturmbauer, C. & Meyer, A. (1993)
40. Sturmbauer, C., Verheyen, E., et al. (1994)
41. Greenwood, P.H. (1984)
42. Gashagaza, M.M. (1991)
43. Yamaoka, K. (1991)
44. Mayr, E. (1984)
45. Sturmbauer, C. & Meyer, A. (1993)
46. Gould, S.J. & Eldredge, N. (1993)
47. Coulter, G.W. (1991)
48. Futuyma, D.J. (1998)
49. Johnson, N.K., Remsen, J.V., Jr., et al. (1998)

50. Mayr, E. (1982)
51. Paterson, H.E.H. (1993)
52. Denton, E.J. & Nicol, J.A.C. (1962)
53. Bouton, N., Witte, F., et al. (1999)
54. Goldschmidt, T., Witte, F., et al. (1990)
55. Kawanabe, H., Hori, M., (1997)
56. Liem, K.F. (1973)
57. Claes, G. & Vree, F.D. (1989)
58. Witte, F. & Oijen, M.J.P.v. (1990)
59. Meyer, A. (1989)
60. Bouton, N., Van Os, N., et al. (1998)
61. Kornfield, I. (1991)
62. Meyer, A. (1987)
63. Yamaoka, K. (1991)
64. Robinson, B.W. & Wilson, D.S. (1998)
65. Seehausen, O. (1996)
66. Seehausen, O. & Bouton, N. (1997)
67. Stiassny, M.L.J. (1997)
68. Rüber, L., Verheyen, E., et al. (1998)
69. Grant, P.R. & Grant, B.R. (1998)
70. Fryer, G. (1991)
71. Coulter, G.W., Allanson, B.R., et al. (1986)
72. Ribbink, A.J. (1994)
73. Loiselle, P. (1971)
74. Greenwood, P.H. (1991)
75. Stauffer, J.R., Jr., Bowers, et al. (1996)
76. Elder, H.Y., Garrod, D.J. & Whitehead, P.J.P. (1971)
77. Albertson, R.C., Markert, J.A., et al. (1999)
78. Snoeks, J., Rüber, L. & Verheyen, E. (1994)
79. Vos, L.d. & Snoeks, J. (1994)
80. Ochi, H., Sato, Y. & Yanagisawa, Y. (1999)
81. Vos, L.d. & Snoeks, J. (1994)
82. Thompson, A.B., Allison, E.H., et al. (1996)
83. Lowe-McConnell, R.H. (1975)
84. Greenwood, P.H. (1991)
85. Seehausen, O., Alphen, J.J.M.v., et al. (1999)
86. Trendall, J. (1988)
87. McKaye, K.R. & Gray, W.N. (1984)
88. Ribbink, A.J., Marsh, B.A., et al. (1983)
89. Brichard, P. (1978b)
90. Lowe-McConnell, R.H. (1987)
91. Trendall, J. (1988)
92. Coulter, G.W. (1991)
93. Ribbink, A.J. (1991)
94. Hasler, A.D., Scholz, A.T., et al. (1978)
95. Fryer, G. (1961)
96. Fryer, G. (1996)
97. Seehausen, O. (1996)
98. Oppen, M.J.H.v. et al. (1997)
99. Kohda, M. et al. (1996)
100. Arnegard, M.E. et al. (1999)
101. Dominey, W.J. (1984)
102. Greenwood, P.H. (1965)
103., 104. Greenwood, P.H. (1984)
105. Lande, R. (1981)
106. Dominey, W.J. (1984)
107. Holzberg, S. (1978)

108. Seehausen, O., Witte, F., et al. (1998)
109. Seehausen, O. & Alphen, J.J.M.v. (1998)
110. Seehausen, O., Alphen, J.J.M.v., et al. (1999)
111. Seehausen, O. & Alphen, J.J.M.v. (1998)
112. Seehausen, O., Alphen, J.J.M.v., et al. (1999)
113. Seehausen, O. & Alphen, J.J.M.v. (1998)
114. Arnegard, M.E. et al. (1999)
115. Mayr, E. (1976)
116. Seehausen, O., Alphen, J.J.M.v., et al. (1999
117. Rundle, H.D. & Schluter, D. (1998)
118. Wood, C.C. & Foote, C.J. (1996)
119. Johnson, T.C., Scholz, C.A., et al. (1996)
120. Greenwood, P.H. (1991)
121. Doorn, v.G.S., Noest, A.J., et al. (1998)
122. Kondrashov, A.S. & Shpak, M. (1998)
123. Turner, G. & Burrows, M.T. (1995)
124. Higashi, M., Takimoto, G., et al. (1999)
125. Dickman, M.C., Annett, C., et al. (1990)
126. Bush, G.L. (1994)
127., 128 Roberts, T.R. & Stewart, D.J. (1976)
129. Konings, A. (1998)
130. Stiassny, M.L.J., Schliewen, U.K., et al. (1992)
132. Schliewen, U.K., Tautz, D., et al. (1994)
133., 134 Maley, J.D. et al. (1990)
135., 136. Stiassny, M.L.J., Schliewen, U.K., et al. (1992)
137. Schliewen, U.K., Tautz, D., et al. (1994)
138. Kling, G.W. (1987)
139. Schliewen, U.K., Tautz, D., et al. (1994)
140. Barlow, G.W. & Munsey, J.W. (1976)
141. Ribbink, A.J., Greenwood, P.H., et al. (1991)
142., 143., Nxomani, C., Ribbink, A.J., et al. (1999).
144. McKaye, K.R. (1980)
145. Meyer, A. (1990)
146. Greenwood, P.H. (1998)
147. Barlow, G.W., Francis, R.C., et al. (1990)
148. Barlow, G.W. & Munsey, J.W. (1976)
149. Barlow, G.W. (1998)

Chapter 13

1. Baumgartner, T.R., Soutar, A., et al. (1992)
2. United Nations Food .. . . (1995)
3., 4., 5. Naylor, R.L. et al. (1998)
6. Heggberget, T.G. et al. (1993)
7. Anonymous (1998)
8. Courtenay, W.R., Jr. (1973)
9. Bailey, M. (1998)
10. Trendall, J. (1988a)
11. Hert, E. (1995)
12. Munthali, S.M. & Ribbink, A.J. (1998)
13., 14., 15. Turner, G.F. (1994)
16. Baskin, Y. (1992)
17. Chimits, P. (1957)

18. Pullin, R.S.V. (1991)
19. Gunther, J. (1996)
20. Leveque, C. (1997)
21. Pullin, R.S.V. (1991)
22. Brem, G., Brenig, B., Hoerstgen-Schwark, G., et al. (1988)
23. Bentsen, H.B. et al. (1998)
24. Rahman, M.A., Mak, R., et al. (1998)
25. Marengoni, N.G., Onoue, Y., et al. (1998)
26. Pullin, R.S.V. (1991)
27. Wohlfarth, G.W. & Wedekind, H. (1991)
28. Wohlfarth, G.W. (1994)
29. Pullin, R.S.V. (1991)
30. Wohlfarth, G.W., Rothbard, S.,et al. (1990)
31. Lobel, P.W. (1980)
32. Incer, J. (1976)
33. McKaye, K.R. Ryan, J.D., et al. (1995)
34., 35., 36 Zaret, T.M. & Paine, R.T. (1973)
37. Shafland, P.L. (1995)
38., 39 Futuyma, D.J. (1998)
40. Baskin, Y. (1992)
41. Pitcher, T.J. & Hart, P.J.B. (1995)
42., 43. Reynolds, J.E., Greboval, D.F., et al. (1995)
44. Witte, F. et al. (1992a)
45. Baskin, Y. (1992)
46. Reynolds, J.E., Greboval, D.F., et al. (1995)
47. Ogutu-Ohwayo, R. (1995)
48. Ogutu-Ohwayo, R., Hecky, R.E., et al. (1997)
49. Riedmiller, S. (1994)
50. Reynolds, J.E., Greboval, D.F., et al. (1995)
51. Riedmiller, S. (1994)
52. Reynolds, J.E., Greboval, D.F., et al. (1995)
53. Ogutu-Ohwayo, R., Hecky, R.E., et al. (1997)
54. Riedmiller, S. (1994)
55. Baskin, Y. (1992)
56. Ochumba, P.B.O. (1995)
57. Ogutu-Ohwayo, R., Hecky, R.E., et al. (1997)
58. Kaufman, L. & Ochumba, P. (1993)
59. Ogutu-Ohwayo, R., Hecky, R.E., et al. (1997)
60. Witte, F., Goldschmidt, T., et al. (1995)
61. Witte, F. et al. (1992b)
62. Goundswaard, K. & Witte, F. (1997)
63. Witte, F. et al. (1992b)
64. Kaufman, L. (1992)
65. Goldschmidt, T., Witte, F., (1993)
66., 67. Witte, F. et al. (1992b)
68. Ogutu-Ohwayo, R., Hecky, R.E., et al. (1997)
69. Hecky, R.E. Bugenyi, F.W.B.,et al. (1994)
70. Kaufman, L. & Ochumba, P. (1993)
71. Ochumba, P.B.O. (1995)
72. Kaufman, L. & Ochumba, P. (1993)
73. Ochumba, P.B.O. (1995)
74. Ogutu-Ohwayo, R., Hecky, R.E., et al. (1997)
75. Seehausen, O. (1996)
76. Witte, F., Goldschmidt, T., et al. (1995)
77. Witte, F. et al. (1992b)

78., 79. Witte, F., Goldschmidt, T., et al. (1995)
80. Seehausen, O., Alphen, J.J.M.v., et al. (1997)
81. Crapon de Caprona, M.-D. (1986)
82., 83. Seehausen, O., Alphen, J.J.M.v., et al. (1997)
84. Seehausen, O. (1996)
85. Witte, F. et al. (1992b)
86. Witte, F., Goldschmidt, T., et al. (1995)
87. Seehausen, O. (1999)
88. Seehausen, O. (1996)
89. Goldschmidt, T. (1996)
90. Seehausen, O. (1996)
91. Harrison, I.J. & Stiassny, M.L.J. (1999)

REFERENCES

Alatalo, R. V., and Ratti, O. (1995). Sexy son hypothesis: Controversial once more. *Trends in Ecology and Evolution, 10*(2), 52–53.

Albertson, R. C., Markert, J. A., Danley, P. D., and Kocher, T. D. (1999). Phylogeny of a rapidly evolving clade: The cichlid fishes of Lake Malawi, East Africa. *Proceedings of the National Academy of Science USA, 96*(9), 5107–5110.

Albrecht, H. (1968). Freiwasserbeobachtungen an Tilapien (Pisces, Cichlidae) in Ostafrika. *Zeitschrift für Tierpsychologie, 25*, 377–394.

Albrecht, H., Apfelbach, R., and Wickler, W. (1968). Über die Eigenstandigkeit der Art *Tilapia grahami* Boulenger, in reinem Süsswasser (Pisces, Cichlidae). *Senckenbergiana biologica, 49*(2), 107–118.

Andersson, M. (1994). *Sexual Selection.* Princeton: Princeton University Press.

Anonymous. (1998). *Aquaculture.* Orlando, Florida: Florida Agricultural Statistics Service.

Apfelbach, R. (1967). Kampfverhalten und Brutpflegeform bei Tilapien (Pisces, Cichlidae). *Die Naturwissenschaften, 54*(3), 1–2.

Arnegard, M. E., Markert, J. A., Danley, P. D., Stauffer, J. R., Jr., Ambali, A. J., and Kocher, T. D. (1999). Population structure and colour variation of the cichlid fish *Labeotropheus fuelleborni* Ahl along a recently formed archipelago of rocky habitat patches in southern Lake Malawi. *Proceedings of the Royal Society of London Series B Biological Sciences, 266*(1415), 119–130.

Aronson, L. R. (1949). An analysis of reproductive behavior in the mouthbreeding cichlid fish, *Tilapia macrocephala* (Bleeker). *Zoologica, 34*, 133–158.

Aronson, L. R. (1951). Factors influencing the spawning frequency in the female cichlid fish *Tilapia macrocephala. American Museum Novitates, (1484)*, 1–26.

Axelrod, R., and Hamilton, W. D. (1981). The evolution of cooperation. *Science* (Washington, D.C.), *211*, 1390–1396.

Baerends, G. P. (1986). On causation and function of the pre-spawning behaviour of cichlid fish. *Journal of Fish Biology, 29*(Suppl. A), 107–121.

Baerends, G. P. (1993). A comparative study of stimulus selection in the filial following response of fry of substrate spawning cichlid fish. *Behaviour, 125*(1–2), 79–155.

Baerends, G. P., and Baerends-van Roon, J. M. (1950). An introduction to the study of the ethology of cichlid fishes. *Behaviour Supplement, 1*, 1–242.

Baerends, G. P., Bennema, B. E., and Vogelzang, A. A. (1960). Über die Änderung der Sehschärfe mit dem Wachstum bei *Aequidens portalagrensis* (Hensel) (Pisces, Cichlidae). *Zoologische Jahrbücher, Abteilung Systematik, Ökologie und Geographie der Tiere, 88*, 65–78.

Baerends, G. P., Wanders, J. B. W., and Vodegel, R. (1986). The relationship between marking patterns and motivational state in the pre-spawning behaviour of the cichlid fish *Chromidotilapia guentheri* (Sauvage). *Netherlands Journal of Zoology, 36*, 88–116.

Bailey, M. (1998). Messing about in boats, Part II. *Cichlid News,* 7(3), 15–20.

Baker, R. R., and Bellis, M. A. (1995). *Human Sperm Competition: Copulation, Masturbation, and Infidelity.* London: Chapman and Hill.

Baldaccini, N. E. (1973). An ethological study of reproductive behaviour including the colour patterns of the cichlid fish *Tilapia mariae* (Boulanger). *Monitore zoologico italiano,* 7, 247–290.

Balmford, A. (1991). Mate choice on leks. *Trends in Ecology and Evolution, 6*, 87–92.

Balon, E. K. (1977). Early ontogeny of *Labeotropheus* Ahl, 1927 (Mbuna, Cichlidae, Lake Malawi), with a discussion on advanced protective styles in fish reproduction and development. *Environmental Biology of Fishes, 2(2),* 147–176.

Balon, E. K. (1981). Saltatory processes and altricial to precocial forms in the ontogeny of fishes. *American Zoologist, 21*, 573–596.

Balon, E. K. (1985). *Early life histories of fishes. New developmental, ecological and evolutionary perspectives.* Dordrecht: Dr. W. Junk Publishers.

Balshine-Earn, S., and Lotem, A. (1998). Individual recognition in a cooperatively breeding cichlid: Evidence from video playback experiments. *Behaviour, 135*(3), 369–386.

Balshine-Earn, S., and McAndrew, B. J. (1995). Sex-role reversal in the black-chinned tilapia, *Sarotherodon melanotheron* (Rüppel) (Cichlidae). *Behaviour, 132*(11–12), 861–874.

Balshine-Earn, S., Neat, F. C., Reid, H., and Taborsky, M. (1998). Paying to stay or paying to breed? Field evidence for direct benefits of helping behavior in a cooperatively breeding fish. *Behavioral Ecology, 9*(5), 432–438.

Banister, K., and Clarke, M. A. (1980). A revision of the large *Barbus* (Pisces, Cyprinidae) of Lake Malawi with a reconstruction of the history of the southern African Rift Valley Lakes. *Journal of Natural History, 14*, 483–542.

Barel, C. D. N. (1983). Towards a constructional morphology of cichlid fishes (Teleostei, Perciformes). *Netherlands Journal of Zoology, 33*(4), 357–424.

Barel, C. D. N., Ligtvoet, W., Goldschmidt, T., Witte, F., and Goudswaard, P. C. (1991). The haplochromine cichlids in Lake Victoria: An assessment of biological and fisheries interests. In M. H. A. Keenleyside (Ed.), *Cichlid Fishes: Behaviour, Ecology, and Evolution* (pp. 258–279). New York: Chapman and Hall.

Barlow, G. W. (1964). Ethology of the Asian teleost *Badis badis.* V. Dynamics of fanning and other parental activities, with comments on the behavior of the larvae and postlarvae. *Zeitschrift für Tierpsychologie, 21*, 99–123.

Barlow, G. W. (1968). Effect of size of mate on courtship in a cichlid fish. *Communications in Behavioral Biology, Part A, 2*(4), 149–160.

Barlow, G. W. (1970). A test of appeasement and arousal hypotheses of courtship behavior in a cichlid fish, *Etroplus maculatus. Zeitschrift für Tierpsychologie, 27*, 779–806.

Barlow, G. W. (1972). The attitude of fish eye-lines in relation to body shape and to stripes and bars. *Copeia, 1972*(1), 4–12.

Barlow, G. W. (1974a). Contrasts in social behavior between Central American cichlid fishes and coral-reef surgeon fishes. *American Zoologist, 14*, 9–34.

Barlow, G. W. (1974b). Extraspecific imposition of social grouping among surgeonfishes (Pisces: Acanthuridae). *Journal of Zoology* (London), *174*, 333–340.

Barlow, G. W. (1974c). Hexagonal territories. *Animal Behaviour, 22*, 876–878.

Barlow, G. W. (1975). On the sociobiology of some hermaphroditic serranid fishes, the hamlets, in Puerto Rico. *Marine Biology, 33*, 295–300.

Barlow, G. W. (1976). The Midas cichlid in Nicaragua. In T. B. Thorson (Ed.), *Investigations of the Ichthyofauna of Nicaraguan Lakes* (pp. 333–358). Lincoln: School of Life Sciences, University of Nebraska.

Barlow, G. W. (1981). Patterns of parental investment, dispersal, and size among coral-reef fishes. *Environmental Biology of Fishes, 6*, 65–85.

Barlow, G. W. (1983). Do gold Midas cichlid fish win fights because of their color, or because they lack normal coloration? A logistic solution. *Behavioral Ecology and Sociobiology, 13*, 197–204.

Barlow, G. W. (1987). Spawning, eggs, and larvae of the longnose filefish *Oxymonacanthus longirostris,* a monogamous coralivore. *Environmental Biology of Fishes, 20*, 183–194.

Barlow, G. W. (1991). Mating systems among cichlid fishes. In M. H. A. Keenleyside (Ed.), *Cichlid Fishes: Behaviour, Ecology, and Evolution* (pp. 173–190). New York: Chapman and Hall.

Barlow, G. W. (1992). Mechanisms of mate choice in monogamy. In P. Bateson and M. Gomendio (Eds.), *Behavioural Mechanisms in Evolutionary Perspective* (pp. 26–29). Madrid: Instituto Juan March de Estudios e Investigaciones.

Barlow, G. W. (1993). The puzzling paucity of feeding territories among freshwater fishes. *Marine Behavior and Physiology, 23*, 155–174.

Barlow, G. W. (1998). Sexual-selection models for exaggerated traits are useful but constraining. *American Zoologist, 38*, 59–69.

Barlow, G. W., Francis, R. C., and Baumgartner, J. V. (1990). Do the colours of parents, companions, and self influence assortative mating in the polychromatic Midas cichlid? *Animal Behaviour, 40*, 713–722.

Barlow, G. W., and McKaye, K. R. (1982). A comparison of feeding, spacing, and aggression in color morphs of the Midas cichlid. II. After 24 hours without food. *Behaviour, 80*, 127–142.

Barlow, G. W., and Munsey, J. W. (1976). The red devil-Midas-arrow cichlid species complex in Nicaragua. In T. B. Thorson (Ed.), *Investigations of the Ichthyofauna of Nicaraguan Lakes* (pp. 359–369). Lincoln: School of Life Sciences, University of Nebraska.

Barlow, G. W., and Rogers, W. (1978). Female Midas cichlid's choice of mate in relation to parents' and to own color. *Biology of Behavior, 3*, 137–146.

Barlow, G. W., Rogers, W., and Cappeto, R. V. (1977). Incompatibility and assortative mating in the Midas cichlid. *Behavioral Ecology and Sociobiology, 2*, 49–59.

Barlow, G. W., Rogers, W., and Fraley, N. (1986). Do Midas cichlids win through prowess or daring? It depends. *Behavioral Ecology and Sociobiology, 19*, 1–8.

Barlow, G. W., and Siri, P. (1994). Polychromatic Midas cichlids respond to dummy opponents: Color, contrast, and context. *Behaviour, 130*(1–2), 77–112.

Barlow, G. W., and Siri, P. (1997). Does sexual selection account for the conspicuous head dimorphism in the Midas cichlid? *Animal Behaviour, 52*(3), 573–584.

Barnett, C. (1981). The role of urine in parent-offspring communication in a cichlid fish. *Zeitschrift für Tierpsychologie, 55*, 173–182.

Baroiller, J., Chourrout, D., Fostier, A., and Jalabert, B. (1995). Temperature and sex chromosomes govern sex ratios of the mouthbrooding cichlid fish *Oreochromis niloticus. Journal of Experimental Zoology, 273*, 216–223.

Baskin, Y. (1992). Africa's troubled waters. *BioScience, 42*(7), 476–481.

Bateson, P. (1994). The dynamics of parent-offspring relationships in mammals. *Trends in Ecology and Evolution, 9*, 399–403.

Baumgartner, T. R., Soutar, A., and Ferreira-Bartrina, V. (1992). Reconstruction of the history of pacific sardine and northern anchovy populations over the past two millennia from sediments of the Santa Barbara Basin, California. *California Cooperative Oceanic Fisheries Investigations, 33*, 24–40.

Baylis, J. R. (1974). The behavior and ecology of *Herotilapia multispinosa* (Teleostei, Cichlidae). *Zeitschrift für Tierpsychologie, 34*, 115–146.

Baylis, J. R. (1976a). A quantitative study of long-term courtship: I. Ethological isolation between sympatric populations of the Midas cichlid, *Cichlasoma citrinellum*, and the arrow cichlid, *C. zaliosum. Behaviour, 59*, 59–69.

Baylis, J. R. (1976b). A quantitative study of long-term courtship: II. A comparative study of the dynamics of courtship in two New World cichlid fishes. *Behaviour, 59*, 117–161.

Baylis, J. R. (1981). The evolution of parental care in fishes, with reference to Darwin's rule of male sexual selection. *Environmental Biology of Fishes, 6*, 223–251.

Beeching, S. C. (1993). Eyespots as visual cues in the intraspecific behavior of the cichlid fish *Astronotus ocellatus. Copeia, 1993*(4), 1154–1157.

Beeching, S. C. (1995). Colour pattern and inhibition of aggression in the cichlid fish *Astronotus ocellatus. Journal of Fish Biology, 47*, 50–58.

Beeching, S. C., Gross, S. H., Bretz, H. S., and Hariatis, E. (1998). Sexual dichromatism in a cichlid fish: The ethological significance of female ventral colouration in the convict cichlid, *Cichlasoma nigrofasciatum. Animal Behaviour, 56*(4), 1021–1026.

Bentsen, H. B., Eknath, A. E., Palada-De Vera, M. S., Danting, J. C., Bolivar, H. L., Reyes, R. A., Dionisio, E. E., Longalong, F. M., Circa, A. V., Tayamen, M. M., and Gjerde, B. (1998). Genetic improvement of farmed tilapias: Growth performance in a complete diallel cross experiment with eight strains of *Oreochromis niloticus*. *Aquaculture, 160*(1–2), 145–173.

Bereczkei, T., Voros, S., Gal, A., and Bernath, L. (1997). Resources, attractiveness, family commitment: Reproductive decisions in human mate choice. *Ethology, 103*(8), 681–699.

Bergmann, H.-H. (1968). Eine descriptive Verhaltensanalyse des Segelflossers (*Pterophyllum scalare* Cuv. and Val., Cichlidae, Pisces). *Zeitschrift für Tierpsychologie, 25*, 559–587.

Bern, O., and Avtalion, R. R. (1990). Some morphological aspects of fertilization in tilapias. *Journal of Fish Biology, 36*(3), 375–382.

Billard, R. (1986). Spermatogenesis and spermatology of some teleost fish species. *Reproduction, Nutrition, Development, 26*, 877–920.

Birkhead, T. R., and Møller, A. P. (1992). *Sperm Competition in Birds: Evolutionary Causes and Consequences.* New York: Academic Press.

Bleick, C. R. (1975). Hormonal control of the nuchal hump in the cichlid fish *Cichlasoma citrinellum. General Comparative Endocrinology, 26*, 198–208.

Blüm, V. (1966). Zur hormonalen Steuerung der Brutpflege einiger Cichliden. *Zoologische Jahrbücher, Abteilung Physiologie*, 72, 264–290.

Blüm, V., and Fiedler, K. (1964). Der Einfluss von Prolactin auf das Brutpflegeverhalten von *Symphysodon aequifasciata axelrodi* L.P. Schultz (Cichlidae, Teleostei). *Die Naturwissenschaften, 51*(6), 149–150.

Blumer, L. S. (1979). Male parental care in the bony fishes. *Quarterly Review of Biology, 54*, 149–161.

Borgerhoff Mulder, M. (1991). Human behavioural ecology. In J. R. Krebs and N. B. Davies (Eds.), *Behavioural Ecology: An Evolutionary Approach,* 3rd ed. (pp. 69–103). Oxford: Blackwell Scientific Publications.

Bouton, N., Van Os, N., and Witte, F. (1998). Feeding performance of Lake Victoria rock cichlids: Testing predictions from morphology. *Journal of Fish Biology, 53*(Supplement A), 118–127.

Bouton, N., Witte, F., van Alphen, J. J. M., Schenk, A., and Seehausen, O. (1999). Local adaptations in populations of rock-dwelling haplochromines (Pisces: Cichlidae) from southern Lake Victoria. *Proceedings of the Royal Society of London Series B Biological Sciences, 266*(1417), 355–360.

Braddock, J. C. (1949). The effect of prior residence upon dominance in the fish *Platypoecilus maculatus. Physiological Zoology, 22*, 161.

Brem, G., Brenig, B., Hoerstgen-Schwark, G., and Winnacker, E. L. (1988). Gene transfer in tilapia *Oreochromis niloticus. Aquaculture, 68*(3), 209–220.

Bremer, H., and Walter, U. (1986). Histologische, ultrastrukturelle und topochemische Untersuchungen zur Brutpflege von *Symphysodon aequifasciatus* Pellegrin 1903. *Gegenbaurs morphologische Jahrbuch,* Leipzig, *132*(2), 183–194.

Brestowsky, M. (1968). Vergleichende Untersuchungen zur Elternbindung von *Tilapia*-Jungfischen (Cichlidae, Pisces). *Zeitschrift für Tierpsychologie, 25*(7), 761–828.

Brichard, P. (1978a). *Fishes of Lake Tanganyika.* Neptune City, New Jersey: T. F. H. Publications.

Brichard, P. (1978b). Un cas d'isolement de substrats rocheux au milieu de fonds de sable dans le nord du lac Tanganyka. *Revue de Zoologie Africaine, 92*, 518–524.

Brooks, J. L. (1950). Speciation in ancient lakes. *Quarterly Review of Biology, 23*, 30–36.

Brooks, R., and Caithness, N. (1995). Female choice in a feral guppy population: Are there multiple cues? *Animal Behaviour, 50*(2), 301–307.

Brown, D. H., and Marshall, J. A. (1978). Reproductive behaviour of the rainbow cichlid *Herotilapia multispinosa* (Pisces, Cichlidae). *Behaviour, 67*(3–4), 299–321.

Brown, M. E. (1957). Experimental studies of growth. In M. E. Brown (Ed.), *Physiology of Fishes,* vol. 1 (pp. 361–400). New York: Academic Press.

Bull, J. J. (1980). Sex determination in reptiles. *Quarterly Review of Biology, 55*, 3–21.

Burchard, J. E. (1965). Family structure in the dwarf cichlid *Apistogramma trifasciatum* Eigenmann and Kennedy. *Zeitschrift für Tierpsychologie, 22*, 150–162.

Bush, G. L. (1994). Sympatric speciation in animals: New wine in old bottles. *Trends in Ecology and Evolution, 9*(8), 285–288.

Buxton, C. D. (1993). Life-history changes in exploited reef fishes on the east coast of South Africa. *Environmental Biology of Fishes, 36*, 47–63.

Carlisle, T. R. (1985). Parental response to brood size in a cichlid fish. *Animal Behaviour, 33*, 234–238.

Chien, A. K., and M. S. Salmon (1972). Reproductive behavior of the angelfish *Pterophyllum scalare*. I. A quantitative analysis of spawning and parental behavior. *Forma et Functio, 5*, 45–74.
Chimits, P. (1957). Tilapia in ancient Egypt. *F.A.O. Fisheries Bulletin, 10*, 1–5.
Claes, G., and Vree, F. D. (1989). Asymmetrical pharyngeal mastication in *Oreochromis niloticus*. *Koninklijk Museum voor Midden-Afrika Tervuren Belgie Annalen Zoologische Wetenschappen, 257*, 69–72.
Claes, G., Vree, F. D., and Vandewalle, P. (1991). Masticatory operations and the functions of pharyngeal jaw movements in cichlids. *Koninklijk Museum voor Midden-Afrika Tervuren Belgie Annalen Zoologische Wetenschappen, 263*, 85–90.
Clarke, G. M. (1998). Developmental stability and fitness: The evidence is not quite so clear. *American Naturalist, 152*(5), 762–766.
Coe, M. J. (1966). The biology of *Tilapia grahami* Boulenger in Lake Magadi, Kenya. *Acta Tropica, 23*, 146–177.
Cole, J. E., and Ward, J. A. (1969). The communication function of pelvic fin-flickering in *Etroplus maculatus* (Pisces: Cichlidae). *Behaviour, 35*, 179–199.
Coleman, R. M. (1993). The Evolution of Parental Investment in Fishes. Unpublished Ph.D. dissertation, University of Toronto, Toronto.
Coleman, R. M., and Galvani, A. P. (1998). Egg size determines offspring size in Neotropical cichlid fishes (Teleostei: Cichlidae). *Copeia, 1998*(1), 209–213.
Conover, D. O. (1984). Adaptive significance of temperature-dependent sex determination in a fish. *American Naturalist, 123*, 297–313.
Coulter, G. W. (1991). The benthic fish community. In G. W. Coulter (Ed.), *Lake Tanganyika and Its Life* (pp. 151–199). Oxford: Oxford University Press.
Coulter, G. W., Allanson, B. R., Bruton, M. N., Greenwood, P. H., Hart, R. C., Jackson, P. B. N., and Ribbink, A. J. (1986). Unique qualities and special problems of the African Great Lakes. *Environmental Biology of Fishes, 17*, 161–183.
Courtenay, S. C., and Keenleyside, M. H. A. (1983). Wriggler-hanging: A response to hypoxia by brood-rearing *Herotilapia multispinosa* (Teleostei, Cichlidae). *Behaviour, 85*(3–4), 183–197.
Courtenay, W. R., Jr. (1973). Exotic aquatic organisms in Florida with emphasis on fishes: A review and recommendations. *Transactions of the American Fisheries Society, 102*, 1–12.
Coyne, J. A., and Sohn, J. J. (1978). Interspecific brood care in fishes: Reciprocal altruism or mistaken identity? *American Naturalist, 112*, 447–450.
Crapon de Caprona, M.-D. (1980). Olfactory communication in a cichlid fish, *Haplochromis burtoni*. *Zeitschrift für Tierpsychologie, 52*(2), 113–134.
Crapon de Caprona, M.-D. (1986). Are "preferences" and "tolerances" in cichlid mate choice important for speciation? *Journal of Fish Biology, 29*(Suppl. A), 151–158.
Crapon de Caprona, M.-D., and Fritzsch, B. (1984). Interspecific fertile hybrids of haplochromine Cichlidae (Teleostei) and their possible importance for speciation. *Netherlands Journal of Zoology, 34*, 503–538.
Dambach, M. (1963). Vergleichende Untersuchungen über das Schwarmverhalten von *Tilapia*-Jungfischen (Cichlidae, Teleostei). *Zeitschrift für Tierpsychologie, 20*, 267–296.
Dambach, M., and Wallert, I. (1966). Das Tilapia-Motiv in der altaägyptischen Kunst. *Chronique d'Egypte, 51*(82), 273–294.
Darwin, C. (1871). *The Descent of Man and Selection in Relation to Sex*, 1st ed. London: John Murray.
Davies, N. B. (1978). Territorial defense in the speckled wood butterfly *(Pararge aegeria)*: The resident always wins. *Animal Behaviour, 26*, 138–147.
Davies, N. B., and Brooke, M. (1991). Coevolution of the cuckoo and its hosts. *American Scientist, 79*(January), 92–98.
De Silva, S. S., Maitipe, P., and Cumaranatunge, R. T. (1984). Aspects of the biology of the euryhaline Asian cichlid, *Etroplus suratensis*. *Environmental Biology of Fishes, 10*(1–2), 77–87.
Denton, E. J., and Nicol, J. A. C. (1962). Why fishes have silvery sides; and a method of measuring reflectivity. *Journal of Physiology, 165*, 12–15.
Desprez, D., and Melard, C. (1998). Effect of ambient water temperature on sex determinism in the blue tilapia *Oreochromis aureus*. *Aquaculture, 162*(1–2), 79–84.
Dickman, M. C., Annett, C., and Barlow, G. W. (1990). Unsuspected cryptic polymorphism in the polychromatic Midas cichlid. *Biological Journal of the Linnean Society, 39*, 239–249.
Dierkes, P., Taborsky, M., and Kohler, U. (1999). Reproductive parasitism of broodcare helpers in a cooperatively breeding fish. *Behavioral Ecology and Sociobiology, 10*(5), 510–555.

Dominey, W. J. (1984). Effects of sexual selection and life history on speciation: Species flocks in African cichlids and Hawaiian *Drosophila.* In A. A. Echelle and I. Kornfield (Eds.), *Evolution of Fish Species Flocks* (pp. 231–249). Orono: University of Maine at Orono Press.

Dominey, W. J., and Snyder, A. M. (1988). Kleptoparasitism of freshwater crabs by cichlid fishes endemic to Lake Barombi Mbo, Cameroon, West Africa. *Environmental Biology of Fishes, 22*(2), 155–160.

Doorn, G. S. v., Noest, A. J., and Hogeweg, P. (1998). Sympatric speciation and extinction driven by environment dependent sexual selection. *Proceedings of the Royal Society of London Series B Biological Sciences, 265*(1408), 1915–1919.

Eccles, D. H. (1986). Is speciation of demersal fishes in Lake Tanganyika restrained by physical limnological conditions? *Biological Journal of the Linnean Society, 29*(2), 115–122.

Eccles, D. H., and Lewis, D. S. C. (1976). A revision of the genus *Docimodus* Boulenger (Pisces: Cichlidae), a group of fishes with unusual feeding habits from Lake Malawi. *Biological Journal of the Linnean Society, 58*, 165–172.

Echelle, A. A., and Kornfield, I. (Eds.). (1984). *Evolution of Fish Species Flocks.* Orono: University of Maine at Orono Press.

Edwards, A. W. F. (1998). Natural selection and the sex ratio: Fisher's sources. *American Naturalist, 151*(6), 564–569.

Elder, H. Y., Garrod, D. J., and Whitehead, P. J. P. (1971). Natural hybrids of the African cichlid fishes *Tilapia spilurus nigra* and *Tilapia leucosticta*: A case of hybrid introgression. *Biological Journal of the Linnean Society, 3*, 103–146.

Emlen, S. T. (1991). Evolution of cooperative breeding in birds and mammals. In J. R. Krebs and N. B. Davies (Eds.), *Behavioural Ecology: An Evolutionary Approach,* 3rd ed. (pp. 3201–337). Boston: Blackwell.

Emlen, S. T. (1995). An evolutionary theory of the family. *Proceedings of the National Academy of Sciences of the United States of America, 92*, 8092–8099.

Emlen, S. T. (1997). Predicting family dynamics in social vertebrates. In J. R. Krebs and N. B. Davies (Eds.), *Behavioural Ecology: An Evolutionary Approach,* 4th ed. (pp. 128–253). Oxford: Blackwell.

Endler, J. A. (1992). Signals, signal conditions, and the direction of evolution. *American Naturalist, 139*, S125-S153.

Enquist, M., Leimar, O., Ljungberg, T., Mallner, Y., and Segerdahl, N. (1990). A test of the sequential assessment game: Fighting in the cichlid fish *Nannacara anomala. Animal Behaviour, 40*(1), 1–14.

Evans, M. R., and Norris, K. (1996). The importance of carotenoids in signaling during aggressive interactions between male firemouth cichlids *(Cichlasoma meeki). Behavioral Ecology,* 7(1), 1–6.

Faaborg, J. (1981). The characteristics and occurrence of cooperative polyandry. *Ibis, 123*, 477–484.

Falter, U., and Charlier, M. (1989). Mate choice in pure-bred and hybrid females of *Oreochromis niloticus* and *Oreochromis mossambicus* based upon visual stimuli (Pisces: Cichlidae). *Biology of Behaviour, 14*(3), 218–228.

Falter, U., and Dolisy, D. (1989). The effect of female sexual pheromones on the behavior of *Oreochromis niloticus, Oreochromis mossambicus,* and hybrid males (Pisces: Cichlidae). *Koninklijk Museum voor Midden-Afrika Tervuren Belgie Annalen Zoologische Wetenschappen, 257*, 35–38.

Farias, I. P., Orti, G., Sampaio, I., Schneider, H., and Meyer, A. (1999). Mitochondrial DNA phylogeny of the family Cichlidae: Monophyly and fast molecular evolution of the Neotropical assemblage. *Journal of Molecular Evolution, 48*(6), 703–711.

Fernandes, M. D. O., and Volpato, G. L. (1993). Heterogeneous growth in the Nile tilapia: Social stress and carbohydrate metabolism. *Physiology and Behavior, 54*(2), 319–323.

Finley, L. (1983). *Synodontis multipunctatus* reproduction and maternal mouthbrooding cichlids: A cuckoo relationship? *Journal of the American Cichlid Association, 98*, 17–18.

Fischer, E. A. (1981). Sexual allocation in a simultaneously hermaphroditic coral reef fish. *American Naturalist, 117*, 64–82.

Fisher, R. A. (1930). *The Genetical Theory of Natural Selection.* Oxford: Clarendon Press.

Forey, P. L. (1995). Agnathans recent and fossil, and the origin of jawed vertebrates. *Reviews in Fish Biology and Fisheries, 5*(3), 267–303.

Fox, H. E., White, S. A., Kao, M. H. F., and Fernald, R. D. (1997). Stress and dominance in a social fish. *Journal of Neuroscience, 17*(16), 6463–6469.

Fox, H. M., and Vevers, G. (1960). *The Nature of Animal Colours.* London: Sidgwick and Jackson.

Francis, R. C. (1990). Temperament in a fish: A longitudinal study of the development of individual differences in aggression and social rank in the Midas cichlid. *Ethology, 86,* 311–325.

Francis, R. C. (1992). Sexual lability in teleosts: Developmental factors. *Quarterly Review of Biology, 67,* 1–18.

Francis, R. C., and Barlow, G. W. (1993). Social control of primary sex differentiation in the Midas cichlid. *Proceedings of the National Academy of Sciences of the United States of America, 90,* 10673–10675.

Francis, R. C., Soma, K., and Fernald, R. D. (1993). Social regulation of the brain-pituitary-gonadal axis. *Proceedings of the National Academy of Sciences of the United States of America, 90,* 7794–7798.

Fraser, S. A. (1996). The influence of predators on adoption behavior in adult convict cichlids *(Cichlasoma nigrofasciatum). Canadian Journal of Zoology, 74*(6), 1165–1173.

Fraser, S. A., Wisenden, B. D., and Keenleyside, M. H. A. (1993). Aggressive behaviour among convict cichlid *(Cichlasoma nigrofasciatum)* fry of different sizes and its importance to brood adoption. *Canadian Journal of Zoology, 71*(12), 2358–2362.

Fryer, G. (1961). Observations on the biology of the cichlid fish *Tilapia variabilis* Boulenger in the northern waters of Lake Victoria (East Africa). *Revue de zoologie et de Botanique Africaines, 64,* 1–33.

Fryer, G. (1991). The evolutionary biology of African cichlid fishes. *Musée Royal de l'Afrique Centrale Tervuren, Belgique, Annales Sciences Zoologiques, 263.*

Fryer, G. (1996). Endemism, speciation, and adaptive radiation in great lakes. *Environmental Biology of Fishes, 45*(2), 109–131.

Fryer, G., and Iles, T. D. (1972). *The Cichlid Fishes of the Great Lakes of Africa: Their Biology and Evolution.* Edinburgh: Oliver and Boyd.

Futuyma, D. J. (1998). *Evolutionary Biology.* Sunderland, Massachusetts: Sinauer Associates.

Gaemers, P. A. M. (1984). Taxonomic position of the Cichlidae (Pisces, Perciformes) as demonstrated by the morphology of their otoliths. *Netherlands Journal of Zoology,* 34, 566–596.

Gaemers, P. A. M. (1986). Recent progress in cichlid taxonomy based on otoliths, and its significance for the phylogeny of tilapiines and haplochromines (Perciformes, Pisces). *Musée Royal de l'Afrique Centrale, Annales Sciences Zoologiques, 251,* 143–150.

Gaines, S. D., and Lubchenco, J. (1982). A unified approach to marine plant-herbivore interactions. II. Biogeography. *Annual Review of Ecology and Systematics, 13,* 111–138.

Gashagaza, M. M. (1991). Diversity of breeding habits in lamprologine cichlids in Lake Tanganyika. *Physiology and Ecology Japan, 28*(1–2), 29–65.

Geerts, M. (1996). The age of cichlid fishes. *The Cichlids Yearbook, 6,* 92–95.

Genner, M. J., Turner, G. F., and Hawkins, S. J. (1999). Resource control by territorial male cichlid fish in Lake Malawi. *Journal of Animal Ecology, 68*(3), 522–529.

Ghiselin, M. T. (1969). The evolution of hermaphroditism among animals. *Quarterly Review of Biology, 44,* 189–208.

Goldschmidt, T. (1991). Egg mimics in haplochromine cichlids (Pisces, Perciformes) from Lake Victoria. *Ethology, 88,* 177–190.

Goldschmidt, T. (1996). *Darwin's Dreampond: Drama on Lake Victoria.* Cambridge, Massachusetts: MIT Press.

Goldschmidt, T., Witte, F., and De Visser, J. (1990). Ecological segregation in zooplanktivorous haplochromine species (Pisces: Cichlidae) from Lake Victoria. *Oikos, 58*(3), 343–355.

Goldschmidt, T., Witte, F., and Wanink, J. (1993). Cascading effects of the introduced Nile perch on the detritivorous/phytoplanktivorous species in the sublittoral areas of Lake Victoria. *Conservation Biology,* 7(3), 686–700.

Goodwin, N. B., Balshine-Earn, S., and Reynolds, J. D. (1998). Evolutionary transitions in parental care in cichlid fish. *Proceedings of the Royal Society of London Series B Biological Sciences, 265*(1412), 2265–2272.

Gottfried, M. D. (1986). Developmental transition in feeding morphology of the Midas cichlid. *Copeia, 1986,* 1028–1030.

Gould, S. J., and Eldredge, N. (1993). Punctuated equilibrium comes of age. *Nature* (London), *366*(6452), 223–227.

Goundswaard, K., and Witte, F. (1997). The catfish fauna of Lake Victoria after the Nile perch upsurge. *Environmental Biology of Fishes, 49*(1), 21–43.

Grant, P. R., and Grant, B. R. (1998). Speciation and hybridization of birds on islands. In P. R. Grant (Ed.), *Evolution on Islands* (pp. 142–162). New York: Oxford University Press.

Grantner, A., and Taborsky, M. (1998). The metabolic rates associated with resting, and with the performance of agonistic, submissive, and digging behaviours in the cichlid fish *Neolamprologus pulcher* (Pisces: Cichlidae). *Journal of Comparative Physiology B Biochemical Systemic and Environmental Physiology, 168*(6), 427–433.

Gray, J. A. B., and Denton, E. J. (1991). Fast pressure pulses and communication between fish. *Journal of the Marine Biological Association U.K., 71*, 83–106.

Greenberg, B. (1961). Spawning and parental behavior in female pairs of the jewel fish, *Hemichromis bimaculatus* Gill. *Behaviour, 18*(1), 44–61.

Greenberg, B. (1963). Parental behavior and imprinting in cichlid fishes. *Behaviour, 21*, 127–144.

Greenberg, B., Zijlstra, J. J., and Baerends, G. P. (1965). A quantitative description of the behaviour changes during the reproductive cycle of the cichlid fish *Aequidens portalegrensis* Hensel. *Proceedings Koninklijke Nederlandse Akademie van Wetenschappen, Series C, 68*(3), 135–149.

Greenwood, P. H. (1965). The cichlid fishes of Lake Nabugabo, Uganda. *Bulletin of the British Museum (Natural History), Zoology, 12*(9), 315–357.

Greenwood, P. H. (1974). The cichlid fishes of Lake Victoria, East Africa: The biology and evolution of a species flock. *Bulletin of the British Museum (Natural History), Zoology, Supplement 6*, 1–134.

Greenwood, P. H. (1984). African cichlids and evolutionary theories. In A. A. Echelle and I. Kornfield (Eds.), *Evolution of Fish Species Flocks* (pp. 141–154). Orono: University of Maine Press.

Greenwood, P. H. (1991). Speciation. In M. H. A. Keenleyside (Ed.), *Cichlid Fishes: Behaviour, Ecology, and Evolution* (pp. 86–102). New York: Chapman and Hall.

Grier, H. J., and Fishelson, L. (1995). Colloidal sperm-packaging in mouthbrooding tilapiine fishes. *Copeia, 1995*(4), 966–970.

Grinnell, J., and McComb, K. (1996). Maternal grouping as a defense against infanticide by males: Evidence from field playback experiments on African lions. *Behavioral Ecology, 7*(1), 55–59.

Gunther, J. (1996). Growth of the jaguar cichlid *Cichlasoma managuense* (Pisces: Cichlidae) raised in intensive culture in earthen ponds. *Revista de Biologia Tropical, 44*(2 Part B), 813–818.

Hale, E., and Stauffer, J. R., Jr. (1997). Lake Malawi fish production is linked to landscape geology. *Cichlid News, 6*(2), 10–12.

Hale, E. A., Stauffer, J. R., Jr., and Megan, D. M. (1998). Exceptions to color being a sexually dimorphic character in *Melanochromis auratus* (Teleostei: Cichlidae). *Ichthyological Exploration of Freshwaters, 9*(3), 263–266.

Harrington, R. W. (1971). How ecological and genetic factors interact to determine when self-fertilizing hermaphrodites of *Rivulus marmoratus* change into functional secondary males, with a reappraisal of the modes of intersexuality among fishes. *Copeia, 1971*(3), 389–432.

Harrison, I. J., and Stiassny, M. L. J. (1999). The quiet crisis: A preliminary listing of the freshwater fishes of the world that are extinct or "missing in action." In MacPhee (Ed.), *Extinctions in Near Time* (pp. 271–331). New York: Kluwer Academic/Plenum.

Hasler, A. D., Scholz, A. T., and Horrall, R. M. (1978). Olfactory imprinting and homing in salmon. *American Scientist, 66*, 347–355.

Haussknecht, T., and Kuenzer, P. (1991). An experimental study of the building behaviour sequence of a shell-breeding cichlid fish from Lake Tanganyika *(Lamprologus ocellatus)*. *Behaviour, 116*, 127–142.

Hecky, R. E., Bugenyi, F. W. B., Ochumba, P., Talling, J. F., Mugidde, R., Gophen, M., and Kaufman, L. (1994). Deoxygenation of the deep water of Lake Victoria, East Africa. *Limnology and Oceanography, 39*(6), 1476–1481.

Heggberget, T. G., Johnsen, B. O., Hindar, K., Johnson, B., Hansen, L. P., Hvidsten, N. A., and Jensen, A. J. (1993). Interactions between wild and cultured Atlantic salmon: A review of the Norwegian experience. *Fisheries Research* (Amsterdam), *18*(1–2), 123–146.

Heijns, W. (1996). Natural history and husbandry of *Cichlasoma (Archcentrus) centrarchus*. *Cichlid News, 5*(2), 13–16.

Heiligenberg, W. (1964). Ein Versuch zur ganzheitsbezogenen Analyse des Instinktverhaltens eines Fisches (*Pelmatochromis subocellatus kribensis* Boul., Cichlidae). *Zeitschrift für Tierpsychologie, 21*(1), 1–52.

Heiligenberg, W. (1965). Color polymorphism in the males of an African cichlid fish. *Journal of Zoology* (London), *146*, 95–97.

Heiligenberg, W., Kramer, U., and Schulz, V. (1972). The angular orientation of the black eye-bar in *Haplochromis burtoni* (Cichlidae, Pisces) and its relevance to aggressivity. *Zeitschrift für vergleichende Physiologie, 76*, 168–176.

Helfman, G. S., Collette, B. B., and Facey, D. E. (1997). *The Diversity of Fishes.* Malden, Massachusetts: Blackwell Science.

Hert, E. (1985). Individual recognition of helpers by the breeders in the cichlid fish *Lamprologus brichardi* (Poll, 1974). *Zeitschrift für Tierpsychologie, 68*, 313–325.

Hert, E. (1986a). Freeze branding in fish, a method for eliminating colour patterns at the skin surface. *Ethology, 72*, 165–167.

Hert, E. (1986b). On the significance of the egg-spots in an African mouth-brooder (*Haplochromis elegans,* Trewavas 1933). *Musée Royal de l'Afrique Centrale, Annales Sciences Zoologiques, 251*, 25–26.

Hert, E. (1989). The function of egg-spots in an African mouth-brooding cichlid fish. *Animal Behaviour, 37*(5), 726–732.

Hert, E. (1991). Female choice based on egg-spots in *Pseudotropheus aurora* Burgess 1976, a rock-dwelling cichlid of Lake Malawi, Africa. *Journal of Fish Biology, 38*, 951–953.

Hert, E. (1995). The impact of intralacustrine introductions with regard to space utilization and competition for territories to a cichlid fish community in Lake Malawi, Africa. *Ecological Research, 10*(2), 117–124.

Higashi, M., Takimoto, G., and Yamaura, N. (1999). Sympatric speciation by sexual selection. *Nature* (London), *402*, 523–526.

Hildemann, W. H. (1959). A cichlid fish, *Symphysodon discus,* with unique nurture habits. *American Zoologist, 93*(1), 27–34.

Hiraiwa-Hasegawa, M. (1988). Adaptive significance of infanticide in primates. *Trends in Ecology and Evolution, 3*, 102–105.

Högland, J., and Alatalo, R. V. (1995). *Leks.* Princeton: Princeton University Press.

Holder, J. L. (1991). The Mechanisms of Mate Choice in the Midas Cichlid *Cichlasoma citrinellum.* Unpublished Ph.D. dissertation, University of California, Berkeley.

Holder, J. L., Barlow, G. W., and Francis, R. C. (1991). Differences in aggressiveness in the Midas cichlid fish *(Cichlasoma citrinellum)* in relation to sex, reproductive state, and the individual. *Ethology, 88*, 297–306.

Holzberg, S. (1978). A field and laboratory study of the behaviour and ecology of *Pseudotropheus zebra* (Boulenger), an endemic cichlid of Lake Malawi (Pisces: Cichlidae). *Zeitschrift für zoologische Systematik und Evolutionsforschung, 16*(3), 171–187.

Hoogland, R. D., Morris, D., and Tinbergen, N. (1957). The spines of sticklebacks *(Gasterosteus* and *Pygosteus)* as means of defense against predators *(Perca* and *Esox). Behaviour, 10*, 205–236.

Hori, M. (1993). Frequency-dependent natural selection in the handedness of scale-eating cichlid fish. *Science* (Washington, D.C.), *260*, 216–219.

Hunt, G. L., Jr., and Hunt, M. W. (1977). Female-female pairing in western gulls *(Larus occidentalis)* in southern California. *Science* (Washington, D.C.), *196*, 1466–1467.

Hurd, P. L. (1997). Cooperative signaling between opponents in fish fights. *Animal Behaviour, 54*(5), 1309–1315.

Incer, J. (1976). Geography of Lake Nicaragua. In T. B. Thorson (Ed.), *Investigations of the Ichthyofauna of Nicaraguan Lakes* (pp. 3–7). Lincoln: School of Life Sciences, University of Nebraska.

Jackson, P. B. N., and Ribbink, T. (1975). *Mbuna, Rock-Dwelling Cichlids of Lake Malawi, Africa.* Neptune City: T. H. Publishing Company.

Jaeger, R. G. (1981). Dear enemy recognition and costs of aggression between salamanders. *American Naturalist, 117*, 962–974.

Jakobsson, S., Brick, O., and Kullberg, C. (1995). Escalated fighting behaviour incurs increased predation risk. *Animal Behaviour, 49*(1), 235–239.

Johnson, N. K., Remsen, J. V., Jr., and Cicero, C. (1998). *Resolution of the debate over species concepts in ornithology: A new comprehensive biologic species concept.* Paper presented at the International Ornithological Conference, Durban, South Africa.

Johnson, T. C., Scholz, C. A., Talbot, M. R., Kelts, K., Ricketts, R. D., Ngobi, G., Beuning, K., Ssemmanda, I., and McGill, J. W. (1996). Late Pleistocene desiccation of Lake Victoria and rapid evolution of cichlid fishes. *Science* (Washington, D.C.), *273*, 1091–1093.

Jones, A. J. (1972). The early development of substrate brooding cichlids (Teleostei: Cichlidae) with a discussion of a new system of staging. *Journal of Morphology, 136*(3), 255–272.

Kalas, K. (1976). Brutpflegehelfer und Polygamie beim afrikanischen Buntbarsch *Lamprologus brichardi. Die Naturwissenschaften, 63*, 94.

Kano, T. (1989). The sexual behavior of pygmy chimpanzees. In P. G. Heltne and L. A. Marquardt (Eds.), *Understanding Chimpanzees.* Cambridge: Harvard University Press.

Karino, K. (1996). Tactic for bower acquisition by male cichlids, *Cyathopharynx furcifer*, in Lake Tanganyika. *Ichthyological Research, 43*(2), 125–132.

Karino, K. (1997a). Female mate preference for males having long and symmetric fins in the bower-holding cichlid *Cyathopharynx furcifer. Ethology, 103*(11), 883–892.

Karino, K. (1997b). Influence of brood size and offspring size on parental investment in a biparental cichlid fish, *Neolamprologus moorii. Journal of Ethology, 15*(1), 39–43.

Kaufman, L. (1992). Catastrophic change in species-rich freshwater ecosystems: The lessons of Lake Victoria. *BioScience, 42*(11), 846–858.

Kaufman, L., and Ochumba, P. (1993). Evolutionary and conservation biology of cichlid fishes as revealed by faunal remnants in northern Lake Victoria. *Conservation Biology, 7*, 719–730.

Kawanabe, H., Hori, M., and Nagoshi, M. (Eds.). (1997). *Fish Communities in Lake Tanganyika.* Kyoto: Kyoto University Press.

Keeley, E. R., and Grant, J. W. A. (1993). Visual information, resource value, and sequential assessment in convict cichlid *(Cichlasoma nigrofasciatum)* contests. *Behavioral Ecology, 4*, 345–349.

Keenleyside, M. H. A. (1985). Bigamy and mate choice in the biparental cichlid fish *Cichlasoma nigrofasciatum. Behavioral Ecology and Sociobiology, 17*, 285–290.

Keenleyside, M. H. A., and Bietz, B. (1981). The reproductive behaviour of *Aequidens vittatus* (Pisces, Cichlidae) in Surinam, South America. *Environmental Biology of Fishes, 6*, 87–94.

Keenleyside, M. H. A., and Prince, C. E. (1976). Spawning-site selection in relation to parental care of eggs in *Aequidens paraguayensis* (Pisces: Cichlidae). *Canadian Journal of Zoology, 54*(12), 2135–2139.

Kellogg, K. A., Markert, J. A., Stauffer, J. R., Jr., and Kocher, T. D. (1995). Microsatellite variation demonstrates multiple paternity in lekking cichlid fishes from Lake Malawi, Africa. *Proceedings of the Royal Society of London Series B Biological Sciences, 260*(1357), 79–84.

Kellogg, K. A., Markert, J. A., Stauffer, J. R., Jr., and Kocher, T. D. (1998). Intraspecific brood mixing and reduced polyandry in a maternal mouth-brooding cichlid. *Behavioral Ecology, 9*(3), 309–312.

Kirchshofer, R. (1953). Aktionssystem des Maulbrüters *Haplochromis desfontainesii. Zeitschrift für Tierpsychologie, 10*, 297–318.

Kirkpatrick, M., and Ryan, M. J. (1991). The evolution of mating preferences and the paradox of the lek. *Nature* (London), *350*, 33–38.

Kleiman, D. G. (1977). Monogamy in mammals. *Quarterly Review of Biology, 52*, 39–69.

Kling, G. W. (1987). Seasonal mixing and catastrophic degassing in tropical lakes, Cameroon, West Africa. *Science* (Washington, D.C.), *237*(4818), 1022–1024.

Knight, M. E., and Turner, G. F. (1999). Reproductive isolation among closely related Lake Malawi cichlids: Can males recognize conspecific females by visual cues? *Animal Behaviour, 58*(4), 761–768.

Knowlton, N. (1979). Reproductive synchrony, parental investment, and the evolutionary dynamics of sexual selection. *Animal Behaviour, 27*, 1022–1033.

Kocher, T., and McKaye, K. R. (1983). Defense of heterospecific cichlids by *Cyrtocara moori* in Lake Malawi, Africa. *Copeia, 1983*, 544–547.

Kocher, T., and Stepien, C. (Eds.). (1997). *Molecular Systematics of Fishes.* San Diego: Academic Press.

Kocher, T. D., Conroy, J. A., McKaye, K. R., and Stauffer, J. R. (1993). Similar morphologies of cichlid fish in Lakes Tanganyika and Malawi are due to convergence. *Molecular Phylogenetics and Evolution, 2*(3), 158–165.

Kodric-Brown, A., and Brown, J. H. (1984). Truth in advertising: The kinds of traits favored by sexual selection. *American Naturalist, 124*, 309–323.

Kohda, M., Hori, M., and Snhombo, M. (1997). Inter-individual variation in foraging behaviour and dimorphism in predatory cichlids fishes. In H. Kawanabe, M. Hori, and M. Nagosh (Eds.), *Fish Communities in Lake Tanganyika* (pp. 123–136). Kyoto: Kyoto University Press.

Kohda, M., and Takemon, Y. (1996). Group foraging by the herbivorous cichlid fish, *Petrochromis fasciolatus*, in Lake Tanganyika. *Ichthyological Research, 43*(1), 55–63.

Kohda, M., Yanagisawa, Y., Sato, T., Nakaya, K., Niimura, Y., Matsumoto, K., and Ochi, H. (1996). Geographical colour variation in cichlid fishes at the southern end of Lake Tanganyika. *Environmental Biology of Fishes, 45*(3), 237–248.

Kondo, T. (1987). Feeding habits of *Lamprologus savoryi* (Teleostei: Cichlidae) with reference to its social behavior. *Physiology and Ecology Japan, 23*(1), 1–16.

Kondrashov, A. S., and Shpak, M. (1998). On the origin of species by means of assortative mating. *Proceedings of the Royal Society of London Series B Biological Sciences, 265*(1412), 2273–2278.

Konings, A. (1988). *Tanganyika Cichlids.* Pijnacker, The Netherlands: Raket, B.V.

Konings, A. (1998). *Tanganyika Cichlids in Their Natural Habitat.* El Paso, Texas: Cichlid Press.

Konings, A. (1999). *Cyathopharynx* featherfins of Lake Tanganyika. *Tropical Fish Hobbyist, 47*(7), 14–22.

Kornfield, I. (1991). Genetics. In M. H. A. Keenleyside (Ed.), *Cichlid Fishes: Behaviour, Ecology, and Evolution* (pp. 103–128). New York: Chapman and Hall.

Kornfield, I., and Parker, A. (1997). Molecular systematics of a rapidly evolving species flock: The mbuna of Lake Malawi and the search for phylogenetic signal. In T. D. Kocher and C. A. Stepien (Eds.), *Molecular Systematics of Fishes* (pp. 25–37). San Diego: Academic Press.

Kosswig, G. (1947). Selective mating as a factor for speciation in cichlid fish of East African lakes. *Nature* (London), *179*, 604–605.

Kraft, A. v., and Peters, H. M. (1963). Vergleichende Studien über die Oogenese in der Gattung *Tilapia* (Cichlidae, Teleostei). *Zeitschrift für Zellforschung und Mikroskopische Anatomie, 61*(3), 434–485.

Kramer, D. L., and Bryant, M. J. (1995). Intestine length in the fishes of a tropical stream: 2. Relationships to diet: The long and short of a convoluted issue. *Environmental Biology of Fishes, 42*(2), 129–141.

Kruijt, J. P., Vos, G. J. d., and Bossema, I. (1972). The arena system of black grouse. *Proceedings of the 15th International Ornithological Congress, 1970*, 399–423.

Krüter, R. (1991). Tanganyikan cichlids: Three beautiful sanddwelling cichlids. *The Cichlids Yearbook, 1*, 7–10.

Kuenzer, E., and P. Kuenzer (1962). Untersuchungen zur Brutpflege der Zwergcichliden *Apistogramma reitzigi* und *A. borellii. Zeitschrift für Tierpsychologie,* 19, 56–83.

Kühme, W. (1963). Chemisch ausgelöste Brutpflege- und Schwarmreaktionen bei *Hemichromis bimaculatus* (Pisces). *Zeitschrift für Tierpsychologie, 20*(6), 688–704.

Kullander, S. O. (1998). A phylogeny and classification of the South American Cichlidae (Teleostei: Perciformes). In L. R. Malabarba, R. E. Reis, R. P. Vari, Z. M. Lucena, and C. A. S. Lucena (Eds.), *Phylogeny and Classification of Neotropical Fishes* (pp. 461–498). Porto Alegre: Edipucrs.

Kuwamura, T. (1986). Parental care and mating systems of cichlid fishes in Lake Tanganyika: A preliminary field survey. *Journal of Ethology, 4*, 129–146.

Kuwamura, T. (1988). Biparental mouthbrooding and guarding in a Tanganyikan cichlid *Haplotaxodon microlepis. Japanese Journal of Ichthyology, 35*, 62–68.

Kuwamura, T. (1997). The evolution of parental care and mating systems among Tanganyikan cichlids. In H. Kawanabe, M. Hori, and M. Nagoshi (Eds.), *Fish Communities in Lake Tanganyika* (pp. 57–86). Kyoto: Kyoto University Press.

Kuwamura, T., Nakashima, Y., and Yogo, Y. (1994). Sex change in either direction by growth-rate advantage in the monogamous coral goby, *Paragobiodon echinocephalus. Behavioral Ecology, 5*(4), 434–438.

Lack, D. (1954). *The Natural Regulation of Animal Numbers.* London: Oxford University Press.

Lamprecht, J. (1973). Mechanismen des Paarzusammenhaltes beim Cichliden *Tilapia mariae* Boulenger 1899 (Cichlidae, Teleostei). *Zeitschrift für Tierpsychologie, 32*, 10–61.

Lande, R. (1981). Models of speciation by sexual selection of polygenic traits. *Proceedings of the National Academy of Sciences of the United States of America, 78*, 3721–3725.

Langlois, J. H., and Roggman, L. A. (1990). Attractive faces are only average. *Psychological Science, 1*(2), 115–121.

Lanzing, W. J. R. (1974). Sound production in the cichlid *Tilapia mossambica* Peters. *Journal of Fish Biology, 6*, 341–347.

Lanzing, W. J. R., and Bower, C. C. (1974). Development of colour patterns in relation to behaviour in *Tilapia mossambica* (Peters). *Journal of Fish Biology, 6*, 29–41.

Lavery, R. J. (1991). Physical factors determining spawning site selection in a Central American hole nester, *Cichlasoma nigrofasciatum. Environmental Biology of Fishes, 31*(2), 203–206.

Lavery, R. J. (1995). Changes in offspring vulnerability account for the increase in convict cichlid defensive behaviour with brood age: Evidence for the nest crypsis hypothesis. *Animal Behaviour, 49*(5), 1177–1184.
Lavery, R. J., and Colgan, P. W. (1991). Brood age and parental defence in the convict cichlid, *Cichlasoma nigrofasciatum* (Pisces: Cichlidae). *Animal Behaviour, 41*, 945–951.
Leibel, W. (1992). Goin' south. Part 7: Cichlids of the Americas. We start with the "demonfish." *Aquarium Fish Magazine* (November), 43–51.
Leibel, W. S. (1998a). The evolution of mouthbrooding in South American geophagine cichlids (grant proposal). Easton, Pennsylvania: Lafayette College.
Leibel, W. S. (1998b). What's in a name? The demonfish "Jurupari." *Tropical Fish Hobbyist, 47*(3), 154–157.
Leveque, C. (1997). *Biodiversity Dynamics and Conservation: The Freshwater Fish of Tropical Africa.* New York: Cambridge University Press.
Li, H. W., and Brocksen, R. W. (1977). Approaches to the analysis for energetic costs of intraspecific competition for space by rainbow trout *(Salmo gairdneri)*. *Journal of Fish Biology, 11*, 329–341.
Liem, K. F. (1973). Evolutionary strategies and morphological innovations: Cichlid pharyngeal jaws. *Systematic Zoology, 22*(4), 425–441.
Liem, K. F. (1979). Modulatory multiplicity in the feeding mechanism in cichlid fishes, as exemplified by the invertebrate pickers of Lake Tanganyika. *Journal of Zoology* (London), *189*, 93–125.
Liem, K. F. (1991). Functional morphology. In M. H. A. Keenleyside (Ed.), *Cichlid Fishes: Behaviour, Ecology, and Evolution* (pp. 129–150). New York: Chapman and Hall.
Liem, K. F., and Stewart, D. J. (1976). Evolution of the scale-eating cichlid fishes of Lake Tanganyika: A generic revision with a description of a new species. *Bulletin of the Museum of Comparative Zoology, 147*, 319–350.
Limberger, D. (1983). Pairs and harems in a cichlid fish, *Lamprologus brichardi. Zeitschrift für Tierpsychologie, 62*, 115–144.
Lobel, P. S. (1998). Possible species specific courtship sounds by two sympatric cichlid fishes in Lake Malawi, Africa. *Environmental Biology of Fishes, 52*(4), 443–452.
Lobel, P. W. (1980). Invasion by the Mozambique tilapia (*Sarotherodon mossambicus*; Pisces; Cichlidae) of a Pacific atoll marine ecosystem. *Micronesica, 16*, 349–355.
Loiselle, P. (1971). Hybridization in cichlids. *Journal of the American Cichlid Association, 27*, 9–18.
Loiselle, P. V. (1985). *The Cichlid Aquarium.* Melle, Germany: Tetra-Press.
Loiselle, P. V., and Barlow, G. W. (1978). Do fishes lek like birds? In E. S. Reese and J. Lighter (Eds.), *Contrasts in Behavior* (pp. 33–75). New York: Wiley.
Long, J. A. (1995). *The Rise of Fishes: 500 Million Years of Evolution.* Baltimore: Johns Hopkins University Press.
Lorenz, K. Z. (1966). *On Aggression.* New York: Harcourt Brace Jovanovich.
Losey, G. S., Jr. (1978). The symbiotic behavior of fishes. In D. I. Mostofsky (Ed.), *The Behavior of Fish and Other Aquatic Animals* (pp. 1–31). New York: Academic Press.
Lovshin, L. L. (1982). *Tilapia* hybridization. In R. S. V. Pullin and R. H. Lowe-McConnell (Eds.), *The Biology and Culture of Tilapias,* vol. 7 (pp. 279–308). Manila: International Center for Living Aquatic Resources Management.
Lowe (McConnell), R. H. (1956). The breeding behaviour of *Tilapia* species (Pisces; Cichlidae) in natural waters: Observations on *T. karomo* and *T. variabilis* Boulenger. *Behaviour, 9*(2–3), 140–163.
Lowe (McConnell), R. H. (1959). Breeding behaviour patterns and ecological differences between *Tilapia* species and their significance for evolution within the genus *Tilapia* (Pisces: Cichlidae). *Proceedings of the Zoological Society of London, 132*(1–30), 1–30.
Lowe-McConnell, R. H. (1975). *Fish Communities in Tropical Freshwaters.* London: Longman.
Lowe-McConnell, R. H. (1987). *Ecological Studies in Tropical Fish Communities.* New York: Cambridge University Press.
Lucanus, O. (1998). Darwin's ponds: Malawi and Tanganyika. *Tropical Fish Hobbyist, 47*(2), 150–154.
Magurran, A. E. (1990). The adaptive significance of schooling as an anti-predator defence in fish. *Annales zoologica Fennici, 27*, 51–66.
Maitland, P. S. (1977). *The Hamlyn Guide to Freshwater Fishes of Britain and Europe.* London: Hamlyn.

Maley, J. D., Livingstone, A., Giresse, P., Thouveny, N., Brenac, P., Kelts, K., Kling, G., Stager, C., Haag, M., Fournier, M., Bandet, Y., Williamson, D., and Zogning, A. (1990). Lithostratigraphy, volcanism, paleomagnetism, and palynology of Quaternary lacustrine deposits from Brombi Mbo (West Cameroon): Preliminary results. *Journal of Volcanology and Geothermal Research, 42*, 319–335.

Marconato, A., Rasotto, M. B., and Mazzoldi, C. (1996). On the mechanism of sperm release in three gobiid fishes (Teleostei: Gobiidae). *Environmental Biology of Fishes, 46*(3), 321–327.

Marengoni, N. G., Onoue, Y., and Oyama, T. (1998). All-male tilapia hybrids of two strains of *Oreochromis niloticus. Journal of the World Aquaculture Society, 29*(1), 108–113.

Marler, P. (1955). Studies of fighting in chaffinches. (2) The effect on dominance relations of disguising females as males. *British Journal of Animal Behaviour, 3*, 137–146.

Marlier, G., and LeLeup, N. (1954). A curious ecological niche among the fishes of Lake Tanganyika. *Nature* (London), *174*, 935–936.

Marsh, A. C., and Ribbink, A. J. (1986). Feeding schools among Lake Malawi cichlid fishes. *Environmental Biology of Fishes, 15*, 75–79.

Marsh, B. A., Marsh, A. C., and Ribbink, A. J. (1986). Reproductive seasonality in a group of rock-frequenting cichlid fishes in Lake Malawi. *Journal of Zoology* (London), *209*, 9–20.

Martin, E., and Taborsky, M. (1997). Alternative male mating tactics in a cichlid, *Pelvicachromis pulcher:* A comparison of reproductive effort and success. *Behavioral Ecology and Sociobiology, 41*(5), 311–319.

Martinez-Palacios, C. A., Ross, L. G., and Sanchez Licea, V. H. (1996). The tolerance to salinity, respiratory characteristics, and potential for aquaculture of the Central American cichlid, *Cichlasoma synspilum* (Hubbs, 1935). *Aquaculture Research, 27*(4), 215–220.

Matsumoto, K., and Kohda, M. (1998). Interpopulation variation in the mating system of a substrate-breeding cichlid in Lake Tanganyika. *Journal of Ethology, 16*(2), 123–127.

Maynard Smith, J. (1974). The theory of games and the evolution of animal conflicts. *Journal of Theoretical Biology, 47*, 209–221.

Maynard Smith, J. (1977). Parental investment: A prospective analysis. *Animal Behaviour, 25*, 1–9.

Maynard Smith, J. (1979). Game theory and the evolution of behaviour. *Proceedings of the Royal Society of London Series B Biological Sciences, 205*, 475–488.

Mayr, E. (1976). Sympatric speciation. In E. Mayr (Ed.), *Evolution and the Diversity of Life: Selected Essays* (pp. 144–175). Cambridge, Massachusetts: Belknap Press.

Mayr, E. (1982). Processes of speciation. In C. Barigozzi (Ed.), *Mechanisms of Speciation* (pp. 1–19). New York: Liss.

Mayr, E. (1984). Evolution of fish species flocks: A commentary. In A. A. Echelle and I. Kornfield (Eds.), *Evolution of Fish Species Flocks* (pp. 3–11). Orono: University of Maine Press.

Mboko, S. K., Kohda, M., and Hori, M. (1998). Asymmetry of mouth-opening of a small herbivorous cichlid fish *Telmatochromis temporalis* in Lake Tanganyika. *Zoological Science* (Tokyo), *15*(3), 405–408.

McCune, A. R. (1987). Lakes as laboratories of evolution: Endemic fishes and environmental cyclicity. *Palaios, 2*(5), 446–454.

McKaye, K. R. (1977a). Competition for breeding sites between the cichlid fishes of Lake Jiloá, Nicaragua. *Ecology, 58*, 293–302.

McKaye, K. R. (1977b). Defense of a predator's young by a herbivorous fish: An unusual strategy. *American Naturalist, 111*, 301–315.

McKaye, K. R. (1980). Seasonality in habitat selection by the gold color morph of *Cichlasoma citrinellum* and its relevance to sympatric speciation in the family Cichlidae. *Environmental Biology of Fishes, 5*(1), 75–78.

McKaye, K. R. (1981). Field observations on death feigning: A unique behavior by the predatory cichlid, *Haplochromis livingstoni*, of Lake Malawi. *Environmental Biology of Fishes, 6*, 361–365.

McKaye, K. R. (1983). Ecology and breeding behavior of a cichlid fish *Cyrtocara eucinostomus* on a large lek in Lake Malawi, Africa. *Environmental Biology of Fishes, 8*, 81–96.

McKaye, K. R. (1986a). Mate choice and size assortative pairing by the cichlid fishes of Lake Jiloá, Nicaragua. *Journal of Fish Biology, 29*(Supplement A), 135–150.

McKaye, K. R. (1986b). A unique form of biparental care by an African catfish: An evolutionary puzzle. *ANIMA Magazine of Natural History, 6*, 96–97.

McKaye, K. R. (1991). Sexual selection and the evolution of the cichlid fishes of Lake Malawi, Africa. In M. H. A. Keenleyside (Ed.), *Cichlid Fishes: Behaviour, Ecology, and Evolution* (pp. 241–257). New York: Chapman and Hall.

McKaye, K. R., and Barlow, G. W. (1976). Competition between color morphs of the Midas cichlid, *Cichlasoma citrinellum*, in Lake Jiloá, Nicaragua. In T. B. Thorson (Ed.), *Investigations of the Ichthyofauna of Nicaraguan Lakes* (pp. 465–475). Lincoln: School of Life Sciences, University of Nebraska.

McKaye, K. R., and Gray, W. N. (1984). Extrinsic barriers to gene flow in rock dwelling cichlids of Lake Malawi. Microhabitat heterogeneity and reef colonization. In A. Echelle and I. Kornfield (Eds.), *Evolution of Fish Species Flocks* (pp. 169–184). Orono: University of Maine Press.

McKaye, K. R., Louda, S. M., and Stauffer, J. R., Jr. (1990). Bower size and male reproductive success in a cichlid fish lek. *American Naturalist, 135*(5), 597–613.

McKaye, K. R., and Marsh, A. (1983). Food switching by two specialized algae-scraping cichlid fishes in Lake Malawi, Africa. *Oecologia* (Berlin), *56*, 245–248.

McKaye, K. R., and McKaye, N. M. (1977). Communal care and kidnaping of young by parental cichlids. *Evolution, 3*, 674–681.

McKaye, K. R., Mughogho, D. E., and Lovullo, T. J. (1992). Formation of the selfish school. *Environmental Biology of Fishes, 35*, 213–218.

McKaye, K. R., and Oliver, M. K. (1980). Geometry of a selfish school: Defence of cichlid young by bagrid catfish in Lake Malawi, Africa. *Animal Behaviour, 28*, 1287–1290.

McKaye, K. R., Ryan, J. D., Stauffer, J. R., Jr., Perez, L. J. L., Vega, G. I., and Berghe, E. P. v. d. (1995). African tilapia in Lake Nicaragua: Ecosystem in transition. *BioScience, 45*(6), 406–411.

Meek, S. E. (1907). Synopsis of the fishes of the Great Lakes of Nicaragua. *Chicago Field Museum of Natural History,* 7(4), 97–132.

Meral, G. H. (1973). The Adaptive Significance of Territoriality in New World Cichlidae. Unpublished Ph.D. dissertation, University of California, Berkeley.

Mertz, J. C. (1967). The Organization and Regulation of the Parental Behavior of *Cichlasoma nigrofasciatum* (Pisces: Cichlidae), with Special Reference to Parental Fanning. Unpublished Ph.D. dissertation, University of Illinois, Urbana, Illinois.

Meyer, A. (1987). Phenotypic plasticity and heterochrony in *Cichlasoma managuense* (Pisces, Cichlidae) and their implications for speciation in cichlid fishes. *Evolution, 41*(6), 1357–1269.

Meyer, A. (1989). Cost of morphological specialization: Feeding performance of the two morphs in the trophically polymorphic cichlid fish, *Cichlasoma citrinellum*. *Oecologia* (Berlin), *80*(3), 431–436.

Meyer, A. (1990). Ecological and evolutionary consequences of the trophic polymorphism in *Cichlasoma citrinellum* (Pisces: Cichlidae). *Biological Journal of the Linnean Society, 39*(3), 279–300.

Meyer, A. (1993a). Evolution of mitochondrial DNA in fishes. In P. W. Hochachka and T. P. Mommsen (Eds.), *Biochemistry and Molecular Biology of Fishes,* vol. 2 (pp. 1–38). New York: Elsevier Science Publisher B.V.

Meyer, A. (1993b). Phylogenetic relationships and evolutionary processes in East African cichlid fishes. *Trends in Ecology and Evolution, 8*, 279–284.

Meyer, A., Kocher, T. D., Basasibwaki, P., and Wilson, A. C. (1990). Monophyletic origin of Lake Victoria cichlid fishes suggested by mitochondrial DNA sequences. *Nature* (London), *347*, 550–553.

Miller, R. R. (1966). Geographical distribution of Central American freshwater fishes. *Copeia, 1966*, 773–802.

Møller, A. P., and Thornhill, R. (1998). Bilateral symmetry and sexual selection: A meta-analysis. *American Naturalist, 151*(2), 174–192.

Montgomery, W. L. (1975). Interspecific associations of seabasses (Serranidae) in the Gulf of California. *Copeia, 1975*(4), 785–787.

Montgomery, W. L. (1977). Diet and gut morphology in fishes, with special reference to the monkeyface prickleback, *Cebedichthys violaceus* (Stichaeidae: Blennioidei). *Copeia, 1977*(1), 178–182.

Moran, P., Kornfield, I., and Reinthal, P. N. (1994). Molecular systematics and radiation of the haplochromine cichlids (Teleostei: Perciformes) of Lake Malawi. *Copeia, 1994*(2), 274–288.

Moyer, J. T., and Nakazono, A. (1978). Protandrous hermaphroditism in six species of the anemonefish genus *Amphiprion* in Japan. *Japanese Journal of Ichthyology, 25*, 101–106.

Mrowka, W. (1987a). Filial cannibalism and reproductive success in the maternal mouthbrooding cichlid fish *Pseudocrenilabrus multicolor. Behavioral Ecology and Sociobiology, 21*(4), 257–266.
Mrowka, W. (1987b). Oral fertilization in a mouthbrooding cichlid fish. *Ethology, 74*(3), 293–296.
Munthali, S. M., and Ribbink, A. J. (1998). Condition and fecundity of translocated rock-dwelling cichlid fish in Lake Malawi. *Journal of Zoology* (London), *244*(3), 347–355.
Myers, G. S. (1960). The endemic fish fauna of Lake Lanao, and the evolution of higher taxonomic categories. *Evolution, 14*(3), 323–333.
Myers, G. S. (1966). Derivation of the freshwater fish fauna of Central America. *Copeia, 1966*(4), 766–773.
Myrberg, A. A. J., Montgomery, W. L., and Fishelson, L. (1988). The reproductive behavior of *Acanthurus nigrofuscus* (Forskal) and other surgeonfishes (Fam. Acanthruidae) off Eilat, Israel (Gulf of Aqaba, Red Sea). *Ethology, 79*, 31–61.
Nagoshi, M. (1983). Distribution, abundance, and parental care of the genus *Lamprologus* (Cichlidae) in Lake Tanganyika. *African Studies Monographs, Kyoto University, 3*, 39–47.
Nagoshi, M., and Yanagisawa, Y. (1997). Parental care patterns and growth and survival of dependent offspring in cichlids. In H. Kawanabe, M. Hori, and M. Nagoshi (Eds.), *Fish Communities in Lake Tanganyika* (pp. 175–192). Kyoto: Kyoto University Press.
Nakai, K., Yanagisawa, Y., Sato, T., and Niimura, Y. (1990). Lunar synchronization of spawning in cichlid fishes of the tribe Lamprologini in Lake Tanganyika. *Journal of Fish Biology, 37*(4), 589–598.
Nakano, S., and Nagoshi, M. (1990). Brood defense and parental roles in a biparental cichlid fish *Lamprologus toae* in Lake Tanganyika [East Africa]. *Japanese Journal of Ichthyology, 36*(4), 468–476.
Naylor, R. L., Goldburg, R. J., Mooney, H., Beveridge, M., Clay, J., Folke, C., Kautsky, N., Lubchenco, J., Primavera, J., and Williams, M. (1998). Nature's subsidies to shrimp and salmon farming. *Science* (Washington, D.C.), *283*(5390), 883–884.
Neat, F. C. (1998). Mouth morphology, testes size, and body size in male *Tilapia zillii:* Implications for fighting and assessment. *Journal of Fish Biology, 53*(4), 890–892.
Neat, F. C., Huntingford, F. A., and Beveridge, M. M. C. (1998). Fighting and assessment in male cichlid fish: The effects of asymmetries in gonadal state and body size. *Animal Behaviour, 55*(4), 883–891.
Neil, E. H. (1964). An analysis of color changes and social behavior of *Tilapia mossambica. University of California Publications in Zoology, 75*(1), 1–58.
Nelissen, M. (1976). Contribution to the ethology of *Tropheus moorii* Boulenger (Pisces, Cichlidae) and a discussion of the significance of its colour patterns. *Revue de Zoologie Africaine, 90*(1), 17–29.
Nelson, C. M. (1995). Male size, spawning pit size, and female mate choice in a lekking cichlid fish. *Animal Behaviour, 50*(6), 1587–1599.
Nelson, J. S. (1994). *Fishes of the World*, 3rd ed. New York: Wiley.
Newman, L. (1993). Maintenance and breeding of the red hump eartheater, *Geophagus steindachneri* Eigenmann and Hildebrand 1910. *Cichlid News, 2*(4), 14–16.
Nicoll, C. S., and Bern, H. A. (1971). On the actions of prolactin among the vertebrates: Is there a common denominator? In G. W. W. Wolstenholme and J. Knight (Eds.), *Lactogenic Hormones* (pp. 299–324). Edinburgh: Churchill and Livingstone.
Nielsen, J. L. (1994). Invasive cohorts: Impacts of hatchery-reared coho salmon on the trophic, developmental, and genetic ecology of wild stocks. In D. J. Stouder, K. L. Fresh, and R. J. Feller (Eds.), *Theory and Application in Fish Feeding Ecology*, vol. 18 (pp. 361–385): Belle W. Baruch Library in Marine Sciences.
Nishida, M. (1997). Phylogenetic relationships and evolution of Tanganyikan cichlids: A molecular perspective. In H. Dawanabe, M. Hori, and M. Nagoshi (Eds.), *Fish Communities in Lake Tanganyika* (pp. 3–23). Kyoto: Kyoto University Press.
Noakes, D. L. G. (1973). Parental behavior and some histological features of scales in *Cichlasoma citrinellum* (Pisces, Cichlidae). *Canadian Journal of Zoology, 51*(6), 619–622.
Noakes, D. L. G., and Barlow, G. W. (1973). Ontogeny of parent-contacting in young *Cichlasoma citrinellum. Behaviour, 46*, 221–255.
Noble, G. K., and Curtis, B. (1939). The social behavior of the jewel fish *Hemichromis bimaculatus* Gill. *Bulletin of the American Museum of Natural History, 76*, 1–46.
Noonan, K. C. (1983). Female mate choice in the cichlid fish *Cichlasoma nigrofasciatum. Animal Behaviour, 31*, 1005–1010.

Norris, K. S., and Prescott, J. H. (1959). Jaw structure and tooth replacement in the opaleye, *Girella nigricans* (Ayres) with notes on other species. *Copeia, 1959*(4), 275–283.

Nshombo, M. (1991). Occasional egg-eating by the scale-eater *Plecodus straeleni* (Cichlidae) of Lake Tanganyika. *Environmental Biology of Fishes, 31*(2), 207–212.

Nuttall, D. B., and Keenleyside, M. H. A. (1993). Mate choice by the male convict cichlid (*Cichlasoma nigrofasciatum;* Pisces, Cichlidae). *Ethology, 95*(3), 247–256.

Nxomani, C., Ribbink, A. J., and Kirby, R. (1999). DNA profiling of *Tilapia guinasana*, a species endemic to a single sinkhole, to determine the genetic divergence between color forms. *Electrophoresis, 20*(8), 1781–1785.

Ochi, H. (1993a). Maintenance of separate territories for mating and feeding by males of a maternal mouthbrooding cichlid, *Gnathochromis pfefferi*, in Lake Tanganyika. *Japanese Journal of Ichthyology, 40*(2), 173–182.

Ochi, H. (1993b). Mate monopolization by a dominant male in a multi-male social group of a mouthbrooding cichlid, *Ctenochromis horei. Japanese Journal of Ichthyology, 40*(2), 209–218.

Ochi, H. (1996). Mating systems of two midwater-spawning cichlids, *Cyprichromis microlepidotus* and *Paracyprichromis brieni*, in Lake Tanganyika. *Ichthyological Research, 43*(3), 239–246.

Ochi, H., Sato, Y., and Yanagisawa, Y. (1999). Obligate feeding of cichlid eggs by *Caecomastacembelus zebratus* in Lake Tanganyika. *Journal of Fish Biology, 54*(2), 450–459.

Ochumba, P. B. O. (1995). Limnological changes in Lake Victoria since the Nile perch introduction. In T. J. Pitcher and P. J. B. Hart (Eds.), *The Impact of Species Changes in African Lakes* (pp. 33–43). New York: Chapman and Hall.

Oehlert, B. (1958). Kampf und Paarbildung einiger Cichliden. *Zeitschrift für Tierpsychologie, 15*(2), 141–171.

Ogutu-Ohwayo, R. (1995). Diversity and stability of fish stocks in Lakes Victoria, Kyoga, and Nabugabo after establishment of introduced species. In T. J. Pitcher and P. J. B. Hart (Eds.), *The Impact of Species Changes in African Lakes* (pp. 59–81). New York: Chapman and Hall.

Ogutu-Ohwayo, R., Hecky, R. E., Cohen, A. S., and Kaufman, L. (1997). Human impacts on the African Great Lakes. *Environmental Biology of Fishes, 50*(2), 117–131.

Ohm, D. (1978). Sexualdimorphismus, Polygamie und Geschlechtswechsel bei *Crenicara punctulata* Gunther 1986 (Cichlidae, Telesotei). *Sitzungsberichte der Gesellschaft für Naturforschung Fruende Berlin (N.), 18*, 90–104.

Oosten, J. v. (1957). The skin and scales. In M. E. Brown (Ed.), *The Physiology of Fishes,* vol. 1 (pp. 207–244). New York: Academic Press.

Oppen, M. J. H. v., Turner, G. F., Rico, C., Deutsch, J. C., Ibrahim, K. M., Robinson, R. L., and Hewitt, G. M. (1997). Unusually fine-scale genetic structuring found in rapidly speciating Malawi cichlid fishes. *Proceedings of the Royal Society of London Series B Biological Sciences, 264*(1389), 1803–1812.

Oppenheimer, J. R. (1970). Mouth breeding in fishes. *Animal Behaviour, 18*, 493–503.

Oppenheimer, J. R., and Barlow, G. W. (1968). Dynamics of parental behavior in the black-chinned mouth-breeder, *Tilapia melanotheron* (Pisces: Cichlidae). *Zeitschrift für Tierpsychologie, 25*(8), 889–914.

Ostrander, G. K., and Ward, J. A. (1985). The function of the pelvic fins during courtship and spawning in the orange chromide, *Etroplus maculatus. Environmental Biology of Fishes, 13*(3), 203–210.

Owen, R. B., Crossley, R., Johnson, T. C., Tweddle, D., Kornfield, I., Davison, S., Eccles, D. H., and Engstrom, D. E. (1990). Major low lake levels of Lake Malawi and their implications for speciation rates in cichlid fishes. *Proceedings of the Royal Society of London Series B Biological Sciences, 240*(1299), 519–553.

Parker, A., and Kornfield, I. (1996). Polygynandry in *Pseudotropheus zebra*, a cichlid fish from Lake Malawi. *Environmental Biology of Fishes, 47*(4), 345–352.

Paterson, H. E. H. (1993). *Evolution and the Recognition Concept of Species: Collected Writings.* Baltimore: Johns Hopkins University Press.

Paulo, J. (1995). Comparative study of the behavioral pattern of *Steatocranus casuarius* Poll, 1939 and *Leptotilapia tinanti* (Poll, 1939). *Journal of the American Cichlid Association, 164* (October), 6–15.

Payne, R. B. (1977). The ecology of brood parasitism in birds. *Annual Review of Ecology and Systematics, 8*, 1–28.

Perrone, M. (1978). The economy of brood defence by parental cichlid fishes *Cichlasoma maculicauda. Oikos, 31*, 137–141.

Peters, H. M. (1957). Über die Regulation der Gelegegrösse bei Fischen. *Zeitschrift für Naturforschung, 12b*, 255–261.

Peters, H. M. (1965). Angeborenes Verhalten bei Buntbarschen. I. Wege der Analyse. *Umschau, 21* (November), 665–669.

Peters, H. M. (1971). Testis weights in *Tilapia* (Pisces: Cichlidae). *Copeia, 1971*(1), 13–17.

Peters, H. M. (1975). Hermaphroditism in cichlid fishes. In R. Reinboth (Ed.), *Intersexuality in the Animal Kingdom* (pp. 228–235). Berlin: Springer Verlag.

Pitcher, T. J., and Hart, P. J. B. (Eds.). (1995). *The Impact of Species Changes in African Lakes*, vol. 18. New York: Chapman and Hall.

Poll, M. (1986). Classification des Cichlidae du lac Tanganika. Tribus, genres et espèces. *Mémoires de la Classes des Sciences. Académie Royale de Belgique, 45*(2), 1–163.

Power, M. E. (1990). Resource enhancement by indirect effects of grazers: Armored catfish, algae, and sediment. *Ecology, 71*, 897–904.

Praetorius, W. (1932). How the "king" lives at home. *The Aquarium, 1*, 119–120, 140.

Pullin, R. S. V. (1991). Cichlids in aquaculture. In M. H. A. Keenleyside (Ed.), *Cichlid Fishes: Behaviour, Ecology, and Evolution* (pp. 280–309). New York: Chapman and Hall.

Queiroz, K. d., and Gauthier, J. (1994). Toward a phylogenetic system of biological nomenclature. *Trends in Ecology and Evolution, 9*, 27–31.

Queller, D. C., Strassmann, J. E., and Hughes, C. R. (1993). Microsatellites and kinship. *Trends in Ecology and Evolution, 8*(8), 285–288.

Quertermus, C. J., Jr., and Ward, J. A. (1969). Development and significance of two motor patterns used in contacting parents by young orange chromides *(Etroplus maculatus). Animal Behaviour, 17*, 624–635.

Quinn, T. P., Adkison, M. D., and Ward, M. B. (1996). Behavioral tactics of male sockeye salmon *(Oncorhynchus nerka)* under varying operational sex ratios. *Ethology, 102*(4), 304–322.

Radesäter, T., and Fernö, A. (1979). On the function of the "eye-spots" in agonistic behaviour in the fire-mouth cichlid *(Cichlasoma meeki). Behavioural Processes, 4*, 5–13.

Rahman, M. A., Mak, R., Ayad, H., Smith, A., and Maclean, N. (1998). Expression of a novel piscine growth hormone gene results in growth enhancement in transgenic tilapia (*Oreochromis niloticus*). *Transgenic Research*, 7(5), 357–369.

Randall, J. E., and Hartman, W. D. (1968). Sponge-feeding fishes of the West Indies. *Marine Biology, 1*(3), 216–225.

Rangeley, R. W., and Godin, J.-G. J. (1992). The effects of a trade-off between foraging and brood defense on parental behaviour in the convict cichlid fish, *Cichlasoma nigrofasciatum. Behaviour, 120*, 123–138.

Rechten, C. (1980). Brood relief behaviour of the cichlid fish *Etroplus maculatus. Zeitschrift für Tierpsychologie, 52*, 77–102.

Reebs, S. G. (1994). Nocturnal mate recognition and nest guarding by female convict cichlids (Pisces, Cichlidae: *Cichlasoma nigrofasciatum*). *Ethology, 96*, 303–312.

Reebs, S. G., and Colgan, P. W. (1991). Nocturnal care of eggs and circadian rhythms of fanning activity in two normally diurnal cichlid fishes, *Cichlasoma nigrofasciatum* and *Herotilapia multispinosa. Animal Behaviour, 41*, 303–311.

Reyer, H. U. (1986). Breeder-helper interactions in the pied kingfisher reflect the costs and benefits of cooperative breeding. *Behaviour, 96*, 277–393.

Reynolds, J. E., Greboval, D. F., and Mannini, P. (1995). Thirty years on: The development of the Nile perch fishery in Lake Victoria. In T. J. Pitcher and P. J. B. Hart (Eds.), *The Impact of Species Changes in African Lakes* (pp. 181–214). New York: Chapman and Hall.

Ribbink, A. J. (1984). The feeding behaviour of a cleaner and scale, skin, and fin eater from Lake Malawi (*Docimodus evelynae;* Pisces, Cichlidae). *Netherlands Journal of Zoology, 34*(2), 182–196.

Ribbink, A. J. (1991). Distribution and ecology of the cichlids of the African Great Lakes. In M. H. A. Keenleyside (Ed.), *Cichlid Fishes: Behaviour, Ecology, and Evolution* (pp. 36–59). New York: Chapman and Hall.

Ribbink, A. J. (1994). Alternative perspectives on some controversial aspects of cichlid fish speciation. *Advances in Limnology, 44*, 101–125.

Ribbink, A. J., and Chan, T.-Y. (1989). Sneaking in *Pseudocrenilabrus philander* and the prevalence of sneaking in lacustrine and riverine haplochromines. *Musée Royal de l'Afrique Centrale, Annales Sciences Zoologiques, 257*, 23–28.

Ribbink, A. J., Greenwood, P. H., Ribbink, A. C., Twentyman-Jones, V., and Zyl, B. v. (1991). Unique polychromatism of *Tilapia guinasana*, an African cichlid fish. *South African Journal of Science, 87*, 608–611.

Ribbink, A. J., and Lewis, D. S. C. (1982). *Melanochromis crabro* sp. nov.: A cichlid fish from Lake Malawi which feeds on ectoparasites and catfish eggs. *Netherlands Journal of Zoology, 32*(1), 72–87.

Ribbink, A. J., Marsh, A. C., and Marsh, B. A. (1981). Nest-building and communal care of young by *Tilapia rendalli* Dumeril (Pisces, Cichlidae) in Lake Malawi. *Environmental Biology of Fishes, 6*, 219–222.

Ribbink, A. J., Marsh, B. A., Marsh, A. C., Ribbink, A. C., and Sharp, B. J. (1983). A preliminary survey of the cichlid fishes of rocky habitats in Lake Malawi. *South African Journal of Zoology, 18*, 149–310.

Ribbink, A. J., and Ribbink, A. C. (1997). Paedophagia among cichlid fishes of Lake Victoria and Lake Malawi-Nyasa. *South African Journal of Science, 93*(11–12), 509–512.

Riedmiller, S. (1994). Lake Victoria fisheries: The Kenyan reality and environmental implications. *Environmental Biology of Fishes, 39*(4), 329–338.

Roberts, T. R. (1970). Scale-eating American characoid fishes, with special reference to *Probolodus heterostomus*. *Proceedings of the California Academy of Sciences, 38*(20), 383–390.

Roberts, T. R., and Stewart, D. J. (1976). An ecological and systematic survey of fishes in the rapids of the lower Zaire or Congo River. *Bulletin of the Museum of Comparative Zoology, 147*, 239–317.

Robinson, B. W., and Wilson, D. S. (1998). Optimal foraging, specialization, and a solution to Liem's paradox. *American Naturalist, 151*(3), 223–235.

Rogers, W. (1987). Sex ratio, monogamy, and breeding success in the Midas cichlid *(Cichlasoma citrinellum)*. *Behavioral Ecology and Sociobiology, 21*, 47–51.

Rogers, W. (1988). Parental investment and division of labor in the Midas cichlid *(Cichlasoma citrinellum)*. *Ethology, 79*, 126–142.

Rogers, W. (1995). Female choice predicts the best father in a biparental fish, the Midas cichlid *(Cichlasoma citrinellum)*. *Ethology, 100*, 230–241.

Rogers, W., and Barlow, G. W. (1991). Sex differences in mate choice in a monogamous biparental fish, the Midas cichlid *(Cichlasoma citrinellum)*. *Ethology, 87*, 249–261.

Römer, U., and Beisenherz, W. (1996). Environmental determination of sex in *Apistogramma* (Cichlidae) and two other freshwater fishes (Teleostei). *Journal of Fish Biology, 48*, 714–725.

Rossiter, A. (1995). The cichlid fish assemblages of Lake Tanganyika: Ecology, behaviour, and evolution of its species flocks. *Advances in Ecological Research, 26*, 187–252.

Rüber, L., Verheyen, E., Sturmbauer, C., and Meyer, A. (1998). Lake level fluctuations and speciation in rock-dwelling cichlid fish in Lake Tanganyika, East Africa. In P. R. Grant (Ed.), *Evolution on Islands* (pp. 225–240). New York: Oxford University Press.

Rundle, H. D., and Schluter, D. (1998). Reinforcement of stickleback mate preferences: Sympatry breeds contempt. *Evolution, 52*(1), 200–208.

Ryan, M. J. (1985). *The Tungara Frog: A Study in Sexual Selection and Communication.* Chicago: University of Chicago Press.

Ryan, M. J. (1994). Mechanisms underlying sexual selection. In L. A. Real (Ed.), *Behavioral Mechanisms in Evolutionary Ecology* (pp. 190–215). Chicago: University of Chicago Press.

Sargent, R. C., and Gross, M. R. (1986). William's principle: An explanation of parental care in teleost fishes. In T. J. Pitcher (Ed.), *The Behaviour of Teleost Fishes* (pp. 275–293). Baltimore: Johns Hopkins University Press.

Sato, T. (1986). A brood parasitic catfish of mouthbrooding cichlid fishes in Lake Tanganyika. *Nature* (London), *323*, 58–59.

Sato, T. (1987). A brood-parasitic catfish [in Japanese]. *Nikkei Science*, 185, 34–42.

Sato, T. (1994). Active accumulation of spawning substrate: A determinant of extreme polygyny in a shell-brooding cichlid fish. *Animal Behaviour, 48*(3), 669–678.

Schliewen, U. K., Tautz, D., and Pääbo, S. (1994). Sympatric speciation suggested by monophyly of crater lake cichlids. *Nature* (London), *368*(6472), 629–632.

Schütz, D., Taborsky, M., Parker, G., and Sato, T. (1997). Sexual dimorphism in a shell brooding cichlid: The importance of sexual and natural selection. *Advances in Ethology, 32*, 238.

Schütz, M. (1995). Muttermilch. Nicht nur be Säugetieren und Menschen? *Sexualmedizin, 17*(12), 333–336.

Schütz, M., and Barlow, G. W. (1997). Young of the Midas cichlid get biologically active nonnutrients by eating mucus from the surface of their parents. *Fish Physiology and Biochemistry, 16*(1), 11–18.

Schwanck, E. (1986). Filial cannibalism in *Tilapia mariae*. *Journal of Applied Ichthyology, 2*, 65–74.

Schwanck, E. (1987). *Female mate choice in Tilapia mariae*. Unpublished Ph.D. dissertation. University of Stockholm, Stockholm.

Schwanck, E., and Rana, K. (1991). Male-female parental roles in *Sarotherodon galilaeus* (Pisces: Cichlidae). *Ethology, 89*(3), 229–243.

Schwarz, A. (1974). The inhibition of aggressive behavior by sound in the cichlid fish, *Cichlasoma centrarchus*. *Zeitschrift für Tierpsychologie, 35*(5), 508–517.

Seehausen, O. (1996). *Lake Victoria Rock Cichlids: Taxonomy, Ecology, and Distribution.* Germany: Verduyn Cichlids.

Seehausen, O. (1997). Biparental mouthbrooding in a Lake Victoria cichlid. Unpublished manuscript.

Seehausen, O. (1999). A reconsideration of the ecological composition of the cichlid species flock in Lake Victoria before and after the Nile perch boom. In W. L. T. v. Densen and M. J. Morris (Eds.), *Fish and Fisheries of Lakes and Reservoirs in Southeast Asia and Africa* (pp. 281–293). Otley, U.K.: Westbury Publishing.

Seehausen, O., and Alphen, J. J. M. v. (1998). The effect of male coloration on female mate choice in closely related Lake Victoria cichlids (*Haplochromis nyererei* complex). *Behavioral Ecology and Sociobiology, 42*(1), 1–8.

Seehausen, O., Alphen, J. J. M. v., and Witte, F. (1997). Cichlid fish diversity threatened by eutrophication that curbs sexual selection. *Science* (Washington, D.C.), *277*, 1808–1811.

Seehausen, O., Alphen, J. J. M. v., and Witte, F. (1999). Can ancient colour polymorphisms explain why some cichlid lineages speciate rapidly under disruptive sexual selection? *Belgian Journal of Zoology, 129*(1), 43–60.

Seehausen, O., and Bouton, N. (1996). Polychromatism in rock-dwelling Lake Victoria cichlids: Types, distribution, and observations on their genetics. *The Cichlids Yearbook, 6*, 36–45.

Seehausen, O., and Bouton, N. (1997). Microdistribution and fluctuations in niche overlap in a rocky shore cichlid community in Lake Victoria. *Ecology of Freshwater Fish, 6*(3), 161–173.

Seehausen, O., Mayhew, P. J., and Alphen, J. J. M. v. (1999). Evolution of colour patterns in East African cichlid fish. *Journal of Evolutionary Biology, 12*(3), 514–534.

Seehausen, O., Witte, F., Alphen, J. J. M. v., and Bouton, N. (1998). Direct mate choice maintains diversity among sympatric cichlids in Lake Victoria. *Journal of Fish Biology, 53*(Supplement A), 37–55.

Shafland, P. L. (1995). Introduction and establishment of a successful butterfly peacock fishery in southeast Florida canals. In H. L. Schramm, Jr. and R. G. Piper (Eds.), *Uses and Effects of Cultured Fishes in Aquatic Ecosystems* (pp. 443–451). Bethesda, Maryland: American Fisheries Society.

Shennan, M. G. C., Waas, J. R., and Lavery, R. J. (1994). The warning signals of parental convict cichlids are socially facilitated. *Animal Behaviour, 47*(4), 974–976.

Sheridan, L., and Pomiankowski, A. (1997). Female choice for spot asymmetry in the Trinidadian guppy. *Animal Behaviour, 54*(6), 1523–1529.

Short, R. V., and Balaban, E. (Eds.). (1994). *The Differences Between the Sexes.* New York: Cambridge University Press.

Siemens, M. v. (1990). Broodcare or egg cannibalism by parents and helpers in *Neolamprologus brichardi* (Poll 1986) (Pisces: Cichlidae): A study on behavioural mechanisms. *Ethology, 84*(1), 60–89.

Sikkel, P. C. (1995). Diel periodicity of spawning activity in a permanently territorial damselfish: A test of adult feeding hypothesis. *Environmental Biology of Fishes, 42*, 241–251.

Simpson, M. J. A. (1968). The display of the Siamese fighting fish, *Betta splendens*. *Animal Behaviour Monograph, 1*, 1–73.

Sjölander, S. (1972). Feldbeobachtungen an einigen westafrikanischen Cichliden. *Deutsche Aquarien- und Terrarien- Zeitschrift, 19*(3), 42–45.

Smith, N. G. (1968). The advantage of being parasitized. *Nature* (London), *219*, 690–694.

Snoeks, J., Rüber, L., and Verheyen, E. (1994). The Tanganyika problem: Taxonomy and distribution patterns of its ichthyofauna. *Advances in Limnology, 44*, 357–374.

Springer, V. G. (1982). Pacific Plate biogeography, with special reference to shorefishes. *Smithsonian Contributions in Zoology,367,* 1–182.

Springer, V. G., and Smith-Vaniz, W. (1972). Mimetic relationships involving fishes of the family Blenniidae. *Smithsonian Contributions in Zoology, 112*, 1–36.

St. Mary, C. M. (1998). Characteristic gonad structure in the gobiid genus *Lythrypnus* with comparisons to other hermaphroditic gobies. *Copeia, 1998*(3), 720–724.

Stauffer, J. R., Jr. (1991). Description of a facultative cleanerfish (Teleostei: Cichlidae) from Lake Malawi, Africa. *Copeia, 1991*, 141–147.

Stauffer, J. R., Jr., Bowers, N. J., Kellogg, K. A., and McKaye, K. R. (1997). A revision of the blue-black *Pseudotropheus zebra* (Teleostei: Cichlidae) complex from Lake Malawi, Africa, with a description of a new genus and ten new species. *Proceedings of the Academy of Natural Sciences of Philadelphia, 148*, 189–230.

Stauffer, J. R., Jr., Bowers, N. J., Kocher, T. D., and McKaye, K. R. (1996). Evidence of hybridization between *Cynotilapia afra* and *Pseudotropheus zebra* (Teleostei: Cichlidae) following an intralacustrine translocation in Lake Malawi. *Copeia, 1996*(1), 203–208.

Stauffer, J. R., Jr., and Kellogg, K. A. (1996). Sexual selection in Lake Malawi cichlids. *The Cichlids Yearbook, 6*, 23–28.

Stauffer, J. R., Jr., LoVullo, T. J., and McKaye, K. R. (1993). Three new sand-dwelling cichlids from Lake Malawi, Africa, with a discussion of the status of the genus *Copadichromis* (Teleostei: Cichlidae). *Copeia, 1993*, 1017–1027.

Stepien, C. A., and Kocher, T. D. (1997). Molecules and morphology in studies of fish evolution. In T. D. Kocher and C. A. Stepien (Eds.), *Molecular Systematics of Fishes* (pp. 1–11). San Diego: Academic Press.

Stewart, D. J., and Roberts, T. R. (1984). A new species of dwarf cichlid fish with reversed sexual dichromatism from Lac Mai-ndombe, Zaire. *Copeia, 1984*, 92–86.

Stiassny, M. L. J. (1991). Phylogenetic intrarelationships of the family Cichlidae. In M. H. A. Keenleyside (Ed.), *Cichlid Fishes: Behaviour, Ecology, and Evolution* (pp. 1–35). New York: Chapman and Hall.

Stiassny, M. L. J. (1994). A tangled web: Cichlids and their relatives. *Cichlids Yearbook, 4*, 91–94.

Stiassny, M. L. J. (1997). A phylogenetic overview of the lamprologine cichlids of Africa (Teleostei, Cichlidae): A morphological perspective. *South African Journal of Science, 93*(11–12), 513–523.

Stiassny, M. L. J., and Mezey, J. G. (1993). Egg attachment systems in the family Cichlidae (Perciformes: Labroidei), with some comments on their significance for phylogenetic studies. *American Museum Novitates, 3058*, 1–11.

Stiassny, M. L. J., and Raminosoa, N. (1994). The fishes of the inland waters of Madagascar. *Musée Royal de l'Afrique Centrale, Annales Sciences Zoologiques, 275*, 133–149.

Stiassny, M. L. J., Schliewen, U. K., and Dominey, W. J. (1992). A new species flock of cichlid fishes from Lake Bermin, Cameroon with a description of eight new species of *Tilapia* (Labroidei: Cichlidae). *Ichthyological Exploration of Freshwaters, 3*(4), 311–346.

Strauss, E. (1999). Can mitochondrial clocks keep time? *Science* (Washington, D.C.), *283*(5407), 1436–1438.

Streelman, J. T., and Karl, S. A. (1997). Reconstructing labroid evolution with single-copy nuclear DNA. *Proceedings of the Royal Society of London Series B Biological Sciences, 264*(1384), 1011–1020.

Struessmann, C. A., Cota, J. C. C., Phonlor, G., Higuchi, H., and F. Takashima. (1996). Temperature effects on sex differentiation of two South American atherinids, *Odontesthes argentinensis* and *Patagonina hatcheri. Environmental Biology of Fishes, 47*(2), 143–154.

Sturmbauer, C. (1998). Explosive speciation in cichlid fishes of the African Great Lakes: A dynamic model of adaptive radiation. *Journal of Fish Biology, 53*(Supplement A), 18–36.

Sturmbauer, C., and Meyer, A. (1993). Mitochondrial phylogeny of the endemic mouthbrooding lineages of cichlid fishes from Lake Tanganyika in Eastern Africa. *Molecular Biology and Evolution, 10*(4), 751–768.

Sturmbauer, C., Verheyen, E., and Meyer, A. (1994). Mitochondrial phylogeny of the Lamprologini, the major substrate spawning lineage of cichlid fishes from Lake Tanganyika in Eastern Africa. *Molecular Biology and Evolution, 11*(4), 691–703.

Symons, D. (1995). Beauty is in the adaptations of the beholder: The evolutionary psychology of human female sexual attractiveness. In P. R. Abramson and S. D. Pinkerton (Eds.), *Sexual Nature/Sexual Culture* (pp. 80–118). Chicago: University of Chicago Press.

Taborsky, M. (1984). Broodcare helpers in the cichlid fish *Lamprologus brichardi:* Their costs and benefits. *Animal Behaviour, 32*(4), 1236–1252.

Taborsky, M. (1985a). Breeder-helper conflict in a cichlid fish with broodcare helpers: An experimental analysis. *Behaviour, 95*, 45–75.

Taborsky, M. (1985b). On optimal parental care. *Zeitschrift für Tierpsychologie, 70*(4), 331–336.

Taborsky, M. (1994). Sneakers, satellites, and helpers: Parasitic and cooperative behavior in fish reproduction. *Advances in the Study of Behavior, 23*, 1–100.

Taborsky, M. (1998). Sperm competition in fish: Bourgeois males and parasitic spawning. *Trends in Ecology and Evolution, 13*(1), 222–227.

Taborsky, M., and Grantner, A. (1998). Behavioural time-energy budgets of cooperatively breeding *Neolamprologus pulcher* (Pisces: Cichlidae). *Animal Behaviour, 56*(6), 1375–1382.

Taborsky, M., and Limberger, D. (1981). Helpers in fish. *Behavioral Ecology and Sociobiology, 8*, 143–145.

Taylor, F. J. R. (1999). Ultrastructure as a control for protistan molecular phylogeny. *American Naturalist, 154*(Supplement), S125-S136.

Taylor, M. I., Turner, G. F., Robinson, R. L., and Strauffer, J. R., Jr. (1998). Sexual selection, parasites, and bower height skew in a bower-building cichlid fish. *Animal Behaviour, 56*(2), 379–384.

Thompson, A. B., Allison, E. H., and Ngatunga, B. P. (1996). Distribution and breeding biology of offshore pelagic cyprinids and catfish in Lake Malawi-Niassa. *Environmental Biology of Fishes, 47*(1), 27–42.

Thornhill, R., and Gangestad, S. W. (1996). The evolution of human sexuality. *Trends in Ecology and Evolution, 11*(2), 98–102.

Timms, A. M., and Keenleyside, M. H. A. (1975). The reproductive behaviour of *Aequidens paraguayensis. Zeitschrift für Tierpsychologie, 39*, 8–23.

Trendall, J. (1988a). The distribution and dispersal of introduced fish at Thumbi West Island in Lake Malawi, Africa. *Journal of Fish Biology, 33*(3), 357–370.

Trendall, J. (1988b). Recruitment of juvenile mbuna (Pisces: Cichlidae) to experimental rock shelters in Lake Malawi, Africa. *Environmental Biology of Fishes, 22*(2), 117–132.

Trewavas, E. (1947). An example of "mimicry" in fishes. *Nature* (London), *160*(4056), 120.

Trewavas, E. (1962). Fishes of the crater lakes of the northwestern Cameroons. *Bonner Zoologische Beiträge, 1/3*, 146–192.

Trewavas, E. (1983). *Tilapiine Fishes of the Genera Sarotherodon, Oreochromis, and Danakilia.* Ithaca: Comstock Publishing.

Trewavas, E., Green, J., and Corbet, S. A. (1972). Ecological studies on crater lakes in West Cameroon: Fishes of Barombi Mbo. *Journal of the Zoological Society of London, 167*, 41–95.

Trivers, R. L. (1972). Parental investment and sexual selection. In B. Campbell (Ed.), *Sexual Selection and the Descent of Man* (pp. 136–179). Chicago: Aldine.

Trivers, R. L. (1974). Parent-offspring conflict. *American Zoologist, 14*, 249–264.

Tuan, P. A., Little, D. C., and Mair, G. C. (1998). Genotypic effects on comparative growth performance of all-male tilapia *Oreochromis niloticus* (L.). *Aquaculture, 159*(3–4), 293–302.

Turner, G., and Burrows, M. T. (1995). A model of sympatric speciation by sexual selection. *Proceedings of the Royal Society of London Series B Biological Sciences, 260*(1359), 287–292.

Turner, G. F. (1994a). Fishing and the conservation of the endemic fishes of Lake Malawi. *Advances in Limnology, 44*, 481–494.

Turner, G. F. (1994b). Speciation mechanisms in Lake Malawi cichlids: A critical review. *Advances in Limnology, 44*, 139–160.

Turner, G. F. (1997). Small fry go big time. *New Scientist* (2 August), 36–39.

Tyler, M. J., and Carter, D. B. (1981). Oral birth of the young of the gastric brooding frog *Rheobatrachus silus. Animal Behaviour, 29*, 280–282.

United Nations Food and Agriculture Organizations. (1995). *The State of World Fisheries and Aquaculture.* Rome: FAO.
Valerio, M., and Barlow, G. W. (1986). Ontogeny of young Midas cichlids: A study of feeding, filial cannibalism, and agonism in relation to differences in size. *Biology of Behavior, 11*, 16–35.
Vehrencamp, S. L., and Bradbury, J. W. (1984). Mating systems and ecology. In J. R. Krebs and N. B. Davies (Eds.), *Behavioural Ecology: An Evolutionary Approach,* 2nd ed. (pp. 251–278). Sunderland, Massachusetts: Sinauer Associates.
Vincent, A., Ahnesjö, I., Berglund, A., and Rosenqvist, G. (1992). Pipefishes and seahorses: Are they all sex role reversed? *Trends in Ecology and Evolution, 7*, 237–241.
Vos, L. d., and Snoeks, J. (1994). The noncichlid fishes of the Lake Tanganyika basin. *Advances in Limnology, 44*, 391–405.
Walter, B., and Trillmich. (1994). Female aggression and male peace-keeping in a cichlid fish harem: Conflict between and within the sexes in *Lamprologus ocellatus. Behavioral Ecology and Sociobiology, 34*(2), 105–112.
Ward, J. A., and Barlow, G. W. (1967). The maturation and regulation of glancing off the parents by young orange chromides (*Etroplus maculatus:* Pisces-Cichlidae). *Behaviour, 29*, 1–56.
Ward, J. A., and Samarakoon, J. I. (1981). Reproductive tactics of the Asian cichlids of the genus *Etroplus* in Sri Lanka. *Environmental Biology of Fishes, 6*, 95–103.
Ward, J. A., and Wyman, R. A. (1975). The cichlids of the resplendent isle. *Oceans, 8*, 42–47.
Ward, J. A., and Wyman, R. L. (1977). Ethology and ecology of cichlid fishes of the genus *Etroplus* in Sri Lanka: Preliminary findings. *Environmental Biology of Fishes, 2*(2), 137–145.
Warner, R. R. (1975). The adaptive significance of sequential hermaphroditism in animals. *American Naturalist, 109*, 61–82.
Warner, R. R. (1978). Patterns of sex and coloration in the Galapagos wrasses. *Noticias de Galapagos, 27*, 16–18.
Warner, R. R. (1984). Mating behavior and hermaphroditism in coral reef fishes. *American Scientist, 72*, 128–136.
Warzel, F. (1996). Variation in *Crenicichla regani. The Cichlids Yearbook, 6*, 74–79.
Watanabe, T. (2000). The nesting site of a piscivorous cichlid *Lepidiolamprologus profundicola* as a safety zone for juveniles of a zooplanktivorous cichlid *Cyprichromis leptosoma* in Lake Tanganyika. *Environmental Biology of Fishes, 57*(2), 171–177.
Webber, R., Barlow, G. W., and Brush, A. H. (1973). Pigments of a color polymorphism in a cichlid fish. *Comparative Biochemistry and Physiology, 44B*, 1127–1135.
Weber, P. G. (1970). Visual aspects of egg care behaviour in *Cichlasoma nigrofasciatum* (Günther). *Animal Behaviour, 18*(4), 688–699.
Weber, P. G., and Weber, S. P. (1971). Choice of spawn site in *Cichlasoma nigrofasciatum* (Günther) (Pisces: Cichlidae). *Zeitschrift für Tierpsychologie, 28*, 475–478.
Wedekind, C., and Furi, S. (1997). Body odour preferences in men and women: Do they aim for specific MHC combinations or simply heterozygosity? *Proceedings of the Royal Society of London Series B Biological Sciences, 264*(1387), 1471–1479.
West-Eberhard, M. J. (1983). Sexual selection, social competition, and speciation. *Quarterly Review of Biology, 58*, 155–183.
Westneat, D. F., Sherman, P. W., and Morton, P. W. (1990). The ecology and evolution of extra-pair copulations in birds. *Current Ornithology, 7*, 331–370.
Whitfield, A. K., and Blaber, S. J. M. (1979). The distribution of the freshwater cichlid *Sarotherodon mossambius* in estuarine systems. *Environmental Biology of Fishes, 4*, 77–81.
Wickler, W. (1956). Der Haftapparat einiger Cichliden-Eier. *Zeitschrift für Zellforschung, 45*, 304–327.
Wickler, W. (1959). *Teleogramma brichardi* Poll 1959. *Deutsche Aquarien- und Terrarien-Zeitschrift, 8*, 228–230.
Wickler, W. (1962a). "Egg-dummies" as natural releasers in mouth-breeding cichlids. *Nature* (London), *194*(4833), 1092–1093.
Wickler, W. (1962b). Zur Stammesgeschichte funktionnell korrelierter Organ- und Verhaltensmerkmale: Eiattrappen und Maulbruten bei afrikanischen Cichliden. *Zeitschrift für Tierpsychologie, 19*, 129–164.

Wickler, W. (1965). Signal value of the genital tassel in the male *Tilapia macrochir* Blgr. (Pisces: Cichlidae). *Nature* (London), *208*(5010), 595–596.
Wickler, W. (1997). Sexually selected genital adornment and sperm packaging in species of *Oreochromis* (Teleostei: Cichlidae). *Copeia, 1997*(1), 188–190.
Williams, G. C. (1966). *Adaptation and Natural Selection: A Critique of Some Current Evolutionary Thought.* Princeton: Princeton University Press.
Williams, G. C. (1975). *Sex and Evolution.* Princeton: Princeton University Press.
Williams, G. C. (1988). Retrospect on sex and kindred topics. In R. E. Michod and B. R. Levin (Eds.), *The Evolution of Sex: An Examination of Current Ideas* (pp. 287–298). Sunderland, Massachusetts: Sinauer Associates.
Williams, G. C. (1989). A sociobiological expansion of *Evolution and Ethics.* In J. Paradis and G. C. Williams (Eds.), *Evolution and Ethics* (pp. 179–214). Princeton: Princeton University Press.
Williams, G. C. (1992). *Natural Selection: Domains, Levels, and Challenges.* Oxford: Oxford University Press.
Williams, N. J. (1972). On the ontogeny of behaviour of the cichlid fish *Cichlasoma nigrofasciatum* (Günther). Unpublished Ph.D. dissertation, Groningen University, Groningen.
Wilson, E. O. (1975). *Sociobiology: The New Synthesis.* Cambridge, Massachusetts: Belknap Press.
Winemiller, K. O., Kelso-Winemiller, L. C., and Brenkert, A. L. (1995). Ecomorphological diversification and convergence in fluvial cichlid fishes. *Environmental Biology of Fishes, 44*(1–3), 235–261.
Wisenden, B. D. (1994). Factors affecting reproductive success in free-ranging convict cichlids *(Cichlasoma nigrofasciatum). Canadian Journal of Zoology, 72*(12), 2177–2185.
Wisenden, B. D. (1999). Alloparental care in fishes. *Reviews in Fish Biology and Fisheries, 9*, 45–70.
Wisenden, B. D., and Keenleyside, M. H. A. (1992). Intraspecific brood adoption in convict cichlids: A mutual benefit. *Behavioral Ecology and Sociobiology, 31*, 263–269.
Wisenden, B. D., and Keenleyside, M. H. A. (1994). The dilution effect and differential predation following brood adoption in free-ranging convict cichlids *(Cichlasoma nigrofasciatum). Ethology, 96*(3), 203–212.
Wisenden, B. D., and Keenleyside, M. H. A. (1995). Brood size and the economy of brood defence: Examining Lack's hypothesis in a biparental cichlid fish. *Environmental Biology of Fishes, 43*(2), 145–151.
Wisenden, B. D., Lanfranconi-Izawa, L., and Keenleyside, M. H. A. (1995). Fin digging and leaf lifting by the convict cichlid, *Cichlasoma nigrofasciatum:* Examples of parental food provisioning. *Animal Behaviour, 49*(3), 623–631.
Witte, F., Goldschmidt, T., Goudswaard, P. C., Ligtvoet, W., Oijen, M. J. P. v., and Wanink, J. H. (1992). Species extinction and concomitant ecological changes in Lake Victoria. *Netherlands Journal of Zoology, 42*(2–3), 214–232.
Witte, F., Goldschmidt, T., Wanink, J., Oijen, M. v., Goudswaard, K., Witte-Maas, E., and Bouton, N. (1992). The destruction of an endemic species flock: Quantitative data on the decline of the haplochromine cichlids of Lake Victoria. *Environmental Biology of Fishes, 34*(1), 1–28.
Witte, F., Goldschmidt, T., and Wanink, J. H. (1995). Dynamics of the haplochromine cichlid fauna and other ecological changes in the Mwanza Gulf of Lake Victoria. In T. J. Pitcher and P. J. B. Hart (Eds.), *The Impact of Species Changes in African Lakes* (pp. 83–110). New York: Chapman and Hall.
Witte, F., and Oijen, M. J. v. (1990). Taxonomy, ecology, and fishes of Lake Victoria haplochromine trophic groups. *Zoologische Verhandlingen Leiden, 262*(15), 1–47.
Witte, K. E., and Schmidt, J. (1992). *Betta brownorum,* a new species of anabantoids (Teleostei: Belontiidae) from northwestern Borneo, with a key to the genus. *Ichthyological Exploration of Freshwaters, 2*(4), 305–330.
Wohlfarth, G. W. (1994). The unexploited potential of tilapia hybrids in aquaculture. *Aquaculture and Fisheries Management, 25*(8), 781–788.
Wohlfarth, G. W., Rothbard, S., Hulata, G., and Szweigman, D. (1990). Inheritance of red body coloration in Taiwanese tilapias and in *Oreochromis mossambicus. Aquaculture, 84*(3–4), 219–234.

Wohlfarth, G. W., and Wedekind, H. (1991). The heredity of sex determination in tilapias. *Aquaculture, 92*(2–3), 143–156.

Wood, C. C., and Foote, C. J. (1996). Evidence for sympatric genetic divergence of anadromous and nonanadromous morphs of sockeye salmon *(Oncorhynchus nerka). Evolution, 50*(3), 1265–1279.

Woolfenden, G. E., and Fitzpatrick, J. W. (1984). *The Florida Scrub Jay: A Demography of a Cooperatively Breeding Bird.* Princeton: Princeton University Press.

Wyman, R. L., and Ward, J. A. (1972). A cleaning symbiosis between the cichlid fishes *Etroplus maculatus* and *Etroplus suratensis.* I. Description and possible evolution. *Copeia, 1972,* 834–383.

Wyman, R. L., and Ward, J. A. (1973). The development of behavior in the cichlid fish *Etroplus maculatus* (Bloch). *Zeitschrift für Tierpsychologie, 33,* 461–491.

Yamagishi, S., and Kohda, M. (1996). Is the cichlid fish *Julidochromis marlieri* polyandrous? *Ichthyological Research, 43*(4), 469–471.

Yamamoto, T. (1975). The medaka, *Oryzias latipes,* and the guppy, *Lebistes reticulatus.* In R. C. King (Ed.), *Vertebrates of Genetic Interest,* vol. 4 (pp. 133–149). New York: Plenum.

Yamaoka, K. (1985). Intestinal coiling pattern in the epilithic algal-feeding cichlids (Pisces, Teleostei) of Lake Tanganyika, and its phylogenetic significance. *Zoological Journal of the Linnean Society, 84,* 235–261.

Yamaoka, K. (1991). Feeding relationships. In M. H. A. Keenleyside (Ed.), *Cichlid Fishes: Behaviour, Ecology, and Evolution* (pp. 151–172). New York: Chapman and Hall.

Yanagisawa, Y. (1985). Parental strategy of the cichlid fish *Perissodus microlepis,* with particular reference to intraspecific brood farming out. *Environmental Biology of Fishes, 12*(4), 241–250.

Yanagisawa, Y. (1986). Parental care in a monogamous mouthbrooding cichlid *Xenotilapia flavipinnis* in Lake Tanganyika. *Japanese Journal of Ichthyology, 33*(3), 249–261.

Yanagisawa, Y. (1987). Social organization of a polygynous cichlid *Lamprologus furcifer* in Lake Tanganyika. *Japanese Journal of Ichthyology, 34*(1), 82–90.

Yanagisawa, Y., and Nishida, M. (1991). The social and mating system of the maternal mouthbrooder *Tropheus moorii* (Cichlidae) in Lake Tanganyika. *Japanese Journal of Ichthyology, 38*(3), 271–282.

Yanagisawa, Y., and Nshombo, M. (1983). Reproduction and parental care of the scale-eating cichlid fish *Perissodus microlepis* in Lake Tanganyika. *Physiology and Ecology Japan, 20*(1), 23–31.

Yanagisawa, Y., and Ochi, H. (1991). Food intake by mouthbrooding females of *Cyphotilapia frontosa* (Cichlidae) to feed both themselves and their young. *Environmental Biology of Fishes, 30,* 353–358.

Yanagisawa, Y., Ochi, H., and Gashagaza, M. M. (1997). Habitat use in cichlid fishes for breeding. In H. Kawanabe, M. Hori, and M. Nagoshi (Eds.), *Fish Communities in Lake Tanganyika* (pp. 151–173). Kyoto: Kyoto University Press.

Yanagisawa, Y., Ochi, H., and Rossiter, A. (1996). Intra-buccal feeding of young in an undescribed Tanganyikan cichlid *Microdontochromis* sp. *Environmental Biology of Fishes, 47*(2), 191–201.

Yanagisawa, Y., and Sato, T. (1990). Active browsing by mouthbrooding females of *Tropheus duboisi* and *Tropheus moorii* (Cichlidae) to feed the young and/or themselves. *Environmental Biology of Fishes, 27*(1), 43–50.

Yuma, M., and Kondo, T. (1997). Interspecific relationships and habitat utilization among benthivorous cichlids. In H. Kawanabe, M. Hori, and M. Nagoshi (Eds.), *Fish Communities in Lake Tanganyika* (pp. 89–103). Kyoto: Kyoto University Press.

Yuma, M., Narita, T., Hori, M., and Kondo, T. (1998). Food resources of shrimp-eating cichlid fishes in Lake Tanganyika. *Environmental Biology of Fishes, 52*(1–3), 371–378.

Zahavi, A. (1975). Mate selection: A selection for a handicap. *Journal of Theoretical Biology, 53,* 205–214.

Zahavi, A. (1979). Ritualisation and evolution of movement signals. *Behaviour, 72,* 77–81.

Zaret, T. M., and Paine, R. T. (1973). Species introduction in a tropical lake. *Science* (Washington, D.C.), *182,* 449–455.

Zihler. (1982). Gross morphology and configuration of digestive tracts of Cichlidae (Teleostei, Perciformes): Phylogenetic and functional significance. *Netherlands Journal of Zoology, 32*(4), 544–571.

Zoran, M. J., and Ward, J. A. (1983). Parental egg care behaviour and fanning activities for the orange chromide, *Etroplus maculatus. Environmental Biology of Fishes, 8,* 301–310.

INDEX

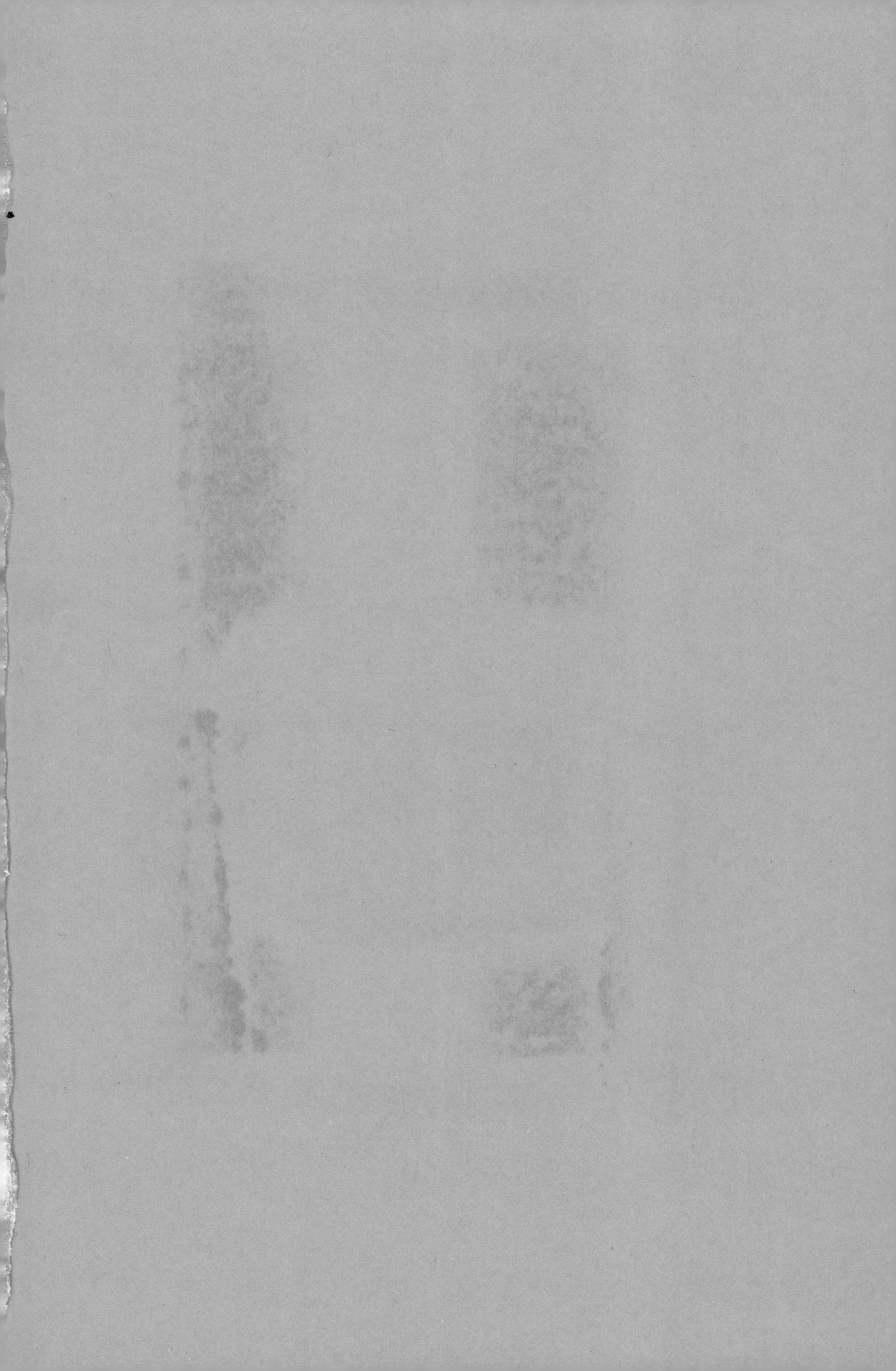

SARATOGA SPRINGS PUBLIC LIBRARY
0 00 02 0272136 1